Handbook of Manufacturing Systems and Design

This book provides a comprehensive overview of manufacturing systems, their role in product/process design, and interconnection with an Industry 4.0 perspective, especially related to design, manufacturing, and operations.

Handbook of Manufacturing Systems and Design: An Industry 4.0 Perspective provides the knowledge related to the theories and concepts of Industry 4.0. It focuses on the different types of manufacturing systems in Industry 4.0 along with associated design, and control strategies. It concentrates on the operations in Industry 4.0 with a particular focus on supply chain, logistics, risk management, and reverse engineering perspectives. Offering basic concepts and applications through to advanced topics, the handbook feeds into the goal of being a single source of knowledge as well as a vehicle to use to explore the future possibilities of design, techniques, methods, and operations associated with Industry 4.0. Concepts with practical applications in the form of case studies are added to each chapter to round out the many attributes this handbook offers.

This handbook targets students, engineers, managers, designers, and manufacturers, and will assist in their understanding of the core concepts of manufacturing systems in connection with Industry 4.0 and optimize alignment between supply and demand in real time for effective implementation of the design concepts.

Handbook of Manufacturing Systems and Design

An Industry 4.0 Perspective

Edited by

Uzair Khaleeq uz Zaman

Ali Siadat

Aamer Ahmed Baqai

Kanwal Naveed

Atal Anil Kumar

CRC Press
Taylor & Francis Group
Boca Raton London New York

CRC Press is an imprint of the
Taylor & Francis Group, an **informa** business

Designed cover image: Shutterstock

First edition published 2024
by CRC Press
6000 Broken Sound Parkway NW, Suite 300, Boca Raton, FL 33487-2742

and by CRC Press
4 Park Square, Milton Park, Abingdon, Oxon, OX14 4RN

CRC Press is an imprint of Taylor & Francis Group, LLC

ISBN: 978-1-032-35321-0 (hbk)
ISBN: 978-1-032-35571-9 (pbk)
ISBN: 978-1-003-32752-3 (ebk)

DOI: 10.1201/9781003327523

Typeset in Times
by MPS Limited, Dehradun

The Open Access version of chapter 4 was funded by Ministry of Science and ICT, and chapters 6 and 8 were funded by University of Luxembourg.

Contents

Forewords

ACADEMIC FOREWORD

Manufacturing paradigms are currently being modified because of the need for resilience, sobriety, agility, and many other issues related to sustainability, including social, economic, and environmental impacts. The COVID-19 pandemic period has shown how it is crucial to be prepared for rapid evolution within the manufacturing industry. Fortunately, more than a decade before, the digitization of industry was proposed based on new human-centric environments, with complete assistance and an essential autonomy of production systems. Hopefully, this transformation is still in progress, and small- and medium-sized companies must align their practices with the ones already in place in large original equipment manufacturer (OEM) facilities.

In addition to CAD/CAM/CAE/PLM approaches, the companies currently use several production management systems, particularly ERP and MES information systems. Even if many companies have adopted a real digital integration, progress is still expected in terms of human assistance for operations and decision-making processes and for the real life cycle management of products. Complete traceability of manufacturing processes is necessary, concerning certifications in many fields like aeronautics or medicine. A mutation is going on to a production-on-demand service when and where consumers need the product at the right cost and quantity. Such evolution will transform physical logistics into digital logistics, with digital transfer of the necessary manufacturing information to local producers.

Definitively, in a context where the different countries move to a certain independence, manufacturing contexts and demands are being transformed. Companies must adapt how they produce products at different levels of management and operation, from production shop floors to extended enterprise logistics.

These major evolutions affect all components/stakeholders of value chains, from essential technologies to management and decision-making methods. Digitization is one of the main vectors supporting these evolutions, with new ways of communication between machines, between companies' departments, between companies, but also between companies and subcontractors, and finally, between companies and end-users. Servitization is becoming the rule concerning these companies-consumers relationships, with a real consideration of the whole life cycle of products, including re- and de-manufacturing, circular economy, and recyclability.

But, even if this seems straightforward, it is of significant concern for companies to have their employees aware of this new context of producing goods. It is always necessary to manage resources and energy consumption when minimizing the impact of production effects on the environment. So, life-long learning is becoming essential through internal training sessions, thanks to a new set of publications, papers, and books. The way from theory to practice is complex, but it is necessary to transform the current practices without losing technological knowledge.

This handbook is to become one of these key publications, very useful for many people in the companies. From operators to general managers, all enterprise functions are impacted and concerned by the transformations of industry in general, and Industry 4.0 already became more or less a standard. Integrating new technologies is becoming a priority in many fields, like IoT helping collect data, which should be analyzed to learn more about the actual production world. Thanks to such an approach, the models of production systems become more reliable. They are integrated with digital twins, which help in particular to simulate what-if scenarios and in decision-making processes.

The three main parts of the handbook give significant content to help understand the basic principles and technologies of smart manufacturing and Industry 4.0 and the way to use them in practice. Many additional issues are also addressed to have a complete overview of the most important concepts, models, and methods.

I congratulate all co-editors and authors for this very significant contribution to sharing knowledge in the future manufacturing systems field based on the complete digitization proposed by Industry 4.0 and beyond.

Alain BERNARD
Emeritus Professor at Ecole Centrale de Nantes
Researcher at Digital Science Laboratory of Nantes (LS2N UMR CNRS 6004)
1, rue de la Noë, BP 92101, 44321 - NANTES Cedex 3 – France
Fellow Member of the French Academy of Technologies
(www.academie-technologies.fr)
Vice President of France Additive (franceadditive.tech)
CIRP Fellow Emeritus (www.cirp.net)
Editor-in-Chief International Journal of Product Life Cycle Management
Associate Editor CIRP Journal of Manufacturing Science and Technology

INDUSTRIAL FOREWORD

Manufacturing systems have undergone a massive transformation in the past decade, and the concept of Industry 4.0 has significantly shaped this evolution. Industry 4.0 refers to integrating advanced technologies such as internet of things (IoT), artificial intelligence (AI), and cyber-physical systems into manufacturing processes to create smart factories capable of autonomous decision-making and optimization.

I am excited to introduce *Handbook of Manufacturing Systems and Design: An Industry 4.0 Perspective*. This handbook provides an insightful and comprehensive guide to the integration of Industry 4.0 into manufacturing systems, their design, operations, and management of products/processes.

One of the key strengths of this handbook is how it brings together leading researchers and practitioners from around the world to share their knowledge and expertise. Their contributions offer valuable insights into smart manufacturing systems' design, monitoring, and control, making this handbook an essential resource for students, engineers, managers, designers, and manufacturers.

In conclusion, *Handbook of Manufacturing Systems and Design: An Industry 4.0 Perspective* is an excellent resource that provides a comprehensive overview of the integration of Industry 4.0 into manufacturing systems. It offers practical advice and guidance for those seeking to optimize their manufacturing processes and align supply and demand in real time. I am confident that this Handbook will be a valuable addition to any industrial manager's or engineer's library and a source of inspiration and innovation for years to come.

Konrad Grohs
Director IoT Business Europe
FANUC

Preface

Handbook of Manufacturing Systems and Design: An Industry 4.0 Perspective is concerned with a comprehensive overview of manufacturing systems, their role in product/process design, and interconnection with the Industry 4.0 perspective, especially concerning design, manufacturing, and operations. Since modern manufacturing is composed of more intrinsic operations compared to the traditional view of constructing the physical product, some of the most important decisions in the manufacturing industry are directly related to the nature of products, the economics of production, fluctuations/changes in consumer demands, and subsequent design changes. Moreover, the advancement of information technologies has paved the way for the evolution of production systems, inviting the concept of Industry 4.0.

This handbook aims to provide up-to-date, convincing answers to the myriad problems arising in manufacturing systems, their design, operations, and management of products/processes. This handbook will help students, engineers, managers, designers, and manufacturers to understand the core concepts of manufacturing systems in connection with Industry 4.0 and optimize alignment between supply and demand in real-time for effective implementation of the design concepts. The handbook will also supply literature with the missing pieces and, thus, a research-enabling platform in the design and control of smart manufacturing systems. Moreover, it will be a source of inspiration and, to some extent, implementation for industrial managers and engineers, particularly those contributing to production control in enhancing the manufacturing processes and increasing profits.

For this handbook, we have secured contributions from leading researchers and practitioners from around the globe who have identified critical elements that could be made part of this handbook. It consists of 17 chapters and follows a process-systems engineering approach to design, monitoring, and control of smart manufacturing systems and covers all aspects from design to supply chain and logistics. The topics covered include computer-aided process planning, Quality 4.0, digital twins, robotic systems, control strategies, sensors, energy harvesters, IoT, Logistics 4.0, cloud manufacturing, risk management, and reverse engineering, among others.

The handbook is planned so that by using the table of contents and the index, the reader can locate references to a topic of concern and easily access information referring to a problem at hand.

Uzair Khaleeq uz Zaman
Ali Siadat
Aamer Ahmed Baqai
Kanwal Naveed
Atal Anil Kumar

Editors' Biographies

Uzair Khaleeq uz Zaman is an assistant professor at the College of Electrical and Mechanical Engineering, National University of Sciences and Technology, Pakistan. Dr. Zaman has a PhD in industrial engineering from Arts et Métiers Institute of Technology, France, and carries a blend of experience in academia and the industry. In the industry, he has eight years' worth of experience in operations, product development, and project management. As part of academic career, Dr. Zaman has taught multiple courses along with postgraduate research at Beihang University (China), Georgia Tech (France), Arts et Métiers Institute of Technology (France), and University of Lorraine (France). His research interests lie in integrated product process design, manufacturing system design, optimization and decision-making of production systems, additive manufacturing, design of experiments, product life cycle assessment, and robotic and navigation systems.

Ali Siadat is a professor and director of Design, Manufacturing and Control Laboratory at Arts et Métiers Institute of Technology, France. With over 25 years of experience, Prof. Siadat has supervised 30 PhDs and more than 65 master's thesis. He is a research scholar of an international repute, and his research interests lie in integrated design, process and manufacturing, intelligent computation for smart industrial systems, integrated human factors, optimization and decision-making of production systems, risk analysis, knowledge and information system for quality, and process control. He serves as associate editor of "International Journal of Production Research" and as Board of Editors for the journal "Engineering Applications of Artificial Intelligence". Siadat is actively involved in Factories of the Future program in France and his research activities are industry-oriented and are driven by international collaborations.

Aamer Ahmed Baqai is a professor at the College of Electrical and Mechanical Engineering, National University of Sciences and Technology, Pakistan. He completed his PhD and master's degrees in mechanical engineering from Arts et Métiers Institute of Technology, France. With over 26 years of professional career, he has a vast experience of academia and industry. Baqai is a researcher of international repute and is a regular research visiting professor at Arts et Métiers Institute of Technology, France. He has also acted as a foreign evaluator for PhD thesis students at Arts et Métiers Institute of Technology. His research interests include advanced manufacturing, manufacturing system design and optimization, reconfigurable manufacturing systems, process modeling and optimization, additive manufacturing, and decision-making in manufacturing.

Kanwal Naveed is an electrical engineer who completed her master's degree in control systems from the National University of Sciences and Technology, Pakistan, in 2014. She is also undergoing her PhD in robotics and artificial intelligence from National University of Sciences and Technology, Pakistan. Currently, Miss Naveed is serving as an assistant professor at Department of Mechatronics Engineering where she is also designated as head of undergraduate program. Her research interests include control of service and field robotics, machine learning, artificial intelligence, and non-linear control systems.

Atal Anil Kumar is currently working as a postdoctoral researcher at University of Luxembourg. His research interests include nonlinear control, collaborative robots, robot kinematics and dynamics, and human-robot interaction. He has a bachelor's degree in mechatronics and an Erasmus master's in advanced robotics (with scholarship) from France and Italy. He was awarded the research internship grant by the Indian Institute of Technology, Bombay (IIT-B) for performing research during his bachelor's degree. Atal has also worked as an R&D Engineer in the automotive and steel manufacturing industries to gain significant hands-on experience to apply his theoretical knowledge.

Contributors

Nayyer Aafaq
College of Aeronautical Engineering
National University of Sciences and Technology
Risalpur, Pakistan

Yasir Ahmad
Department of Engineering Management
College of Electrical and Mechanical Engineering
National University of Sciences and Technology
Islamabad, Pakistan

Atal Anil Kumar
Department of Engineering
Faculty of Science, Technology and Medicine
University of Luxembourg
Kirchberg, Luxembourg

Anas Bin Aqeel
Department of Mechatronics Engineering
College of Electrical and Mechanical Engineering
National University of Sciences and Technology
Islamabad, Pakistan

Muhammad Irfan Aziz
College of Aeronautical Engineering
National University of Sciences and Technology
Risalpur, Pakistan

Shahid Aziz
Jeju National University
Jeju City, South Korea

Aamer Ahmed Baqai
Department of Mechanical Engineering
College of Electrical and Mechanical Engineering
National University of Sciences and Technology
Islamabad, Pakistan

Alain Etienne
Arts et Métiers Institute of Technology
Campus de Metz
Metz, France

Stéphane Hubac
STMicroelectronics
Crolles, France

Dong Won Jung
Jeju National University
Jeju City, South Korea

Muhammad Asad Ullah Khalid
Chung-Ang University
Seoul, South Korea

Sri Kolla
Department of Engineering
Faculty of Science, Technology and Medicine
University of Luxembourg
Kirchberg, Luxembourg

Mathias Kühn
Technische Universitat Dresden
Dresden, Germany

Marc Lassagne
Arts et Métiers Institute of Technology
Campus de Metz
Metz, France

Afshan Naseem
Department of Engineering Management
College of Electrical and Mechanical Engineering
National University of Sciences and Technology
Islamabad, Pakistan

Rashid Naseer
Department of Mechanical Engineering
College of Electrical and Mechanical Engineering
National University of Sciences and Technology
Islamabad, Pakistan

Kanwal Naveed
Department of Mechatronics Engineering
College of Electrical and Mechanical Engineering
National University of Sciences and Technology
Islamabad, Pakistan

Jelena Petronijevic
Arts et Métiers Institute of Technology
Campus de Metz
Metz, France

Peter Plapper
Department of Engineering
Faculty of Science, Technology and Medicine
University of Luxembourg
Kirchberg, Luxembourg

Usman Qamar
Department of Computer and Software Engineering
College of Electrical and Mechanical Engineering
National University of Sciences and Technology
Islamabad, Pakistan

Thorsten Schmidt
Technische Universitat Dresden
Dresden, Germany

Julia Schwemmer
Technische Universitat Dresden
Dresden, Germany

Taiba Zahid
NUST Business School
National University of Science and Technology
Islamabad, Pakistan

Uzair Khaleeq uz Zaman
Department of Mechatronics Engineering
College of Electrical and Mechanical Engineering
National University of Sciences and Technology
Islamabad, Pakistan

List of Key Abbreviations

Technical and scientific terms

ACO	Ant colony optimization
AEC	Advanced equipment control
AGV	Automated guided vehicle
AHP	Analytical hierarchy process
AI	Artificial intelligence
AM	Additive manufacturing
ANN	Artificial neural network
APC	Advanced process control
API	Application programming interface
AQMS	Autonomous quality management system
AR	Augmented reality
ASP	Application service provider
BOL	Beginning of life
BOM	Bill of materials
CAD	Computer-aided design
CAE	Computer-aided engineering
CAGR	Compound annual growth rate
CAPP	Computer-aided process planning
CAM	Computer-aided manufacturing
CAQ	Computer-aided quality
CBA	Cost-benefit analysis
CCD	Charged couple device
CEA	Cost-effectiveness analysis
CFMS	Complex flexible manufacturing system
CIM	Computer-integrated manufacturing
CM	Cloud manufacturing
CMM	Contact-measuring machine
CNC	Computer numerically controlled
CONWIP	Constant work in progress
CPPS	Cyber-physical production system
CPS	Cyber-physical system
DCPS	Decentrally controlled production system
DFM	Design for manufacturing
DFMEA	Design failure mode and effect analysis
DIK	Data, information, and knowledge
DNC	Direct numerical control
DNN	Deep neural network
DO	Dissolved oxygen
DoE	Design of experiments
DPC	Decentralized production control

DSS	Decision support system
EIP	Equipment installation process
EMF	Elementary machining features
EOL	End of life
ERP	Enterprise resource planning
ES	Expert system
ETA	Event tree analysis
FCM	Fuzzy cognitive map
FE	Forward engineering
FIV	Flow-induced vibration
FMA	Failure mode avoidance
FMEA	Failure mode and effects analysis
FMS	Flexible manufacturing system
FoF	Factory of future
FoV	Field of view
FT	Feature technology
FTA	Fault tree analysis
GA	Genetic algorithm
GFP	Green fluorescent protein
GT	Group technology
HHD	Handheld device
HMD	Head-mounted device
HMI	Human-machine interface
HOQ	House of quality
HPLC	High-performance liquid chromatography
ICT	Information and communication technology
IIoT	Industrial Internet of things
IM	Intelligent manufacturing
IoS	Internet of service
IoT	Internet of things
ISM	In-situ microscopy
ISPC	Intelligent statistical process control
IT	Information technology
JSSP	Job shop scheduling problems
LCA	Life cycle analysis
LCM	Life cycle management
LiDAR	Light Detection and Ranging
LOOR	Load-oriented order release
MAUT	Multi-attribute utility theory
MEMS	Micro-electromechanical system
MES	Manufacturing execution system
MFD	Manufacturing for design
MIS	Management information system
ML	Machine learning
MOL	Middle of life
MR	Mixed reality

MRP-I	Material requirement planning
MRP-II	Manufacturing resource planning
NC	Numerical control
N/MIR	Near and mid-infrared
NN	Neural network
NURBS	Non-uniform rational basis spline
OCAP	Out of control action plan
OEM	Original equipment manufacturer
OOAC	Organ-on-a-chip
OT	Operation technology
PCA	Principal component analysis
PDA	Production data acquisition
PFMEA	Process failure mode and effect analysis
PI	Proportional-integral
PLM	Product life cycle management
PLS	Partial least squares
PPS	Production planning system
PRM	Probabilistic roadmap
PSO	Particle swarm optimization
PU	Processing unit
PV	Photovoltaic
QA	Quality assurance
QC	Quality control
QFD	Quality function deployment
QMS	Quality management system
RACI	Realize, accountable, consulted, informed
RE	Reverse engineering
RF	Radio frequency
RFID	Radio frequency identification
RNN	Recurrent neural network
ROI	Return on investment
ROS	Robot operating system
RRT	Rapidly exploring random trees
SBO	Simulation-based optimization
SDK	Software development kit
SM	Smart manufacturing
SME	Small and medium enterprises
SPC	Statistical process control
SMO	Smart manufacturing object
SoA	Service-oriented architecture
SQC	Statistical quality control
SVM	Support vector machine
TEER	Transepithelial electrical resistant
TOF	Time of flight
TOPSIS	Technique for order of preference by similarity to ideal solution
TQC	Total quality control

TQM	Total quality management
UI	User interface
UR	Universal robot
VI	Virtual instrument
VIV	Vortex-induced vibration
VR	Virtual reality
WIP	Work in progress
WSN	Wireless sensor network
XR	Extended reality
ZDM	Zero-defect manufacturing

Organizations, institutions, and companies

ASQ	American Society for Quality
EFQM	European Foundation for Quality Management
EU	European Union
IATF	International Automotive Task Force
IFR	International Federation of Robotics
IIC	Industrial Internet Consortium
ISO	International organization for standardization
NI	National Instruments
NIST	National Institute of Standards and Technology
SFEG	Scott Fetzer Electrical Group
SMLC	Smart Manufacturing Leadership Coalition
UNEP	United Nations Environmental Program

Part I

Theory and Concepts in Manufacturing Systems

1 Introduction to Industry 4.0 and Smart Manufacturing

1.1 INTRODUCTION TO INDUSTRY 4.0

Ever since the beginning of industrialization, technological leaps have led to paradigm shifts (industrial revolutions) from water- and steam-powered machines to electrical and digital automated production (Lasi et al., 2014; Vaidya et al., 2018). Industry 1.0 (the first paradigm shift) converted the focus from agriculture to an industrial society, wherein product volume was the sole objective, and product variety was very low. Industry 2.0 (second paradigm shift) brought major technological innovations like electricity, cars, and electronic/mechanical devices. The demand during Industry 2.0 was influenced by two dimensions, i.e., variety and volume. Next up was Industry 3.0, the third paradigm shift, which was characterized by technological innovations such as the change from analog to digital and a significant reduction in average product lifecycles. The demand during Industry 3.0 had three dimensions, i.e., variety, volume, and delivery time (Yin et al., 2018).

Post Industry 3.0, information and communication technology (ICT) is undergoing rapid development. Many disruptive technologies, such as the internet of things (IoT), cloud computing, big data analytics, and artificial intelligence (AI), have emerged, thereby making the manufacturing industry capable and smart in terms of addressing ongoing challenges such as improved quality, increase in customized requirements, and reduction in time to market (Zheng et al., 2018). Moreover, an increasing number of sensors are being used to enable the equipment to sense, act, and communicate with one another (Zhang et al., 2015). As a result, real-time production data can be obtained and shared to accelerate accurate and quick decision-making. The connection of physical manufacturing equipment and devices over the Internet, together with big data analytics in the digital world (e.g., the cloud), has resulted in the emergence of a revolutionary means of production, namely, cyber-physical systems (CPSs) or cyber-physical production systems (CPPs). The widespread application of CPS (or CPPs) has ushered into the fourth paradigm shift, namely, Industry 4.0 (Kagermann et al., 2013). Figure 1.1 shows the four stages of the industrial revolution.

1.2 SMART MANUFACTURING/INTELLIGENT MANUFACTURING IN INDUSTRY 4.0

In the last few decades, the introduction of intelligence into manufacturing has become an important topic for industries and researchers. Academic and industry

DOI: 10.1201/9781003327523-2

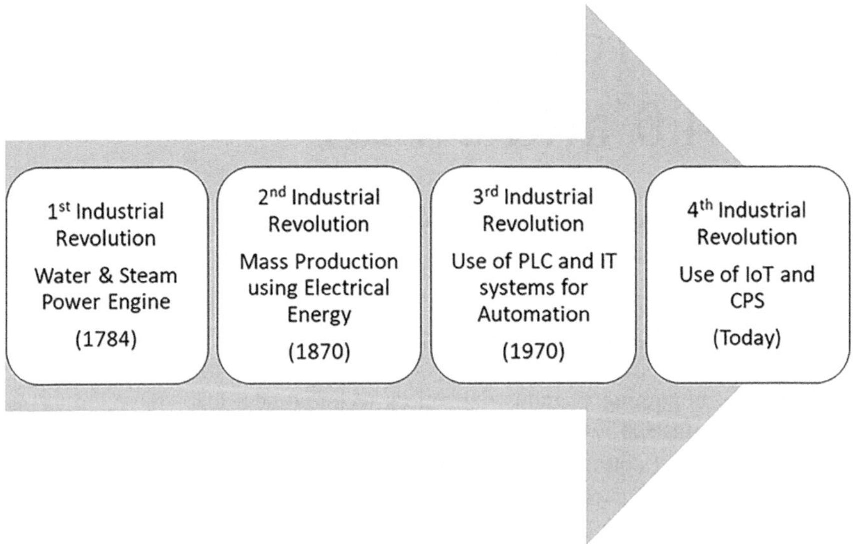

FIGURE 1.1 Four industrial revolutions.

Source: Adapted from Vaidya et al. (2018).

researchers have been using terminologies such as smart manufacturing (SM) and intelligent manufacturing (IM) in the context of Industry 4.0 as manufacturing systems are updated to an intelligent level (Kusiak, 2018; Liang et al., 2018; Zhou et al., 2018). IM (also known as SM) is a broad concept of manufacturing to optimize production and product transactions by fully using advanced information and manufacturing technologies (Kusiak, 1990). The complete lifecycle of a product and the associated value chain is covered by SM, starting from the conceptual design to production, delivery, and eventually recycling/disposing of. SM/IM also integrates customer/user input/feedback in real-time, making the manufacturing process adaptive, agile, and intelligent (IEC, 2023). The interconnection of every step in the manufacturing process is hence the core objective associated with SM/IM.

Zhou et al. (2018) split SM/IM evolution into three stages. The digital manufacturing era before 2000 is defined as the first stage that supported machines and system-level operations using computers. The SM era after 2000 is termed the second stage, where dynamic environments and customer needs are addressed through digital manufacturing. Next-generation IM (NGIM) era after 2020 is termed the third stage, which integrates human and machine systems using the IoT and machine learning (ML). Thoben et al. (2017) stated the use of IM and SM equitably, with the former targeting the technologies to organizational concepts and the latter targeting analytics and controls. Numerous researchers defined SM as an imminent form of IM with smart technologies such as IoT, big data, and cloud computing, enabling Industry 4.0. Zheng et al. (2018) proposed a conceptual framework for Industry 4.0 SM systems, as shown in Figure 1.2.

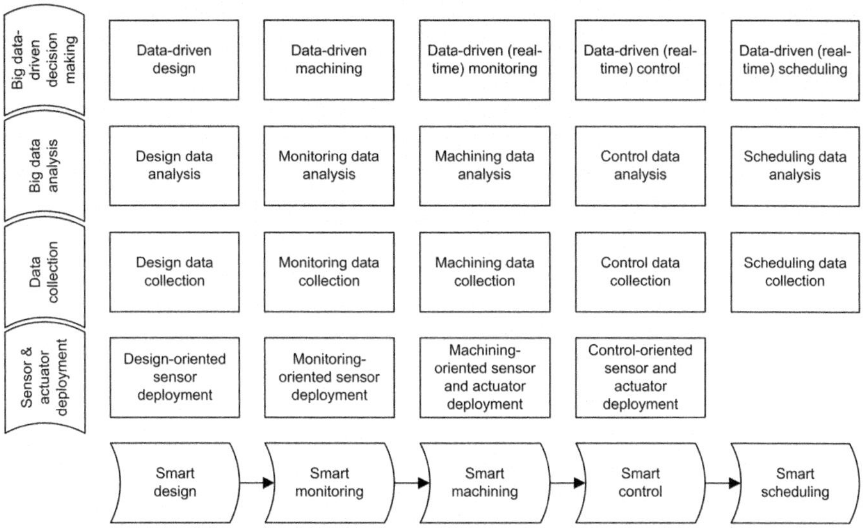

FIGURE 1.2 Conceptual framework of Industry 4.0 smart manufacturing systems.

Source: Adapted from Zheng et al. (2018).

1.2.1 Origins of SM/IM

The idea of SM appeared in the late 1980s through its association with AI, a tool to enhance SM. This was followed by the book titled *Smart Manufacturing and Artificial Intelligence* in 1987, which elaborated on the importance of AI in improving production and profits associated with manufacturing (Krakauer, 1987). The concept of SM re-emerged after two decades with its development in Industry 4.0. The National Institute of Standards and Technology (NIST) (Lu et al., 2016) and Smart Manufacturing Leadership Coalition (SMLC, 2020) defined the core definitions of SM. Similarly, the concept of IM originated from artificial and manufacturing intelligence, with its earlier publication dating back to 1988–1995 (Wang et al., 2021). The Japanese did pioneering work on IM (Okabe, 2013) when they established Intelligent Manufacturing System (IMS). Later, IMS was utilized by US and European researchers as well.

1.2.2 Factory of the Future

Since the early twentieth century, researchers and practitioners have been idealizing what the factory of the future (FoF) will look like. Questions such as how the future production environments will be and how machines and humans will communicate with each other are repeatedly asked. Today, products have shorter lifecycles, markets are dynamic, safety and reliability concerns are increasing, lifecycles are becoming more concise, and customers' awareness regarding quality is at its peak. In such scenarios, modern manufacturing systems, environments, and processes are affected directly and indirectly. Moreover, with the complex interaction of machines, processes, and components, knowing how humans will come to terms with such an interaction is imperative. Nevertheless, such a scenario requires optimized human-machine operations and the interconnection of the 'digital' and 'real' worlds to get a bird's eye view of the complete value chain for increased production with greater efficiency using fewer resources. FoF is, therefore, the true application of the paradigms of SM and IM wherein the production facilities and manufacturing plants are striving towards systems that are not only fully connected but also adaptive, autonomous, dependable, intelligent, scalable, collaborative, agile, and sustainable (Moghaddam & Nof, 2017).

Furthermore, the SM traits in the paragraphs mentioned above, along with digitization, bring forth new decision support systems (DSS) that utilize real-time production data and assist in revamping the entire product lifecycle. Concepts such as open value chain, flexible production, human-centered manufacturing, innovative business models (such as crowdsourcing, anything-as-a-service, and symbiotic ecosystems), and local initiatives [from Smart manufacturing leadership coalition (SMLC), Industrial internet consortium (IIC), Japan's e-factory, German Industrie 4.0, intelligent manufacturing (China)] have become the backbone of the FoF (IEC, 2023). Last but not least, with driving technologies of Industry 4.0, such as connectivity and interoperability, FoF will always be prone to security and safety issues. Figure 1.3 shows the total concept for the manufacturing system security for FoF.

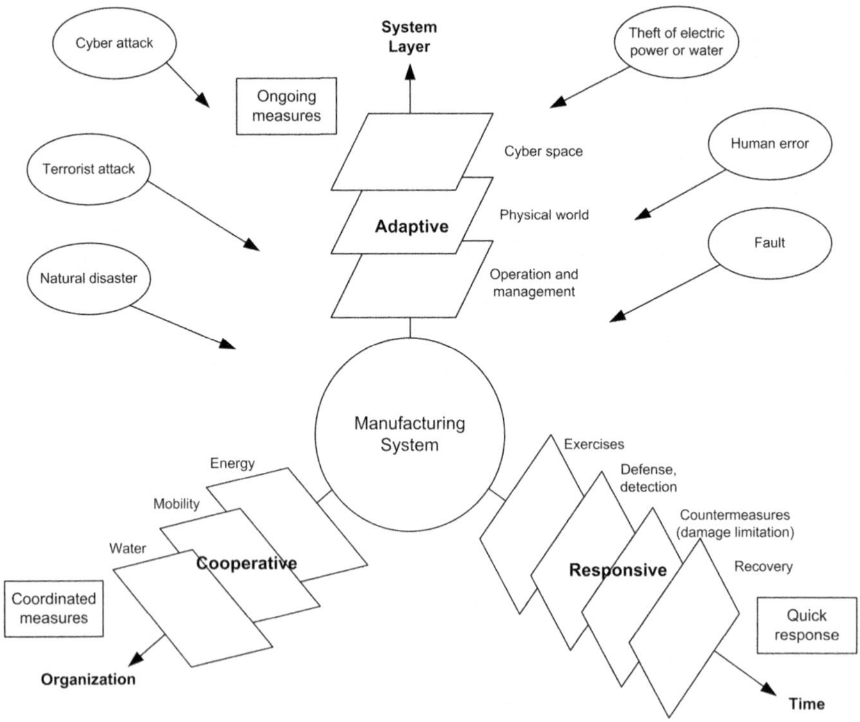

FIGURE 1.3 Total concept for manufacturing system security.

Source: Adapted from IEC (2023).

1.2.3 Key Pillars of Industry 4.0 and SM/IM

Technological advancements common to SM and IM are viewed as a new generation of information technology (IT) comprising Industry 4.0, IoT, big data, cloud and fog computing, AI, ML, digital twins, and CPS (Wang et al., 2021). Vaidya et al. (2018) emphasized nine pillars of Industry 4.0 that are making the manufacturing paradigm smarter and more intelligent. The pillars are listed in Table 1.1.

TABLE 1.1
Nine Pillars of Industry 4.0

Ser	Type of Pillar	Description
1	Big Data and Analytics	Big data and analytics enable the collection and comprehensive evaluation of data from many different sources to support real-time decision-making, optimization of production quality, energy saving, and improvement in equipment service.
2	Autonomous Robots	An autonomous robot is used to perform autonomous production method more precisely and also work in the places where human workers are restricted to work. Autonomous robots can complete given task precisely and intelligently within the given time limit and also focus on safety, flexibility, and versatility.
3	Simulation	Simulations will be used more extensively in plant operations to leverage real-time data to mirror the physical world in a virtual model, which can include machines, products, and humans, thereby reducing machine setup times and increasing quality.
4	System Integration: Horizontal and Vertical System Integration	The paradigm of Industry 4.0 is essentially outlined by three dimensions of integration: (a) horizontal integration across the entire value creation network, (b) vertical integration and networked manufacturing systems, and (c) end-to-end engineering across the entire product life cycle.
5	Industrial IoT (IIoT)	IoT is also known as Internet of Everything (IoE) which consists of Internet of Service (IoS), Internet of Manufacturing Services (IoMs), Internet of People (IoP), an embedded system and Integration of Information and Communication technology (IICT). Collectively, it is called IIoT.
6	Cyber security and CPS	Reliable communications as well as sophisticated identity and access management of machines and users is important for Industry 4.0 to address the issue of cybersecurity threats which increases dramatically with the increased connectivity and use of standard communication protocols.
7	Cloud	Cloud-based IT platform serves as a technical backbone for the connection and communication of manifold elements of the Application Centre Industry 4.0. 'Digital production' is a concept of having the connections of different devices to same cloud to share information to one another and can be extended to set of machines from a shop floor as well as the entire plant.

(Continued)

TABLE 1.1 (Continued)
Nine Pillars of Industry 4.0

Ser	Type of Pillar	Description
8	Additive Manufacturing	Additive-manufacturing methods will be widely used to produce small batches of customized products that offer construction advantages, such as complex, lightweight designs. High-performance, decentralized additive manufacturing systems will reduce transport distances and stock on hand.
9	Augmented Reality	Augmented-reality-based systems can support a variety of services, such as selecting parts in a warehouse and sending repair instructions over mobile devices.

Source: Adapted from Vaidya et al. (2018).

1.2.4 Challenges and Opportunities in SM/IM

A new era leaping towards Industry 4.0 comes with its own challenges creating more research, production, and industry-wide expansion opportunities. A not-so-technical challenge is keeping up with evolving technology. The industrial setups must ensure the employees are capable enough to handle the constantly changing technology, which gives rise to the need for the right workforce in an evolving industrial arrangement. Moreover, work environment safety while dealing with smart machines is also a significant concern. When dealing with a cobot in real-time, the bot must be appropriately trained to maintain a safe working zone with the human interacting with it. Such scenarios have given rise to digital twins and many other AI-based learning techniques to facilitate a safe work environment.

AI, cloud computing, and data analytics have also taken the productivity capabilities of SM by storm. However, it is very hard to judge these concepts' impacts on industrial sustainability ideologies which have given rise to cutting-edge research opportunities to review SM from a sustainability point of view and shift the paradigm towards an automated and sustainability-integrated manufacturing setup (Abubakr et al., 2020). In addition, data handling, security, and sharing are also a concern where many industrial setups are not only faced with the challenge of dataset collection but the smart use of this data to train a system. Cyber security has hence become a significant concern.

This handbook, therefore, attempts to present a comprehensive picture to our readers by discussing the challenges related to manufacturing system design in context of Industry 4.0 highlighting the research opportunities, case studies, and expected future trends.

1.3 FUTURE PERSPECTIVES OF SM/IM IN INDUSTRY 4.0

The future research perspectives for SM/IM in Industry 4.0 will be the core focus of this handbook in chapters to follow with the integration of manufacturing science, sensor technology, and ICT. For a future Industry 4.0 IM system, the perspectives can be many. However, to simplify, the future perspectives can be categorized into 'smart machines' (like smart robots and CPS-enabled smart machine tools), 'smart design' [such as augmenting design with virtual reality (VR)/augmented reality (AR), and coupling computer-aided design (CAD)/computer-aided manufacturing (CAM) software with smart physical prototype systems], 'smart monitoring' via sensors, 'smart control' using cyber-physical production-control systems, 'smart scheduling' through big data analytics, 'cloud-based solutions' [using cloud computing and service-oriented architecture (SOA)], and 'agent-based implementations' utilizing multi-agent technologies.

1.4 CONCLUSION

SM/IM are important paradigms in Industry 4.0. With increasing attention to Industry 4.0, SM/IM is becoming more important in advancing modern industry and economy. SM/IM is considered a critical future perspective in research and application. It provides added value to various products and systems by applying cutting-edge technologies to traditional products in manufacturing and services. In the chapters to follow in the handbook, each pillar of Industry 4.0 is discussed in detail, focusing on SM/IM.

REFERENCES

Abubakr, M., Abbas, A. T., Tomaz, I., Soliman, M. S., Luqman, M., Hegab, H. 2020. Sustainable and smart manufacturing: An integrated approach. *Sustainability*, 12(6): 2280.

International Electrotechnical Commission (IEC). 2023. Smart manufacturing. https://www.iec.ch/smart-manufacturing (accessed on 23 January 2023).

Kagermann, H., Wahlster, W., Helbig J. 2013. Recommendations for implementing the strategic initiative INDUSTRIE 4.0: Securing the future of German manufacturing industry. Final report of the Industrie 4.0 Working Group, Forschungsunion. https://www.din.de/blob/76902/e8cac883f42bf28536e7e8165993f1fd/recommendations-for--implementing-industry-4-0-data.pdf (accessed on 5 June 2023).

Krakauer, J. 1987. *Smart manufacturing with artificial intelligence*. Dearborn: Computer and Automated Systems Association of the Society of Manufacturing Engineers.

Kusiak, A. 1990. *Intelligent manufacturing systems*. Old Tappan: Prentice Hall Press.

Kusiak, A. 2018. Smart manufacturing. *International Journal of Production Research*, 56(1–2): 508–517.

Lasi, H., Fettke, P., Feld, T., Hoffmann, M. 2014. Industry 4.0. *Business and Information Systems Engineering*, 6: 239–242.

Liang, S., Rajora, M., Liu, X., Yue, C., Zou, P., Wang, L. 2018. Intelligent manufacturing systems: A review. *International Journal of Mechanical Engineering and Robotics Research*, 7(3): 324–330.

Lu, Y., Morris, K. C., Frechette, S. 2016. Current standards landscape for smart manufacturing systems. Technical paper. Gaithersburg: National Institute of Standards and Technology. https://www.nist.gov/publications/current-standards-landscape-smart-manufacturing--systems (accessed on 5 June 2023).

Moghaddam, M., Nof, S. Y. 2017. The collaborative factory of the future. *International Journal of Computer Integrated Manufacturing*, 30(1): 23–43.

Okabe, T., Bunce, P., Limoges, R. 2013. Next generation manufacturing systems (NGMS) in the IMS program. In: Okino, N., Tamura, H., Fuji, S. (eds) *Advances in production management systems*, pp. 43–54. Boston: Springer Nature.

Smart Manufacturing Leadership Coalition (SMLC). 2020. Implementing 21st century smart manufacturing. *Schaumburg: Control Global; c2004–2020* [cited 2020 Aug 22]. Available from: https://www.controlglobal.com/whitepapers/2011/110621-smlc-smart-manufacturing/

Thoben, K. D., Wiesner, S., Wuest, T. 2017. Industrie 4.0 and smart manufacturing—A review of research issues and application examples. *International Journal of Automation Technology*, 11(1): 4–16.

Vaidya, S., Ambad, P., Bhosle, S. 2018. Industry 4.0 – A Glimpse. *Procedia Manufacturing*, 20: 233–238.

Wang, B., Tao, F., Fang, X., Liu, C., Liu, Y., Freiheit, T. 2021. Smart manufacturing and Intelligent manufacturing: A comparative review. *Engineering*, 7: 737–757.

Yin, Y., Stecke, K. E., Li, D. 2018. The evolution of production systems from Industry 2.0 through Industry 4.0. *International Journal of Production Research*, 56(1–2): 848–861.

Zhang, Y. F., Zhang, G., Wang, J., Sun, S., Si, S., Yang, T. 2015. Real-time information capturing and integration framework of the Internet of manufacturing things. *International Journal of Computer Integrated Manufacturing*, 28(8): 811–822.

Zheng, P., Wang, H., Sang, Z., Zhong, R. Y., Liu, Y., Liu, C., Mubarok, K., Yu, S., Xu, X. 2018. Smart manufacturing systems for Industry 4.0: Conceptual framework, scenarios, and future perspectives. *Frontiers of Mechanical Engineering*, 13(2): 137–150.

Zhou, J., Li, P., Zhou, Y., Wang, B., Zang, J., Meng, L. 2018. Toward new-generation intelligent manufacturing. *Engineering*, 4(1): 11–20.

2 Computer-Aided Design, Computer-Aided Process Planning, and Computer-Aided Manufacturing

Anas Bin Aqeel and Uzair Khaleeq uz Zaman
Department of Mechatronics Engineering, College of Electrical and Mechanical Engineering, National University of Sciences and Technology, Islamabad, Pakistan

Shahid Aziz
Jeju National University, South Korea

Aamer Ahmed Baqai
Department of Mechanical Engineering, College of Electrical and Mechanical Engineering, National University of Sciences and Technology, Islamabad, Pakistan

2.1 INTRODUCTION TO CAD/CAPP/CAM

Designing and manufacturing a product is customers' demand for better daily life. The process of understanding from basic demand to delivery of any product is generally described through a product life cycle. The product life cycle is typically a complex mechanism, as illustrated in Figure 2.1. The requirement to create a product starts from a customer's demand to market analysis. Once the gap is found, product design, process planning, and product manufacturing occur. Finally, product sales and use/discard stages occur. Even though the initial and final stages of the market, sales, and services are essential management fields, but are outside the technical aspect of this chapter. Therefore, only product design, process planning, and product manufacturing stages will be elaborated in detail. The development of computer technologies has helped engineers in different phases of product development. Hence, computer-aided design (CAD), computer-aided process planning (CAPP), and computer-aided manufacturing (CAM) are the automation capabilities of product design, process planning, and product manufacturing, respectively.

CAD/CAPP/CAM generally represents the integration of CAD, CAPP, and CAM. With the demand for better quality, low cost, and reduced lead time, manufacturers' requirements and competition to introduce new products are on the rise.

DOI: 10.1201/9781003327523-3

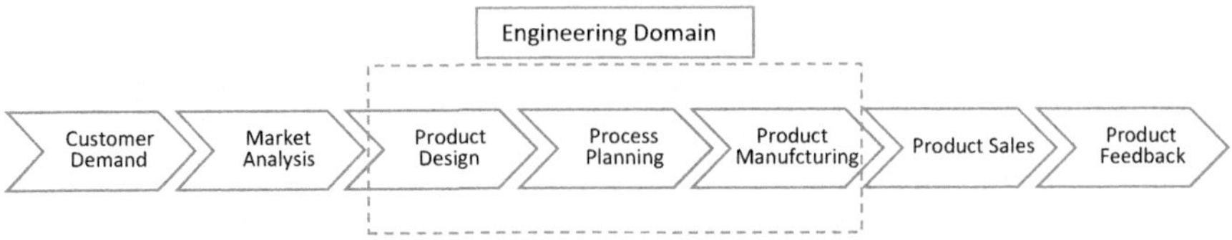

FIGURE 2.1 Schematic of product life cycle (modified from Xue, 2018a).

Product Design CAD	Process Planning CAPP	Product Manufacturing CAM
➢ Geometric modelling ➢ Engineering analysis ➢ Simulation ➢ Scientific computing ➢ Graphics ➢ Engineering database ➢ Data communication ➢ Reverse engineering	➢ Rough cast design ➢ Machining process selection ➢ Operation design ➢ Routing design ➢ Ratio of man hour ➢ Tooling ➢ Fixture and jigs	➢ NC programming ➢ Tool path planning ➢ Cutting data file ➢ Simulation of tool path ➢ NC code verification ➢ Check out and trial manufacture ➢ 3D printing

FIGURE 2.2 Tasks for product design, process planning and product manufacturing stages (modified from Xue, 2018a).

Human labor is getting replaced by robots, and with the introduction of Industry 4.0, a seamless integration using CAPP bridges the two essential tools of design and manufacturing known as CAD and CAM. Numerous tasks are oriented to accomplish in each stage of the engineering domain, as shown in Figure 2.2.

2.2 COMPUTER-AIDED DESIGN

Since 1980, computer-aided designs have been primarily used to illustrate 2D engineering drawings and 3D modeling of physical components. Moreover, it is also used to define engineering processes from conceptual designs to product blueprints, along with static and dynamic analysis of assemblies for better manufacturing (Xue, 2018a). The use of computers by designers in the 1960s brought in the 3D construction and numerical control (NC) programming areas within the aerospace and automotive industries. Since computers were overpriced, the first commercial CAD applications were utilized in the automotive and aerospace industry, which later commercialized to personal computers in the 1970s. CAD implementation evolved from 2D drawings (identical to hand-made drawings) in the 1970s to solid modeling in the 1980s. A few decisive products related to solid modeling packages in 1981 were Romulus (ShapeData), Uni-Solid (Unigraphics), and surface modeler CATIA (Dassault Systemes). Autodesk founded a 2D system titled AutoCAD in 1982, with the next breakthrough of Pro/Engineer using feature-based modeling method in 1988. In the late 1980s and early 1990s, the development of geometrically and topologically consistent 3D object systems, such as B-rep solid modeling kernel, ACIS (Spatial

Technology Inc.), and Parasolid (Shape Data), was obtained. SolidWorks in 1995, Solid Edge in 1996, and Iron CAD in 1998 were then introduced. Subsequently, at the beginning of the 21st century, CAD packages introduced functional modules (Kasik et al., 2005).

2.2.1 Mathematical Models in 3D CAD Systems

In product development, applications required representations of geometric entities using mathematical models. Each application is represented based on different geometric orientations, with few requiring simple entities such as lines and arcs and others requiring curves and surfaces. The few significant geometric modeling systems are detailed as follows.

2.2.1.1 Wireframe

A wireframe is a geometric model system derived from designers representing the 3D shape of solid objects using metal wires. Surfaces are outlined by solid lines, including opposite and normally hidden views. The wireframe is the least complex method in 3D representation in comparison to solid and surface modeling, as depicted in Figure 2.3.

Wireframe models are known for defining complex solid objects. They are created by indicating each edge where the object's vertices through lines or curves are joined, or multiple mathematical surfaces meet. The straight line and arcs are represented through four unique curves in matrix form: B-Spline curve, Hermilton and Ferguson curve, Bezier curve, and NURBS curve, as shown in Equations 2.1–2.4 (Xue, 2018d).

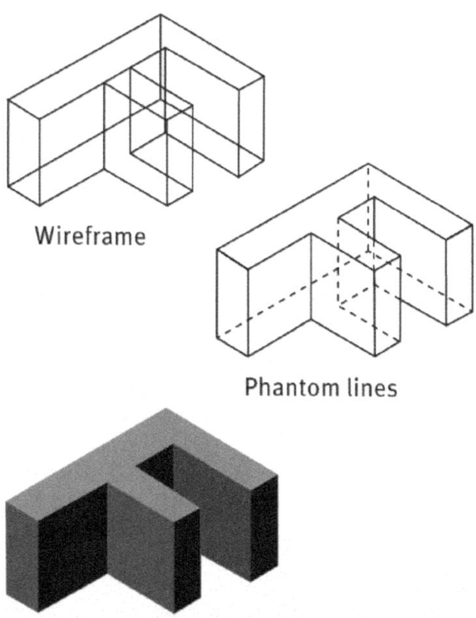

FIGURE 2.3 Wireframe model, phantom lines and surface/solid model.

$$r_i(u) = \begin{bmatrix}1 & u & u^2 & u^3\end{bmatrix} \frac{1}{6}\begin{bmatrix}1 & 4 & 1 & 0\\ -3 & 0 & 3 & 0\\ 3 & -6 & 3 & 0\\ -1 & 3 & -3 & 1\end{bmatrix}\begin{bmatrix}V_i\\ V_{i+1}\\ V_{i+2}\\ V_{i+3}\end{bmatrix} \tag{2.1}$$

where $0 \le u \le 1$

$$r(u) = \begin{bmatrix}1 & u & u^2 & u^3\end{bmatrix}\begin{bmatrix}1 & 0 & 0 & 0\\ 0 & 0 & 1 & 0\\ -3 & 3 & 2 & -1\\ 2 & -2 & 1 & 1\end{bmatrix}\begin{bmatrix}r(0)\\ r(1)\\ r'(0)\\ r'(1)\end{bmatrix} \tag{2.2}$$

where $0 \le u \le 1$

$$r_i(u) = \begin{bmatrix}1 & u & u^2 & u^3\end{bmatrix}\begin{bmatrix}1 & 0 & 0 & 0\\ -3 & 3 & 0 & 0\\ 3 & -6 & 3 & 0\\ -1 & 3 & -3 & 1\end{bmatrix}\begin{bmatrix}V_0\\ V_1\\ V_2\\ V_3\end{bmatrix} \tag{2.3}$$

where $0 \le u \le 1$

$$\begin{aligned} R(u) &= \sum_{i=0}^{n} V_i^{w} \cdot N_{i,k}(u) \\ r(u) = H\{R(u)\} &= \frac{\sum_{i=0}^{n} W_i \cdot V_i \cdot N_{i,k}(u)}{\sum_{i=0}^{n} W_i N_{i,k}(u)} \end{aligned} \tag{2.4}$$

where $0 \le u \le 1$

2.2.1.2 Surface

Surface modeling is a mathematical technique to represent solid objects and is widely used for CAD illustrations and architectural renderings. It's more complex than a wireframe for object representation. However, on the contrary, it is less sophisticated than solid modeling. Free-form surfaces linked to the surface modeling in matrix mode are represented as Coons surface, Bezier surface, B-Spline surface, and NURBS surface, as shown in Equations 2.5–2.8 (Xue, 2018d).

$$\begin{aligned} r(u, w) &= r_1 + r_2 + r_3 \\ r(u, w) &= [-1 \quad 1 \quad -u \quad u]\begin{bmatrix}0 & r(u, 0) & r(u, 1)\\ r(0, w) & r(0, w) & r(0, 1)\\ r(1, w) & r(1, w) & r(1, 1)\end{bmatrix}\begin{bmatrix}-1\\ 1-w\\ w\end{bmatrix} \\ r(u, w) &= [-1 \quad 1 \quad -u \quad u]\begin{bmatrix}0 & r(u, 0) & r(u, 1)\\ r(0, w) & r(0, 0) & r(0, 1)\\ r(1, w) & r(1, 0) & r(1, 1)\end{bmatrix}\begin{bmatrix}-1\\ 1-w\\ w\end{bmatrix} \end{aligned} \tag{2.5}$$

where $0 \le u \le 1$ and $0 \le w \le 1$

$$V = \begin{bmatrix} V_{0,0} & V_{0,1} & V_{0,2} & V_{0,3} \\ V_{1,0} & V_{1,1} & V_{1,2} & V_{1,3} \\ V_{2,0} & V_{2,1} & V_{2,2} & V_{2,3} \\ V_{3,0} & V_{3,1} & V_{3,2} & V_{3,3} \end{bmatrix}$$

$$r(u,w) = \begin{bmatrix} 1 & u & u^2 & u^3 \end{bmatrix} \begin{bmatrix} 1 & 0 & 0 & 0 \\ -3 & 3 & 0 & 0 \\ 3 & -6 & 3 & 0 \\ -1 & 3 & -3 & 1 \end{bmatrix} \begin{bmatrix} V_{0,0} & V_{0,1} & V_{0,2} & V_{0,3} \\ V_{1,0} & V_{1,1} & V_{1,2} & V_{1,3} \\ V_{2,0} & V_{2,1} & V_{2,2} & V_{2,3} \\ V_{3,0} & V_{3,1} & V_{3,2} & V_{3,3} \end{bmatrix} \begin{bmatrix} 1 & -3 & 3 & -1 \\ 0 & 3 & -6 & 3 \\ 0 & 0 & 3 & -3 \\ 0 & 0 & 0 & 1 \end{bmatrix} \begin{bmatrix} 1 \\ w \\ w^2 \\ w^3 \end{bmatrix} \tag{2.6}$$

where $0 \le u \le 1$ and $0 \le w \le 1$

$$V = \begin{bmatrix} V_{0,0} & V_{0,1} & V_{0,2} & V_{0,3} \\ V_{1,0} & V_{1,1} & V_{1,2} & V_{1,3} \\ V_{2,0} & V_{2,1} & V_{2,2} & V_{2,3} \\ V_{3,0} & V_{3,1} & V_{3,2} & V_{3,3} \end{bmatrix}$$

$$r(u,w) = \begin{bmatrix} 1 & u & u^2 & u^3 \end{bmatrix} \frac{1}{6} \begin{bmatrix} 1 & 4 & 1 & 0 \\ -3 & 0 & 3 & 0 \\ 3 & -6 & 3 & 0 \\ -1 & 3 & -3 & 1 \end{bmatrix} \begin{bmatrix} V_{0,0} & V_{0,1} & V_{0,2} & V_{0,3} \\ V_{1,0} & V_{1,1} & V_{1,2} & V_{1,3} \\ V_{2,0} & V_{2,1} & V_{2,2} & V_{2,3} \\ V_{3,0} & V_{3,1} & V_{3,2} & V_{3,3} \end{bmatrix} \begin{bmatrix} 1 & -3 & 3 & -1 \\ 4 & 0 & -6 & 3 \\ 1 & 3 & 3 & -3 \\ 0 & 0 & 0 & 1 \end{bmatrix} \begin{bmatrix} 1 \\ w \\ w^2 \\ w^3 \end{bmatrix} \tag{2.7}$$

where $0 \le u \le 1$ and $0 \le w \le 1$

$$r(u,w) = \frac{\sum_{i=0}^{n_u} \sum_{j=0}^{n_w} W_{i,j} \cdot V_{i,j} \cdot N_{i,k_u}(u) \cdot N_{j,k_w}(w)}{\sum_{i=0}^{n_u} \sum_{j=0}^{n_w} V_{i,j} \cdot N_{i,k_u}(u) \cdot N_{j,k_w}(w)}$$
$$r(u,w) = \frac{\sum_{i=0}^{n_u} \sum_{j=0}^{n_w} W_{i,j} \cdot V_{i,j} \cdot N_{i,k_u}(u) \cdot N_{j,k_w}(w)}{\sum_{i=0}^{n_u} \sum_{j=0}^{n_w} W_{i,j} \cdot N_{i,k_u}(u) \cdot N_{j,k_w}(w)} \tag{2.8}$$

where $0 \le u \le 1$ and $0 \le w \le 1$

2.2.2 Software Packages in CAD Systems

The development of CAD technology established numerous CAD packages available in the market for engineers and designers for various products. Here, only a few most frequently used CAD software packages are briefly elaborated.

2.2.2.1 AutoCAD

AutoCAD was released initially in 1979 with the name "Interact CAD" and is referred to as the documented name of MicroCAD by Autodesk company. This commercial software was developed for 2D and 3D CAD and drafting and was commercially introduced in 1982 on microcomputers. Its original file format is .dwg, with .dxf as an interchangeable format. For 2D drawing exchange, the DXF file format has developed into an interoperable CAD data standard.

2.2.2.2 SolidWorks

SolidWorks Corporation, founded in 1993 by an MIT graduate, is a solid modeling software that Dassault later acquired in 1997. It utilizes parametric feature-based modeling to create models and assemblies with the software established using Parasolid-kernel.

In SolidWorks, parameters relate to constraints to capture shape or geometry standards. These constraints can be numeric, such as circle diameters and lines, or geometric parameters, like a concentric, parallel, horizontal, tangent, etc. Relations could be used here amongst the parameters to specify design intent. Likewise, features relate to building blocks, such as shapes and operations to construct a part. Shape-based features relate to 2D or 3D sketches like holes, slots, etc., which are later extruded or cut to add or delete the part—however, operation-based features include chamfer, shells, fillets, etc. In SolidWorks, dimensions are adapted to define the size and locations of the shape and relations such as concentric, parallel, horizontal, and tangent. Likewise, the assembly model uses mates to determine similar conditions and their restrictions in movement.

SolidWorks files use Microsoft structured storage files such as SLDDRW known drawing files, SLDPRT known part files, and SLDASM known assembly files.

2.2.2.3 PTC Creo or Pro/Engineer

PTC Creo, previously known as Pro/Engineer, is a feature-based solid modeling software with ten collaborative application provider suites. It's the first rule-based constraint, parametric 3D modeler in the industry. Optimization of design and product development can be achieved using this approach as it captures product behavior using dimensions, parameters, features, and their relationship. CREO provides a complete platform for product design, analysis, and manufacturing. Its data is interchangeable without conversion between CAD, Computer Aided Engineering (CAE), and CAM and can monitor the Bill of Materials (BOM) of the product.

2.2.2.4 Siemens-NX or Unigraphics UG

NX, earlier known as NX Unigraphics or UG, is a leading sophisticated CAD/CAM/CAE software package developed by Unigraphics at first and later secured by Siemens PLM Software. A few essential functions of its CAD module include parametric solid modeling, free-form surface modeling, product, and manufacturing information (PMI), reverse engineering, and sheet metal design. Likewise, the key functions in CAE and CAM modules comprise stress analysis, kinematic analysis, computational fluid dynamics (CFD) analysis, thermal analysis, and numerical control programming.

2.2.2.5 CAXA

CAXA is a Chinese industrial software that provides 2D and 3D design software, MES, product life cycle management (PLM), and a platform of industrial cloud service. Its offices in Beijing, Nanjing, and United States are part of many CAD and CAPP technical standards. They are famous for providing PLM solutions to digital design and manufacturing, such as CAD designs, process planning, data management, and manufacturing processes [CAM, Manufacturing Execution System (MES), and Direct Numerical Control (DNC)].

2.3 PROCESS PLANNING

'Process planning' is the term accepted as a bridge connecting the CAD designs to its manufacturing and is pivotal for the effective integration of CAD into CAM. According to the Society of Manufacturing Engineering (SME), "Process planning is a systematic determination of the methods by which a product is to be manufactured economically and competitively" (Xu, 2009b). Process planning is the most viable option for selecting manufacturing processes and machines to obtain the required operations to manufacture a component. Figure 2.4 shows the diagram of process planning (Xue, 2018b). Traditionally, based on design specifications and available resources, manufacturing process experts were utilized for process planning. However, their different experiences have led to inconsistencies in process planning. The knowledge of detailed manufacturing processes is required to acquire consistent and optimized planning, which has led to the development and evolution of CAPP (Park, 2003; Vidal et al., 2005).

Singh (1996) defined the basic steps in developing a process plan as listed below. However, the sequence to follow amongst the steps is not strictly linear, as these could be interdependent.

- Analysis of part requirements.
- Workpiece/material selection.
- Determination of manufacturing operations and sequences.
- Machine tool selection.
- Tools, jigs, and inspection equipment selection.
- Machining (cutting speed, feed, and depth of cut) and manufacturing time (setup time, lead time, and process time) selection conditions.

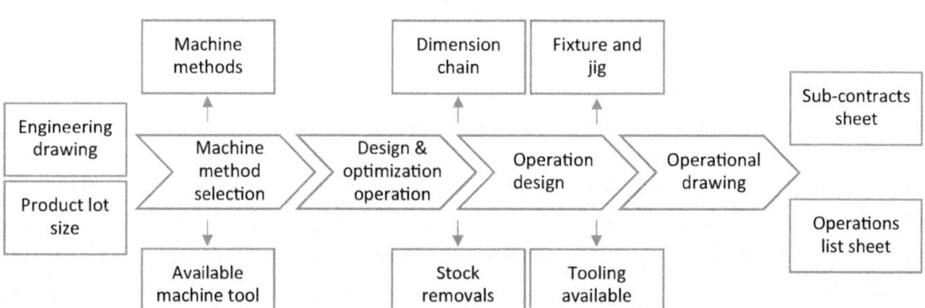

FIGURE 2.4 Diagram of process planning (modified from Xue, 2018b).

During the process, primary CAPP operations relate to the selection of machine tools, cutting tools, machining operations, and time and cost associated with the part manufacturing. Other important parameters that work as a function of CAPP operations to achieve an optimized algorithm are the available facilities and the quality standard required for production.

The concept of developing computerized process planning dates to the mid-1960s. In 1965, Niebel presented the idea to implement process planning through computers with the first CAPP system, AUTO PROS, launched by Norway in 1969 and later commercialized in 1973 (Niebel, 1967). US company named Computer-Aided Manufacturing International (CAM-I) established the most significant CAM-I automated process planning system in 1976 (Butterfield et al., 1986). Later, the area was covered by numerous research and review articles (Alting & Zhang, 1989; Cay & Chassapis, 1997; Zhang & Xie, 2007).

2.4 PRINCIPAL PROCESS PLANNING APPROACHES

Process planning is categorized into two basic principal approaches: manual experience-based planning and computer-aided process planning.

2.4.1 Manual Experience-Based Planning Method

Manual experience-based planning is established on the planner's experience and knowledge of production capabilities, equipment, tools, and manufacturing processes. It varies with the planner's approach to solving the problem and consumes more time. Generally, to detail a plan for producing parts from raw materials, planners use part lists or bill of materials (BOMs) to communicate manufacturing requirements and the order in which the product should be built. The activities in the development of process planning include the following steps:

- Reading engineering drawings to analyze machining requirements.
- Selection of machining methods and optimizing routing plan.
- Selection of machine equipment and design fixture.
- Selection of machine parameters (speed, feed, and depth of cut).
- Calculation of dimensions and tolerances.
- Calculation of manufacturing time (process time, lead time, and setup time).
- Filling out required forms.

A routing lineup of the operations is required for sequenced product manufacturing. Each operation identifies specific information such as work center, machine run, tools and inspection, labor working hours, and setup time standards. Manufacturing routings are established to reduce lead time with BOM to determine how, where, and when the quantities are required for work order completion.

2.4.2 Computer-Aided Process Planning Method

As briefly explained, the CAPP method has become an optimal planning system in a synchronous engineering environment. It is essential to integrate a process planning system into an organizational flow. Suppose the design changes are made. In that case, the CAPP system should be able to update the cost estimates, or if a machine breakdown occurs on the floor, the CAPP system must provide the most advanced alternative economical solution to the situation. In comparison to the experience-based planning method, CAPP helps in achieving the following:

- Production of consistent process plans.
- Reduction in cost and lead-time
- Reduction in planner's skillset requirement.
- Increase in production capacity.

During the early years, the variant approach was the basis for developing CAPP systems. Since 1980, semi generative and generative approaches have been utilized for CAPP. Later, artificial intelligence (AI), also termed a knowledge-based or expert system, was proposed. Jianbin Xue (2018c) defined the evolutionary path of computerized process planning into six stages as given below:

- Manual classification: standard process planning
- Computer-maintained process planning
- Variant CAPP
- Generative CAPP
- Dynamic, generative CAPP
- Hybrid CAPP

Manufacturers classify the parts into families in the first stage of manual process planning and develop some standardized process plans. During this manual system, parts are retrieved, marked up, and retyped. This manual stage results in a productivity increase but doesn't improve the production processes. The second stage entails to initial evolution of the CAPP system by storing a process plan electronically. Initially, CAPP evolves into a variant CAPP system, a group technology (GT) coding and classification approach to determine part attributes or parameters. Around 90% of the baseline process plan selection for a part family is entered manually and stored in the system, with 10% effort added by the planner to fine-tune the process. In the fourth stage of generative CAPP, process planning decision rules are built, which operate on feature-based technology coding resulting in nominal manual interaction. With the evolution of the CAPP system being more generative to obtain automatic part classification and other design parameters, AI is utilized to fully integrate CAPP into Computer Integrated Manufacturing (CIM) environment. In this fifth stage of dynamic generative CAPP, numerous parameters in the development of process plans, such as plant capacity, tools availability, equipment loadings, maintenance timings, etc., are considered. This dynamic stage varies depending on time variation and technical requirements. The sixth stage of hybrid CAPP systems is designed to optimize the process planning by implementing AI, neural networks, and other advanced algorithms. This CAPP system helps integrate CAPP into CAD/CAM intelligently by incorporating multiple algorithms.

2.5 CLASSIFICATION OF COMPUTER-AIDED PROCESS PLANNING SYSTEMS

The CAPP can be classified into five types: variant CAPP method, generative CAPP method, expert-system-based CAPP method, neural-network-based CAPP method, and hybrid CAPP method (Xu, 2009a), with variant and generative as the significant types.

2.5.1 Variant CAPP

The variant CAPP approach is relatively straightforward based on a computerized extension of the manual approach. A process plan is created for a new part through changes made to the existing plan of a similar part by recalling, identifying, and retrieving the data/code. Generally, these process plans are associated with families of parts and are known as "master parts". Part classification and coding schemes are utilized to group part families with a developed standard process plan. The variant CAPP system allows flexibility to edit and change according to requirements as well. Initially, the challenge is related to the selection of classification and coding structure standards and the development of baseline standards for part families. Numerous methods have been utilized, with the most used as GT. Another significant method is termed "Feature Technology" (FT).

2.5.1.1 Group Technology

Group technology (GT) method capitalizes to obtain part similarities based on design and manufacturing for its family classification. Suppose one company disassembles around 2000 parts into individual components and identifies that 90% of the parts can be divided into five major families. Moreover, the basic components of a pump, for instance, can be divided into seals, motors, housing, flanges, and shafts. The major industrial coding systems used for GT are Metal Institute Classification (MICLASS) System, KK-3 System, and Optiz Coding System.

a. **MICLASS System:** In the 1970s, TNO, a Netherlands-based organization, developed a MICLASS system, which is now renamed MUTICLASS. The system was developed to automate and standardize designs, production, and management functions involving 30 digits. It's a survey-based interactive session with a computer to assign part code with codes for the machine part, as shown in Table 2.1.
b. **The KK-3 System** is a 21-digit decimal system that generates code for classifying dimensions and dimensional ratios. It is used to classify rotary components as listed in Table 2.2.

c. **Optiz Coding System:** It is the first extensive hybrid coding system named after a German professor (Zehtaban et al., 2016). It consists of a maximum of fourteen digits and is categorized into three sections. The first section with the first five digits was termed "from code" describing part shape and topology of the part. Later four digits were termed as "supplementary code" representing the dimension, precision, raw material shape, and accuracy of part. And the third section, "secondary code", consists of four alphabets, A, B, C, and D, that define the organization customization with organization-specific information. The structure of the Optiz coding system is shown in Figure 2.5 (MACCONNELL, 1970). An example of a rotating and non-rotating part for Optiz code is shown in Figure 2.6.

TABLE 2.1
MICLASS Code for Machined Part (Xue, 2018b)

Pos.	Code	Items	Meaning
0	1	Prefix code	Machine parts/sheet metal code
1	2		Round part, all diameters visible from one end
2	7	Basic form	One machine outside diameter visible from each end and a groove
3	0		No inside diameter
4	0	General Manufacturing operations	No secondary elements (holes, slots, curved faces etc.)
5	4	Function	Axle
6	1		
7	1		Outside diameter range: 0.251–0.371 in
8	0		
9	5	Dimensions	Length: 12.51–13.00 in
10	7		
11	0		Inside diameter: non
12	0		
13	8	Tolerances	Geometric concentricity tolerance < 0.01
14	2	Material chemistry	High strength steel
15	8		
16	2	Material shape	Round bar stock
17	1	Production quality	26–57
18	0	Tool axis complexity	No secondary element orientation
19	3		Thread diameter 5/16 in
20	5	User defined	Thread per inch: 18
….	….		….
30	….		….

TABLE 2.2
Structure of KK-3 System for Rotational Components (Xue, 2018b)

Digit	Items (Rotational Component)		
1	Part name		General classification
2			Detail classification
3	Materials		General classification
4			Detail classification
5	Major dimensions		Length
6			Diameter
7	Primary shapes and ratio of major dimensions		
8	Shape details and kind processes	External surfaces	External surface and outer primary shape
9			Concentric screw-threaded parts
10			Functional cut-off parts
11			Extraordinary shape parts
12			Forming
13			Cylindrical surface
14		Internal surface	Internal primary shape
15			Internal curved shape
16			Internal flat and cylindrical shape
17		End surface	
18		Nonconcentric holes	Regularly located holes
19			Special holes
20		Noncutting process	
21	Accuracy		

In GT, part classification can be done using various coding methods. For the selected part feature, an alphanumeric value for each part is adopted and classified through part grouping in agreement with the code value. The code structures used are:

- Hierarchical: Interpretation of each symbol
- Chain: The interpretation of each symbol is fixed.
- Hybrid: Consolidation of hierarchical and chain structures.

Besides the GT method, the variant approach also capitalizes on FT, fuzzy clustering method, and production flow to obtain a part classification.

2.5.1.2 Feature Technology

For the classification of the part family, variant CAPP also utilizes FT, fuzzy clustering, and production flow methods. For manufacturing parts, a feature-based system uses form features to describe its component geometry and functionality related to process planning. A part's geometry is outlined by the user through its shape and features to be

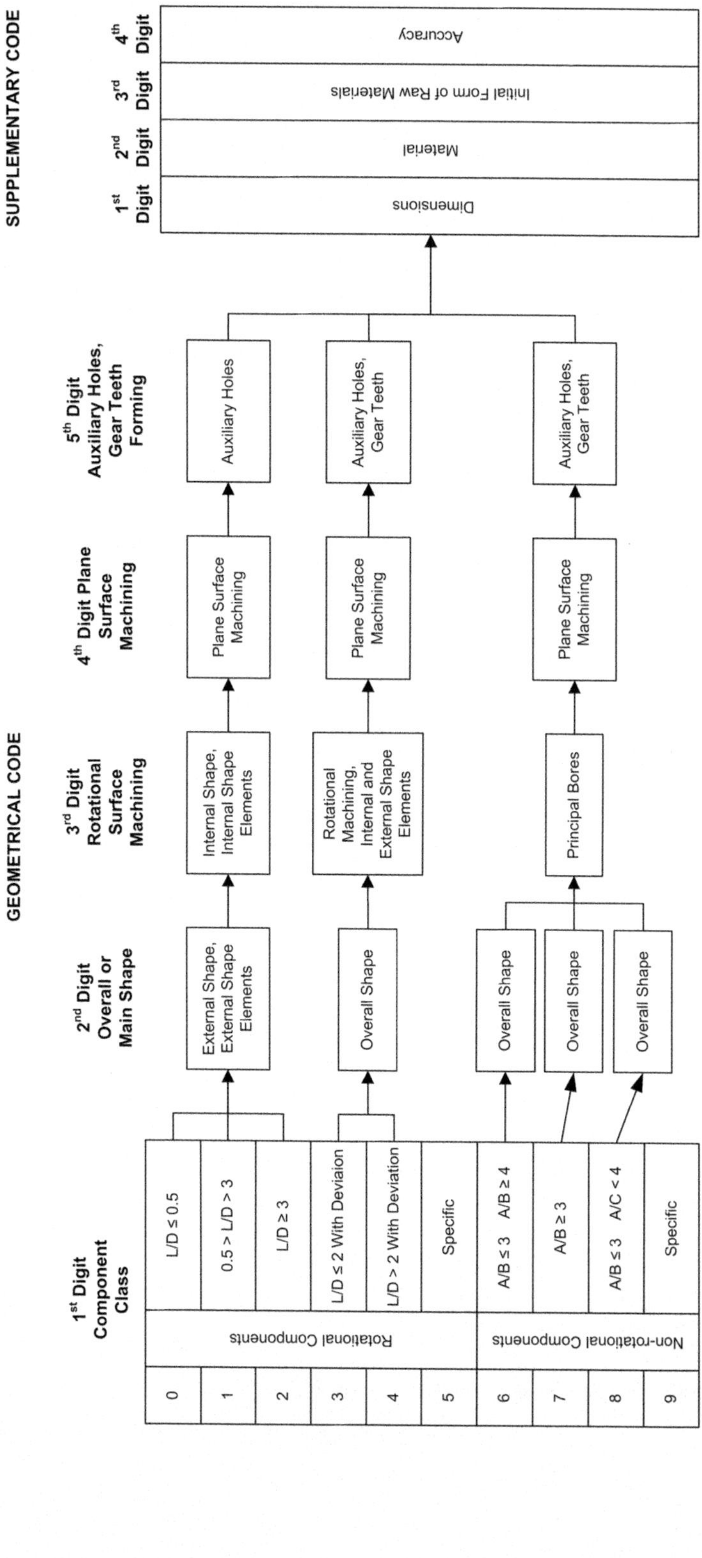

FIGURE 2.5 Optiz coding system structure (adapted from Holland et al., 2002).

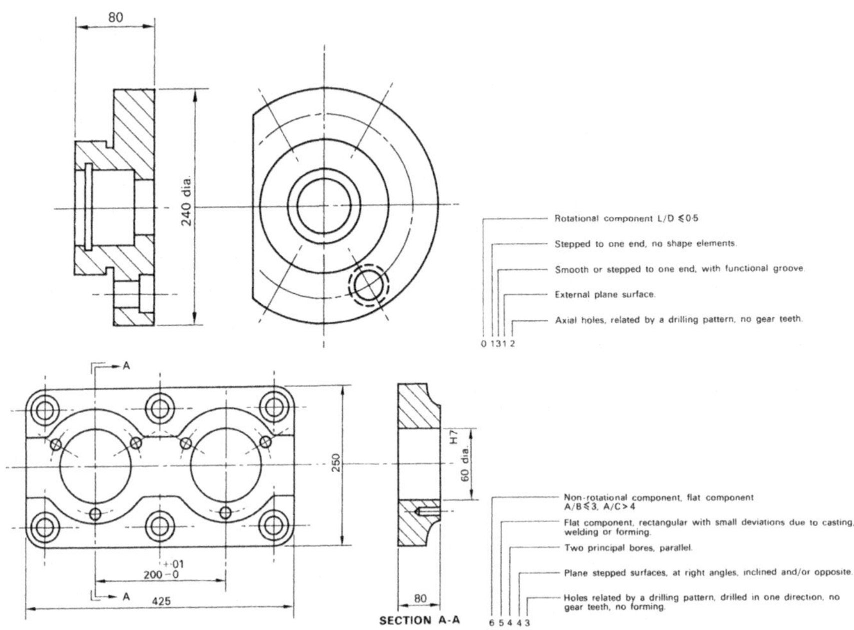

FIGURE 2.6 Rotating (013124279) and non-rotating part codes (654425078) (adapted from Macconnell, 1970).

machined to generate the component's database. The process of selection includes operation sequencing, tool selection, and planning. Moreover, feature interaction is used during the operation sequencing to acquire optimized and consistent process plans. FT allows CAD/CAM systems to bridge the missing gap, which further allows the CAM-I system to categorize part features for three major components: rotational, sheet metal, and prismatic. This technology benefits engineers and manufacturers by enabling them to perform tasks separately using the feature language, such as holes, pockets, etc.

Despite the similarities between variant process planning and manual experience-based planning, variant process planning's ability to utilize information management due to computational capabilities is an edge. Its advantages include:

- Reduction in labor and time demand due to dynamic processing and evaluation of intricate activities and decisions.
- Beneficial to small and medium scale factories due to low development cost and time.
- Constructive to company needs by virtue of standardized procedure utilized by planner's manufacturing knowledge.

The disadvantages include the following:

- Challenging to endure flexibility in editing.

- Challenging to sufficiently accommodate numerous combinations of materials, precision, sizes, machine loading, and different processing sequences.

2.5.2 Generative CAPP

The generative CAPP system is an automatic process planned using algorithms, decision logic, geometry-based data, and formulas. Numerous CAPP systems are developed using generative CAPP approaches, such as knowledge-based CAPP expert systems, AI systems, neural network-based CAPP systems, etc.

Knowledge-based CAPP system is the most frequently used system that stores the knowledge of a computer program and utilizes that knowledge to solve the problem intelligently. Such computer programs act as human experts by simulating the decision process, and hence, decision rule formulation is the first key point for generative CAPP implementation. These decision-based rules are formulated through decision trees, "if-then" programming statements, or AI approaches. Since the nature of part affects the level of complexity in decision rule formulation and its success rate, sheet metal part fabrication and simpler parts have been the area of focus for generative CAPP systems with recent significant progress made in assembly systems such as Printed Circuit Board (PCB)'s assembly.

A generative CAPP system is an automatic plan generated from scratch, utilizing computer programs without expert assistance. The optimal process plan is created by a computer program based on technical and logical decisions. However, to achieve these optimal parameters, one must enter numerous input parameters such as part geometry, material, etc. These operations constitute of the following steps:

- Input of part information
- Selection of machining methods
- Creation, modification, and expansion of the process plan
- Operation of design

The most used knowledge-based approach includes two main steps: knowledge representation and inference mechanism. Knowledge representation is to perform a task that incorporates the real-world problem to computer implementation, for example, sorting a hole in a part with the second step defining the attributes of the hole such as depth, length, and diameter. The inference mechanism, on the other hand, allows the computer to find a solution based on the IF-THEN statement. For instance, IF the hole is found, THEN a drill action is done on the part.

2.6 COMPUTER-AIDED MANUFACTURING

Like computational assistance is involved in designing and process planning areas, computational power is a source of assistance in manufacturing activities. CAM can be elaborated in both general and narrow viewpoints. In general, computational applications are used to counter manufacturing problems such as machine control, planning, storage, plant communication, mechanical and electrical testing, machine monitoring, etc. Hence, CAM is a direct or indirect computer application implementation of modern manufacturing plants such as machine control, material control, production scheduling, managing schedules, inventory control, etc. On the contrary, computer numerically controlled (CNC) machining is the narrow term used for CAM in manufacturing plants producing mechanical and electrical equipment. CAM is also termed computer-aided machining. Using a computer software, machine tools and machinery are controlled in CAM for faster production and precise tooling for dimensionality and material consistency.

Floor Automation evolution dates to the introduction of numerical control (NC), which resulted in a significant increase in production through the development of direct numerical control (DNC) (Gao et al., 2004). For the purpose of sophisticated manufacturing and saving costly tools from damaging, NC computer software packages (G-code) were introduced to avoid machine and tool damage. These are currently used in most manufacturing industries for high-quality production.

2.7 INTEGRATED CAD/CAPP/CAM—CASE STUDIES

This section highlights the case studies associated with integrating CAD/CAPP/CAM processes. The case study selection is based on the evolution of the integration process between CAD/CAPP/CAM systems, with the oldest study from 2002 and the latest published recently.

2.7.1 Case Study—Generative CAPP System for Aerospace Prismatic Component

The development of CAPP has been highlighted as a critical link between CAD and CAM. CAPP system in the past was divided into two basic approaches, with the variant approach termed as a new process plan generation technique using existing standard process plan using stored data. Whereas the generative approach devised a new process plan from scratch. The basic requirement of the generative approach is a feature recognition technology, which interprets the model in manufacturable features. However, with the lack of feature recognition technology, a third-category semi-generative approach has been the main research highlight for the past few decades.

Sadaiah et al. (2002) designed and developed a generative CAPP system for prismatic components and termed it PSG-CAPP. PSG-CAPP system was categorized into three modules, with the first related to feature extraction (Gindy et al., 1993; Han et al., 2000; J. Y. Lee & Kim, 1998; Subrahmanyam & Wonzy, 1995) and the second and third dealt with planning the setup, cutting tool and parameter selection, machine selection and process plan sheet generation. The authors devised the CAPP system with Visual Basic 6.0 and Oracle 7.3 at the front and back ends, respectively. The system was further modelled using SolidWorks with the Visual Basic program for feature extraction. The characteristics that highlight this CAPP system were its linkage to the CAD module and the automatic extraction of extensive features preceding its process planning. The information flow sequence of the proposed CAPP system is shown in Figure 2.7. The developed system was tested for aerospace components and its ability to handle a range of prismatic components, and their process plan generation was also exhibited. Figure 2.8 shows the sample component of 2D and 3D drawings.

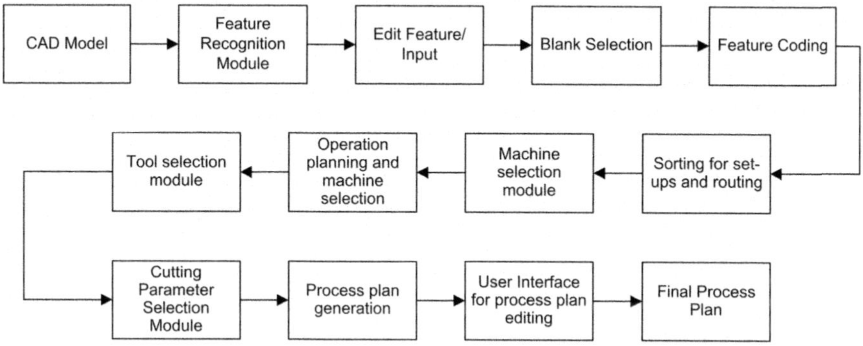

FIGURE 2.7 Information flow sequence of proposed CAPP system (regenerated from Sadaiah et al., 2002).

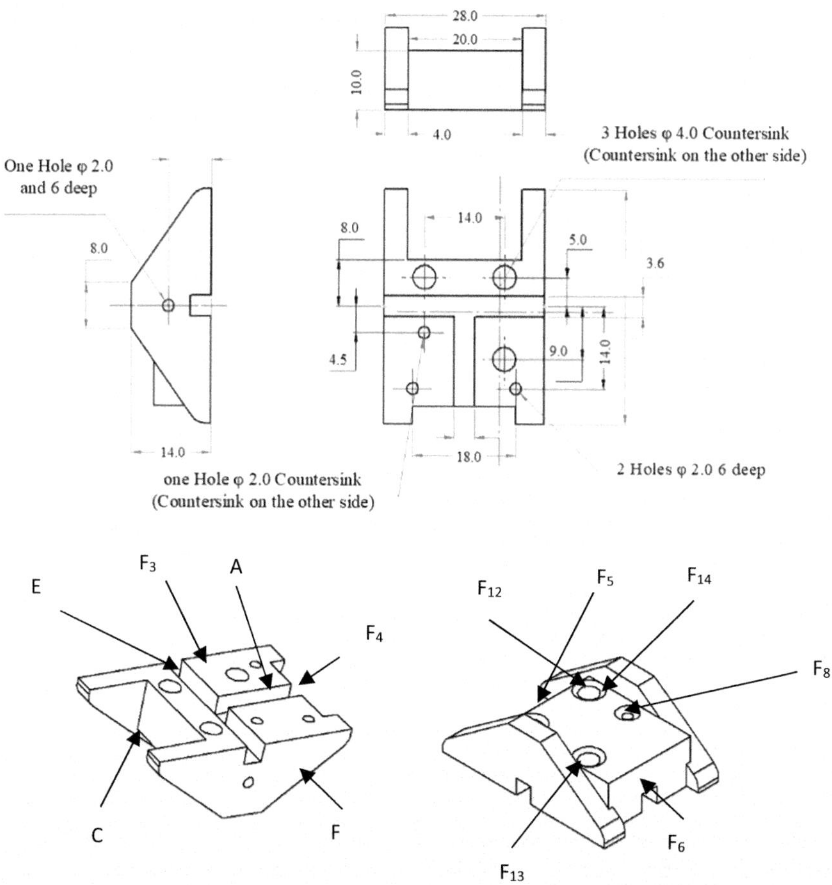

FIGURE 2.8 Sample component 2D and 3D drawings (adapted from Sadaiah et al., 2002).

2.7.2 Case Study—Intelligent CAD/CAPP System for Hydraulic Press Brake

Previously, numerous studies have shown different integration techniques between CAD/CAPP. This case study presented a platform based on Elementary Machining Features (EMF) and a smart approach for setup planning and operation sequence (Borojević et al., 2022). An EMF-based platform as an extension of the SolidWorks Program, which is based on an Application Programming Interface (API), was developed. The structure of the integrated intelligent CAD/CAPP platform based on EMF and Genetic Algorithm (GA) is shown in Figure 2.9. In this new approach, EMF is made intelligent, allowing the inclusion of critical information to EMF, which was later utilized in operation sequencing. This further established a link between part design and process planning phases. Elementary machining operations' list based on production rules is then created from existing and new knowledge gained from the EMFs design phase. The latest information gained is novel in obtaining automatic identification of information for the process planning phase. This comprehensive approach is verified using body of a hydraulic cylinder associated with a hydraulic press brake. Its design is made through the developed structure and verified by new program tools based on API. Furthermore, the solid model was generated after selecting the generic (parent) EMF and its child EMF with appropriate parameters, as shown in Figure 2.10. For the model, among 52 child EMFs, 39 and 13 were used for drilling and milling, respectively. Moreover, the body of the hydraulic cylinder was designed using a solid model comprising 3 and 7 generic EMFs for drilling and milling, respectively.

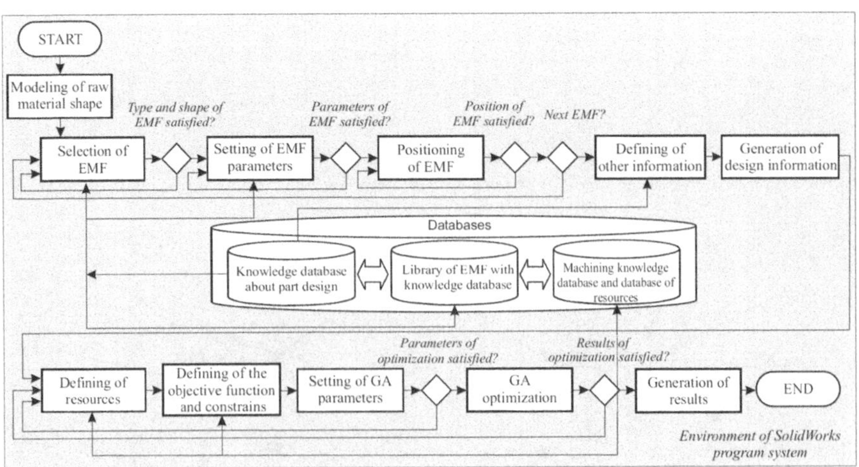

FIGURE 2.9 Structure of integrated intelligent CAD/CAPP platform based on EMF and GA (adapted from Borojević et al., 2022).

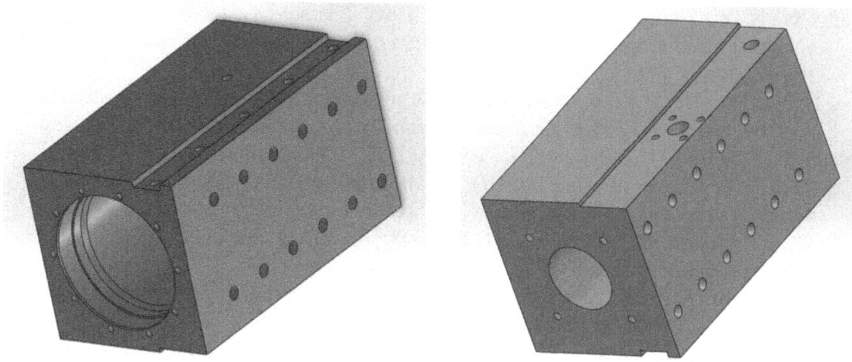

FIGURE 2.10 3D model of the body of hydraulic cylinder (adapted from Borojević et al., 2022).

2.8 DISCUSSION AND CONCLUSION

CAD/CAPP and CAM have been revolutionized for decades. This chapter initially highlights the basic concepts of CAD/CAPP/CAM and their individual roles in the whole product life cycle. After an elaborative discussion on the product life cycle, the CAD section is explained. For the said purpose general product design process and details about its mathematical models have been explained. Later, different computational software and their respective modeling approaches were explored. Once the basic understanding of CAD was developed, process planning is explained in general with both manual and computer-aided approaches. Process planning is used for transforming raw material into its finished part. Moreover, two approaches to bridge CAM to CAPP were explained in detail for the CAPP system: the variant CAPP system and the generative CAPP system. Also, the coding systems, such as group and feature-based, are introduced. At the end of this section, benefits of CAPP implementations are detailed. After CAPP, CAM and its basic terminologies are understood based on both general and narrow point of view. Finally, a few industrial examples regarding CAD/CAPP/CAM integration are added to this chapter.

REFERENCES

Alting, L., Zhang, H. 1989. Computer aided process planning: The state-of-the-art survey. *The International Journal of Production Research*, 27(4): 553–585.

Borojević, S., Milošević, M., Lukić, D., Borojević, B. 2022. An integrated intelligent CAD/CAPP platform: Part I-Product design based on elementary machining features. *Tehnički Vjesnik*, 29(4): 1301–1309.

Butterfield, W. R., Green, M. K., Scott, D. C., Stoker, W. J. 1986. Part features for process planning. *Computer Aided Manufacturing - International Inc.*

Cay, F., Chassapis, C. 1997. An IT view on perspectives of computer aided process planning research. *Computers in Industry*, 34(3): 307–337.

Gao, J., Zheng, D. T., Gindy, N. (2004). Extraction of machining features for CAD/CAM integration. *The International Journal of Advanced Manufacturing Technology*, 24(7), 573–581.

Gindy, N. N. Z., Huang, X., Ratchev, T. M. 1993. Feature-based component model for computer-aided process planning systems. *International Journal of Computer Integrated Manufacturing*, 6(1–2): 20–26.

Han, J., Pratt, M., Regli, W. C. 2000. Manufacturing feature recognition from solid models: A status report. *IEEE Transactions on Robotics and Automation*, 16(6): 782–796.

Holand, P. Standring, P. M. Long, H. Mynors, D. J. 2002. Feature extraction from STEP (ISO 10303) CAD drawing files for metalforming process selection in an integrated design system. *Journal of Materials Processing Technology*, 125–126, 446–455.

Kasik, D. J. Buxton, W. Ferguson, D. R. 2005. Ten CAD challenges. *IEEE Computer Graphics and Applications*, 25(2): 81–92.

Lee, J. Y., Kim, K. 1998. A feature-based approach to extracting machining features. *Computer-Aided Design*, 30(13): 1019–1035.

Macconnell, W. R. (Ed.). (1970). The principles of coding. In *A classification system to describe workpieces*. Pergamon Press, 1–4.

Niebel, B. W. 1967. *Motion and time study*, 2. RD Irwin.

Park, S. C. 2003. Knowledge capturing methodology in process planning. *Computer-Aided Design*, 35(12): 1109–1117.

Sadaiah, M., Yadav, D. R., Mohanram, P. V., Radhakrishnan, P. 2002. A generative computer-aided process planning system for prismatic components. *The International Journal of Advanced Manufacturing Technology*, 20(10): 709–719.

Singh, N. 1996. *Systems approach to computer-integrated design and manufacturing*. Wiley. ISBN-13: 978-0471585176.

Subrahmanyam, S., Wozny, M. 1995. An overview of automatic feature recognition techniques for computer-aided process planning. *Computers in Industry*, 26(1): 1–21.

Vidal, A., Alberti, M., Ciurana, J., Casadesus, M. 2005. A decision support system for optimising the selection of parameters when planning milling operations. *International Journal of Machine Tools and Manufacture*, 45(2): 201–210.

Xu, X. 2009a. Computer-aided process planning and manufacturing. In *Integrating advanced computer-aided design, manufacturing, and numerical control*. IGI Global, 54–74.

Xu, X. 2009b. Integrating advanced computer-aided design, manufacturing, and numerical control: Principles and implementations. IGI Global. 10.4018/978-1-59904-714-0

Xue, J. 2018a. Introduction. *In Integration of CAD/CAPP/CAM*. De Gruyter Science Press: Beijing, China, 1–13.

Xue, J. 2018b. Computer-aided process planning. *In Integration of CAD/CAPP/CAM*. De Gruyter Science Press: Beijing, China, 49–79.

Xue, J. 2018c. *Integration of CAD/CAPP/CAM*. De Gruyter. 10.1515/9783110573091.

Xue, J. 2018d. Computer-aided design. *In Integration of CAD/CAPP/CAM*. De Gruyter Science Press: Beijing, China, 14–48.

Zehtaban, L., Elazhary, O., Roller, D. 2016. A framework for similarity recognition of CAD models. *Journal of Computational Design and Engineering*, 3(3): 274–285.

Zhang, W. J., Xie, S. Q. 2007. Agent technology for collaborative process planning: A review. *The International Journal of Advanced Manufacturing Technology*, 32(3): 315–325.

3 Quality 4.0

Uzair Khaleeq uz Zaman
Department of Mechatronics Engineering, College of Electrical and Mechanical Engineering, National University of Sciences and Technology, Islamabad, Pakistan

3.1 INTRODUCTION TO QUALITY AND QUALITY 4.0

Quality has a broad spectrum, and no single definition encompasses the whole gamut of the word (Drury, 2000). From being the characteristics and features to satisfy particular needs to fitness for use or meeting requirements, quality has been perceived in varying ways, such as quality control (QC), quality assurance (QA), quality improvement, etc. Sebastianelli and Tamimi (2002) suggested various approaches which define quality vis-a-vis product-based, user-based, manufacturing-based, and value-based. The 'product-based approach' re-emphasized the eight dimensions of quality given by Garvin (1984) to measure product quality, i.e., performance, reliability, features, serviceability, durability, perceived quality, aesthetics, and conformance. Moreover, the 'user-based approach' was more marketing-based and defined quality as the limit to which a service or product accomplishes and exceeds the expectations of customers by following both the quality of the design as well as the quality of conformance. The 'manufacturing-based approach' defined quality as the conformance to specifications and laid the basis for statistical QC (SQC). Last but not least, the 'value-based approach' described quality as conformance at an acceptable cost or alternatively as performance at an acceptable price.

Each of the approaches mentioned above has changed significantly within organizations throughout history. To maintain market share and competitiveness, organizations must constantly change with the current technological advances (Merce, 2014). From the first industrial revolution (IR) in the eighteenth century that governed the dawn of the steam engine (Taifa & Vhora, 2019) to the current fourth IR that is guided by cyber-physical systems (CPS) (Maganga & Taifa, 2022), the variations in productivity and quality have been significantly impacted (Sony et al., 2021) and as a result, quality has also undergone several revolutions. For instance, during the first IR or the inspection era, quality went through Quality 1.0, which was characterized by QC and checked if a product was faulty (ASQ, 2021; Broday, 2022). This was followed by Quality 2.0 (in the second IR era), distinguished by quality assurance, and control charts were implemented to control variability. Moreover, during the third IR, Quality 3.0 involved quality management as a strategy with continual improvement and more focus on customer satisfaction (ASQ, 2021). Several tools from total quality management (TQM), such as six

DOI: 10.1201/9781003327523-4

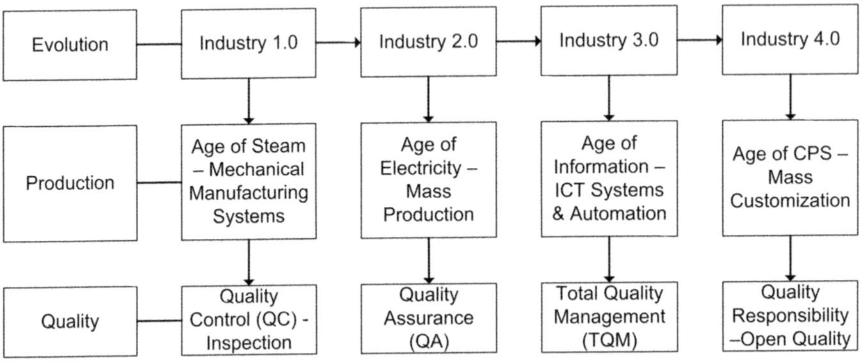

FIGURE 3.1 Evolution of concepts (adapted from Park et al., 2017).

sigma and lean six sigma, were adopted to improve process performance (ASQ, 2021; Broday, 2022). Finally, during the fourth IR era, which is also referred to as Industry 4.0, Quality 4.0 is described by digitalization including concepts such as cyber-CPS, cloud computing, robotics, big data, augmented/virtual reality (AR/VR), artificial intelligence (AI), machine learning (ML), and internet of things (IoT) (Li, 2018; Sony et al., 2021). Quality 4.0 further invites the concept of 'open quality', which helps to attain mass customization by using various approaches and tools to achieve excellence in quality (Broday, 2022). Figure 3.1 shows the evolution of concepts in quality with respect to the industrial revolution (Park et al., 2017).

3.1.1 Background of Quality 4.0

To understand the background for Quality 4.0, it is imperative to realize the role of quality in each era of IR. Inspection and measurement was used to seek uniformity of product in the era of Industry 1.0 wherein quality professionals inspected and repaired by counting and classification. Since detecting non-compliance was the prime objective, methods to solve problems were not considered (Carpinetti, 2012; Garvin, 2002). Next up, control charts were employed to monitor the 'variability' in process in the second IR with the strategy being proactive rather than reactive (Carpinetti, 2012). In addition to the charts, the era was governed by the quantification of quality costs proposed by Juran in 1951, the concept of total quality control (TQC) by Feigenbaum in 1956 and zero-defect program by Crosby in 1961 (Broday, 2022). It was not until the mid-1980s that the term of TQM emerged as an adaptation of TQC. As globalization increased and quality practices were improved, TQM became a global phenomenon. Although TQM has many benefits, a survey conducted by Bernal and Aléson (2015) related to finding the effects of TQM on a company's improved performance revealed that only 'operational performance' was impacted by the adoption of TQM. Also, the traditional quality management systems (QMS) are slower in information processing and handling machine tool rejections, thereby, being ineffective for online inspection and measurements. Years down the lane, companies are still facing issues related to quality management with profit & losses being incurred due to consumer dissatisfaction and elevated costs of poor quality.

Now, in the era of Industry 4.0, the concept of quality has transitioned from quality in mass production to quality in personalized product (Park et al., 2017). As the main Industry 4.0 assumptions are service orientation, real-time capability, interoperability, modularity, decentralisation, and virtualisation (Hermann et al., 2016), it is necessary to incorporate eight pre-requisite aims vis-a-vis standardization of the system, efficient initiatives by the management, deployment of extensive and reliable broadband infrastructure, measures for security and safety, work organization and design, improved resource use efficiency, establishment of regulatory framework, and continuing professional development (Liao et al., 2017).

Moreover, to attain complete control over the complete value chain of the product lifecycle and in light of the above-mentioned Industry 4.0 assumptions and aims, 'digital transformation' is required which ensures that all relevant information is available in real time by connecting all instances involved in the value chain. According to Elshennawy (2004), global manufacturing operations have been digitalized by Industry 4.0 which have subsequently resulted in advances in manufacturing efficiency, supply chain management, business models, and product innovation. Furthermore, manufacturing systems are relying heavily on data and its rapid flow to meet and forecast customer demands at a low cost without sacrificing quality (Constantinescu et al., 2014) thereby implying 'continuous improvement' as the only way to overcome non-value-added costs and reduce manufacturing rejection. Consequently, Quality 4.0 aims to align Industry 4.0 with quality management attributes to improve product quality, customer satisfaction, product innovation, overall enterprise performance, service, and work efficiency (Singh et al., 2022).

3.2 DEFINITIONS OF QUALITY 4.0

Quality 4.0, still a new term in the market, has been defined by various authors in the literature. However, there are two broad approaches through which Sader et al. (2022) defined Quality 4.0. The first approach references Quality 4.0 to Industry 4.0, while the second approach considers Quality 4.0 as the fourth quality management independent evolution phase.

3.2.1 Quality 4.0 as an Implication of Industry 4.0

According to Aldag and Eker (2018, p. 31), "Quality 4.0 is referenced to industry 4.0". This implies that Quality 4.0 is the Industry 4.0-resulted advancement wherein traditional quality management practices, such as control charts, are blended with Industry 4.0 technologies like big data, ML, cloud technologies, IoT, connectivity devices, and AI for real-time quality monitoring to validate products/processes, prevent production errors, and rectify the overall system (Ngo & Schmitt, 2016). Moreover, Jacob (2017, p. 4) concluded that "Quality 4.0 is closely aligning quality management to enable enterprise efficiencies, performance, innovation, and business models". Finally, Küpper et al. (2019, p. 4) considered Quality 4.0 as "the application of Industry 4.0 digital technologies to quality management". All the above definitions imply that Quality 4.0 is an implication of Industry 4.0.

3.2.2 Quality 4.0 as the Fourth Quality Management Independent Evolution Phase

American Society for Quality (ASQ, 2018) differs from the approach in Section 3.2.1 and concludes that the transition from Industry 3.0 to Industry 4.0 and Quality 3.0 to Quality 4.0 happened simultaneously due to the advancements in manufacturing. While Quality 3.0 focuses on traditional practices of quality management, Quality 4.0 can predict the quality issues of tomorrow, maintain credible performance, and reach greater levels of stability and excellence. Moreover, Quality 4.0 can produce higher-quality products at a lower cost by driving the production process, accomplishing greater competitiveness, and increasing responsiveness (Allcock, 2018). Consequently, we can say that although Industry 4.0 is an IT-concentrated approach to manufacturing utilizing networks, sensors, CPS, etc., Quality 4.0 requires greater professional competencies from the quality personnel who can determine not only the required data but also how to use it (ASQ, 2018).

In addition to the above two approaches, other definitions of Quality 4.0 are summarized in Table 3.1.

TABLE 3.1
Definitions for Quality 4.0 (Broday, 2022)

Author(s)	Definition
Jacob (2017)	"Quality 4.0 certainly includes the digitalization of quality management" (p. 8) "Quality 4.0 does not replace traditional quality methods, but rather builds and improves upon them" (p. 8)
Radziwill (2018)	"Quality 4.0 is the name given to the pursuit of performance excellence during these times of potentially disruptive digital transformation" (p. 24)
Salimova et al. (2020)	"An adaptive ability of an object at all stages of the life cycle to meet the needs of a particular consumer on the basis of partnership with stakeholders and digital management of the value chain" (p. 486)
Zonnenshain and Kenett (2020)	"A framework for a quality discipline supporting the fourth industrial revolution. We propose to call it Quality 4.0" (p. 1)
Ramezani and Jassbi (2020)	"Quality '4.0' is a branch of the industry 4.0 movement associated with the digital transformation process connected with emerging technologies" (p. 5)
Sony et al. (2020)	"Quality 4.0 as such is so much more than technology. It is a new method by which digital tools can be used so that organizations' ability to consistently deliver high quality products can be improved" (p. 781)
Escobar et al. (2021)	"Quality 4.0 is founded on a new paradigm based on empirical learning, empirical knowledge discovery, and real-time data generation, collection, and analysis to enable smart decisions" (p. 2320)

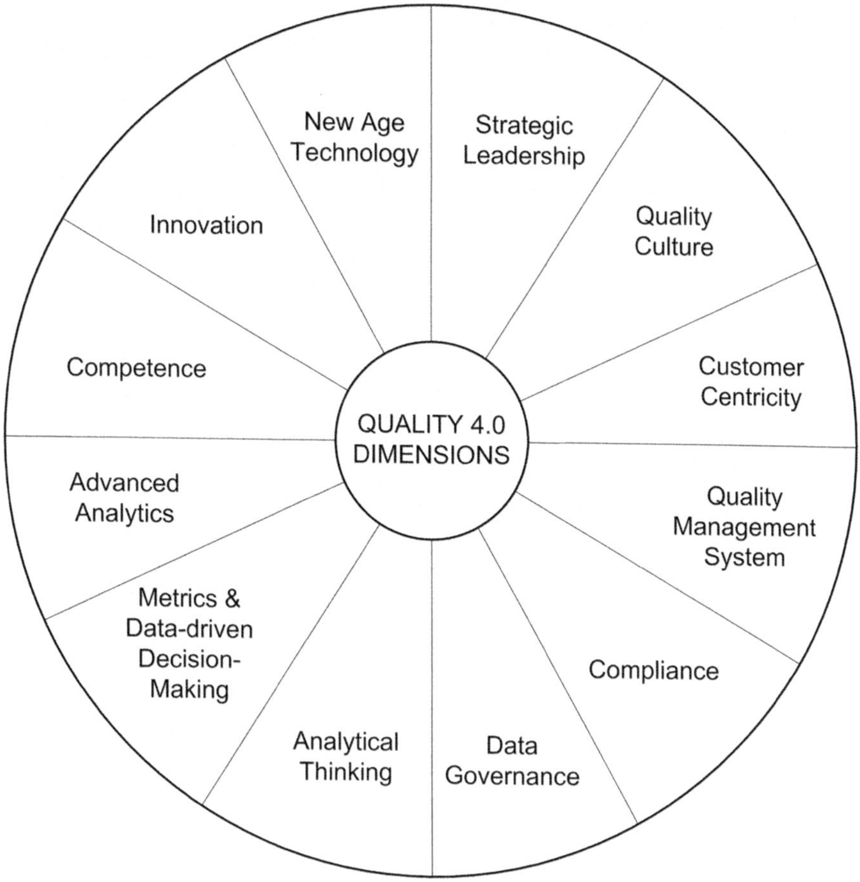

FIGURE 3.2 Dimensions of Quality 4.0 (modified from Sureshchandar, 2022).

3.3 DIMENSIONS OF QUALITY 4.0

Various dimensions of quality are relevant and familiar to Quality 4.0 and have been used to implement quality initiatives effectively. Such dimensions include strategic leadership and commitment from top management, customer focus, quality culture, QMS, capabilities and competence of employees, compliance to quality standards, and analytical thinking. Not only these dimensions assist in building any quality model or system, but also govern operational/quality excellence. Sureshchandar (2022) documented the most recent dimensions of Quality 4.0, as shown in Figure 3.2.

3.4 MODELS OF QUALITY 4.0

3.4.1 European Foundation for Quality Management 2020 Model

The European Foundation for Quality Management (EFQM) 2020 model is an updated and comprehensive business model that incorporates Industry 4.0 elements

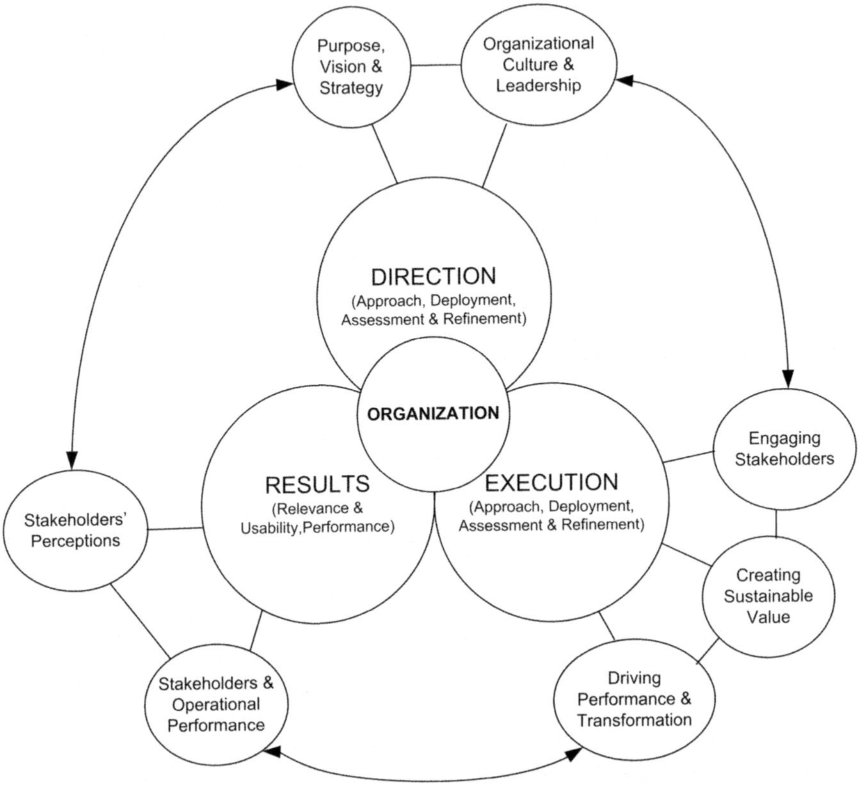

FIGURE 3.3 EFQM 2020 model (modified from Fonseca et al., 2021).

and contributes to Quality 4.0 concepts by encompassing sustainability, improving organizational performance, and emphasizing transformation (Fonseca et al., 2021). Since its inception in 1991, many organizations have been helped by the EFQM model to manage change and improve functioning. Moreover, the EFQM 2020 model incorporates the United Nations Sustainable Development Goals (SDGs) and is supported by European values and business ethics.

The EFQM 2020 model comprehends three different dimensions: direction (why), execution (how), and results (what), with a total of seven criteria. Also, EFQM 2020 model approach outlines creativity, innovation, and disruptive thinking, which are considered necessary elements for Industry 4.0 and factories of the future (FoF) attainment (Fonseca et al., 2021). Figure 3.3 shows the EFQM 2020 model.

3.4.2 TQM and Industry 4.0 Model

Souza et al. (2022) presented the TQM 4.0 ecosystem that can combine technology, quality, and people under three main entities; Industry 4.0, TQM, and QC (see Figure 3.4). For the TQM 4.0 ecosystem to function effectively, creating an environment of interconnection, integration, and collaboration is necessary. For

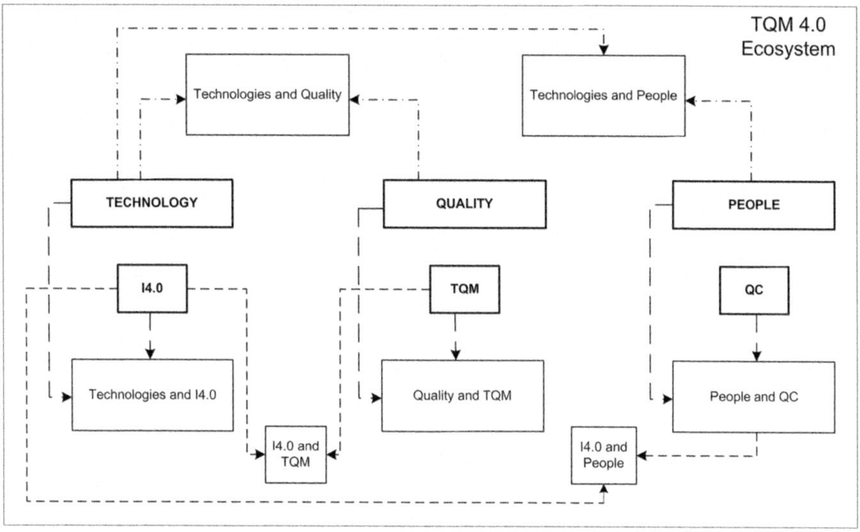

FIGURE 3.4 TQM 4.0 Ecosystem (adapted from Souza et al., 2022).

instance, the integration of 'technology and quality' through Industry 4.0 technologies, such as big data, can make monitoring quality indicators through platforms much easier (Chiarini, 2020). Moreover, combining 'technology and people', data analysis can combine previous scenarios with human perceptions (Bagozi et al., 2019) for better decision-making.

3.4.3 Quality 4.0 and Information Technology/Operation Technology

Quality 4.0 combines information technology (IT) and operation technology (OT) through digital transformation with human intervention (Maganga & Taifa, 2022). Such an integration can improve real-time big data usage and analysis (see Figure 3.5).

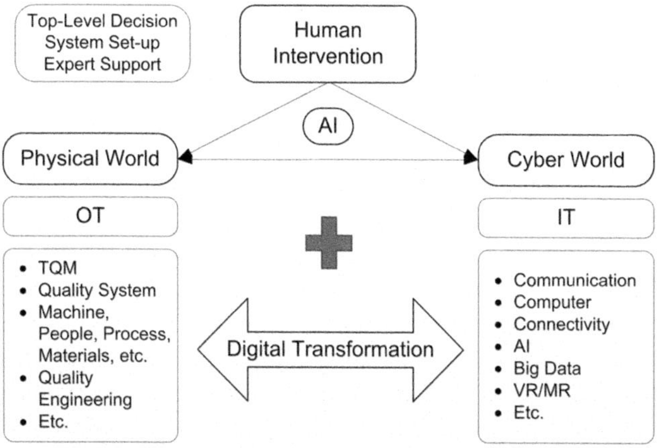

FIGURE 3.5 Quality 4.0 model combining IT and OT through digital transformation (modified from Maganga & Taifa, 2022).

TABLE 3.2
Quality 4.0 Tools (adapted from ASQ, 2021)

Quality 4.0 Tool	Example Implementations in Industry
Artificial Intelligence (AI)	Computer vision, language processing, chatbots, navigation, personal assistants, robotics, complex decision-making
Big Data	Easier access to data sources, tools for analyzing and managing large datasets, infrastructure (e.g., Hadoop, Hive, NoSQL databases)
Blockchain	Enhanced auditability and transparency of transactions (of information and assets), monitoring quality conditions for the processing of transactions
Deep Learning	Pattern recognition for complex systems, image classification, time series forecasting, creating art and sounds, text generation, etc.
Enabling Technologies	5G networks, IoT, affordable actuators and sensors, cloud computing, augmented and mixed reality, virtual reality, etc.
Machine Learning	Email spam filters, fraud detection, forecasting, and text analysis
Data Science	Finding patterns in large datasets, making predictions based on heterogenous datasets, etc.

3.5 QUALITY 4.0 TOOLS

Every industry has some common challenges associated with digital maturity levels and other shortcomings specific to the organization. In this regard, deploying a 'digital strategy' for any organization will not be easy and without hiccups. Although there are well-known quality tools and principles that are rigorously followed by the industries, such as gantt charts, quality function deployment (QFD), house of quality (HOQ), check sheets, scatter diagrams, decision matrix, design of experiments (DoE), value stream mapping, five S (5S), etc., the Quality 4.0 tools listed in Table 3.2 should be utilized for the successful implementation and deployment of the digital transformation as well as alleviating the challenges at-hand for the industries (ASQ, 2021).

3.6 QUALITY 4.0 – CASE STUDIES

3.6.1 Case Study – Autonomous Quality Management System in Automotive Manufacturing

Traditional QMS tools are ineffective for measurement and online inspection and slow in information processing and handling rejection of machine tools. By implementing the concept of Quality 4.0, QC and improvement can be effectively maintained by aligning quality management attributes with Industry 4.0 to improve overall product innovation, enterprise performance, and work efficiency (Elshennawy, 2004; Radziwill, 2018; Singh et al., 2022). Moreover, Quality 4.0 incorporates a CPS that can provide real-time machine/data monitoring and remote supervision/diagnosis/rectification of manufacturing processes/machine tools (Bowers & Pickerel, 2019; Müller et al., 2018; Radziwill, 2018).

Although many researchers have focused on advanced techniques for improving productivity and machining processes, it is imperative to study whether product qualities meet the original equipment manufacturers' (OEM) requirements, i.e., to produce high-quality products at the lowest possible manufacturing cost. Recent research (Chiarini, 2020; Emblemsvag, 2020; Zonnenshain & Kenett, 2020) has developed Autonomous Quality Management System (AQMS) that can automatically inspect and maintain part quality, but in recent times (of COVID-19), there are issues in manufacturing setups related to lack of skilled labour or in some instances 'no labor' at all due to social distance being maintained at workplaces. Such dynamic environments require the digitization of the traditional QMS using Quality 4.0 techniques.

In lieu of the above, the authors (Singh et al., 2022) proposed developing and implementing an AQMS using IoT. The developed system was digitalized using Quality 4.0, and the gage repeatability and reproducibility (gage R&R) of autonomous and traditional QMS systems were evaluated. The methodology followed by the authors is shown in Figure 3.6. The case study was conducted at G.S. Auto International Ltd. (India) and 'hub bolt' was chosen as the product under study. The critical machining processes that contributed most to the rejection of the hub bolt were 'turning operations'. Since defects in manufactured products are mostly associated with machining process capability and human measurement errors, the gage R&R method was used to measure error, and process capability was assessed via six-sigma analysis for the traditional QMS.

All the above processes developed the need for an AQMS that is capable of automatic gauging, detection, correction, segregation, and quality data recording. The developed AQMS system is shown in Figure 3.7. The auto-gauging system consists of a DC supply, an auto-inspection system, a user interface (HDMI or any other medium), a device for memory storage, LVDT gauges, and a logic board containing ESP32 WIFI for communication between gauge interface and machine, and data transmission to the cloud. The developed system was compatible with CNC, SPM, and hydraulic machines.

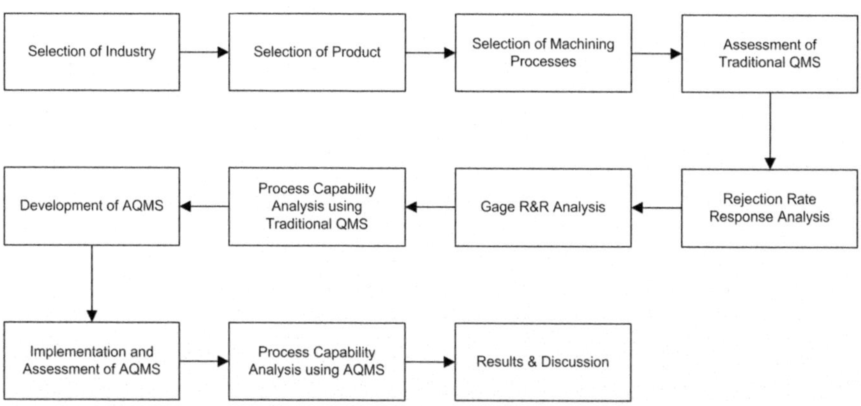

FIGURE 3.6 Research methodology for development and implementation of AQMS in automotive manufacturing (adapted from Singh et al., 2022).

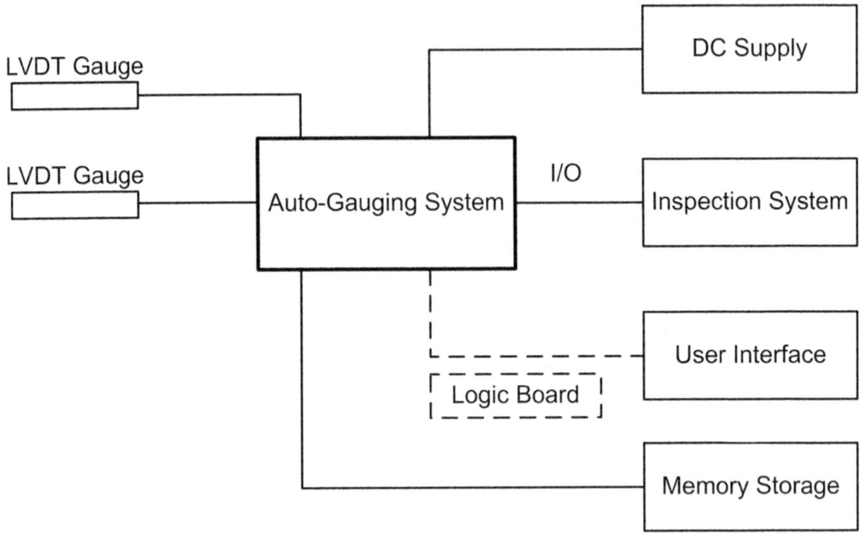

FIGURE 3.7 Components of auto-gauging system in AQMS (modified from Singh et al., 2022).

3.6.2 Case Study – End-to-End Industrial IoT Platform for Quality 4.0 Applications

The authors (Christou et al., 2022) of this case study have introduced a digital platform for predictive maintenance and quality management that can contribute towards implementing zero-defect manufacturing (ZDM). Since the traditional quality improvement methods, along with their barriers, enabling factors, and characteristics, have started to become outdated (Psarommatis et al., 2020), Industry 4.0 technologies, like AI, machine learning, and the considerable drop in the prices of sensors coupled with market requirements, have originated the concept of ZDM. Moreover, ZDM focuses on product and process quality, thereby minimizing product defects by utilizing an optimal combination of predictive, preventive, and corrective techniques.

The platform developed by Christou et al. (2022) is called INTRASOFT Quality Management (INTRA-QM) which uses a cloud/edge computing model and has various components viz. a viz. physical elements [sensors, IoT, cyber-physical production system (CPPS)], data routing & pre-processing (for initialization of CPPS instances), data bus infrastructure (to circulate information to various components), processor engine (to execute data processing logic), device registry (an information directory to access various data sources and devices), a quality management toolkit (for integration and execution of QM and CPPS using various ML algorithms), end user dashboards (for real-time visualization and tracking of data analytics), and field actuation module (to provide the interface of the dashboard with the industrial environment). The platform is shown in Figure 3.8.

The above platform was further deployed on a virtual machine in an industrial setting's shopfloor to validate the proposed framework. The deployment was

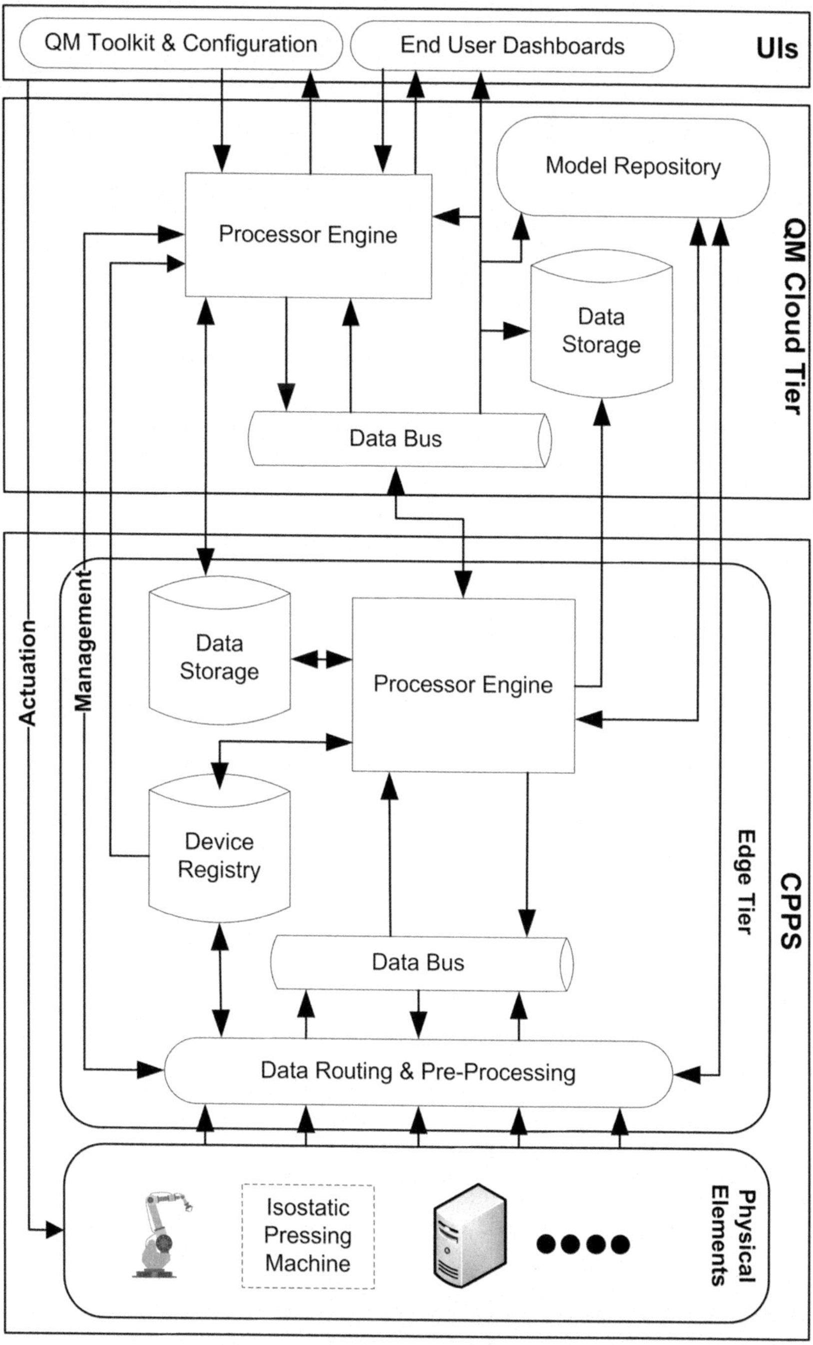

FIGURE 3.8 QM and predictive maintenance platform (modified from Christou et al., 2022).

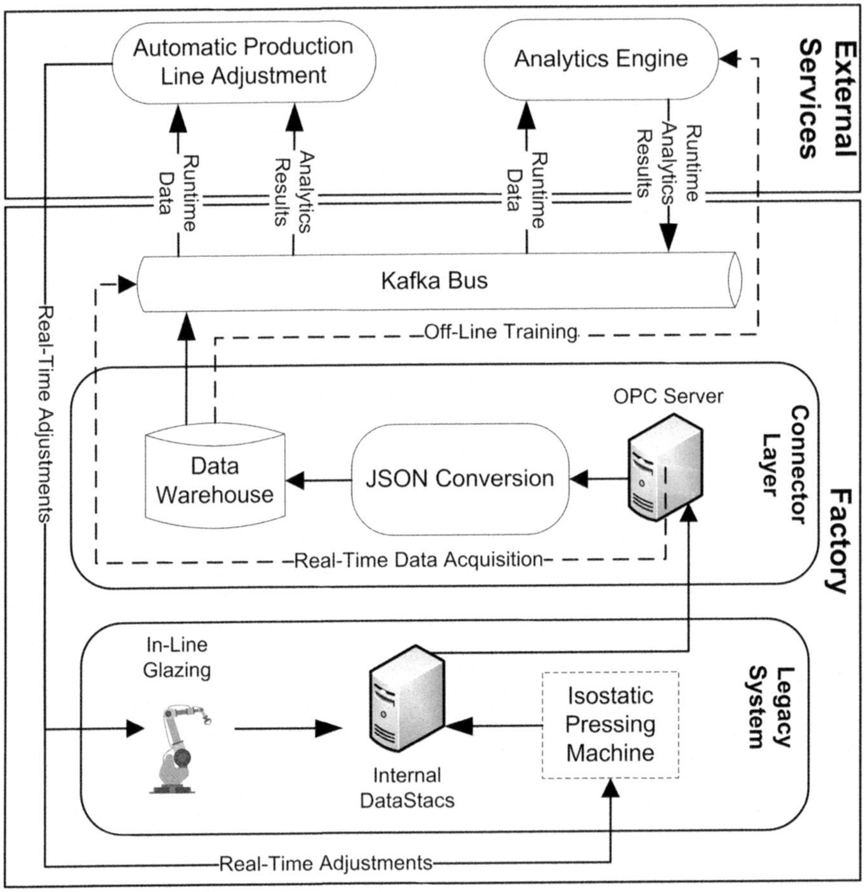

FIGURE 3.9 QM and predictive maintenance framework deployment on a virtual machine (modified from Christou et al., 2022).

composed of two sub-systems: factory and external services (containing a third-party automatic production line adjustment service and an analytics engine). The QM and predictive maintenance deployment is given in Figure 3.9. Not only did the deployment address management and data integration challenges, but it also provided quantifiable benefits to the manufacturer.

3.6.3 Case Study – Quality 4.0: Smart Hybrid Fault Diagnosis System in Plaster Production

Considering the era of emerging technologies of Industry 4.0, QC at the shop floor level has always been a challenging task in smart manufacturing. Although many authors have proposed conceptual and theoretical models associating quality with Industry 4.0, only some have applied the concept in manufacturing setups. The authors of this case study (Ramezani & Jassbi, 2020) have proposed a hybrid model

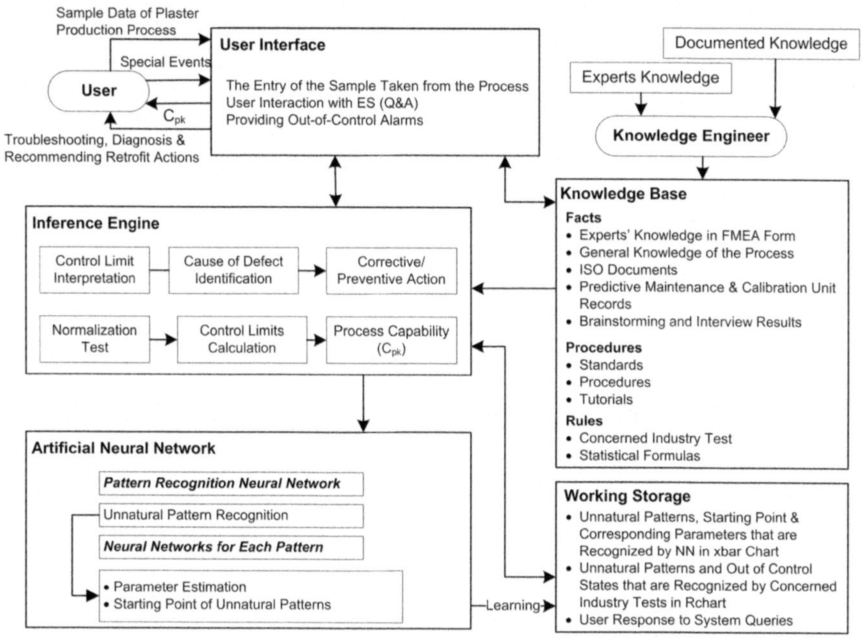

FIGURE 3.10 Hybrid fault diagnosis system (modified from Ramezani & Jassbi, 2020).

to support the troubleshooting in a complex manufacturing system, i.e., the plaster production process. Since any fault detection/diagnosis relies heavily on observing variations within quality characteristics and subsequent recognition of unnatural patterns, a fast and intelligent control chart pattern (CCP) recognition system is required (Hassan, 2014). To accommodate both prescriptive and descriptive approaches, neural networks (NN) and expert systems (ES) were integrated into this case study wherein NN determined fault areas, while ES provided the recommendations for retrofit measures. Figure 3.10 shows the hybrid fault diagnosis system used for intelligent statistical process control (ISPC) for a plaster production process.

3.6.4 Case Study – Industry 4.0 Adoption in QC Processes for Small and Medium Enterprises

Digitalization is the driving force for Quality 4.0. With elements such as real-time monitoring, system integration, data visualization, and integration of design and manufacturing frameworks, quality, from a mere concept, has advanced to become integrated and digitalized, thereby assisting manufacturers/small and medium enterprises (SMEs) to achieve customer satisfaction & operational excellence, conform to quality standards, and minimize cost with the help of digital business systems like enterprise resource planning (ERP) and product lifecycle management (PLM). Figure 3.11 shows how QC processes are integrated with IT and OT systems, from raw materials to finished goods, for SMEs (Dutta et al., 2021).

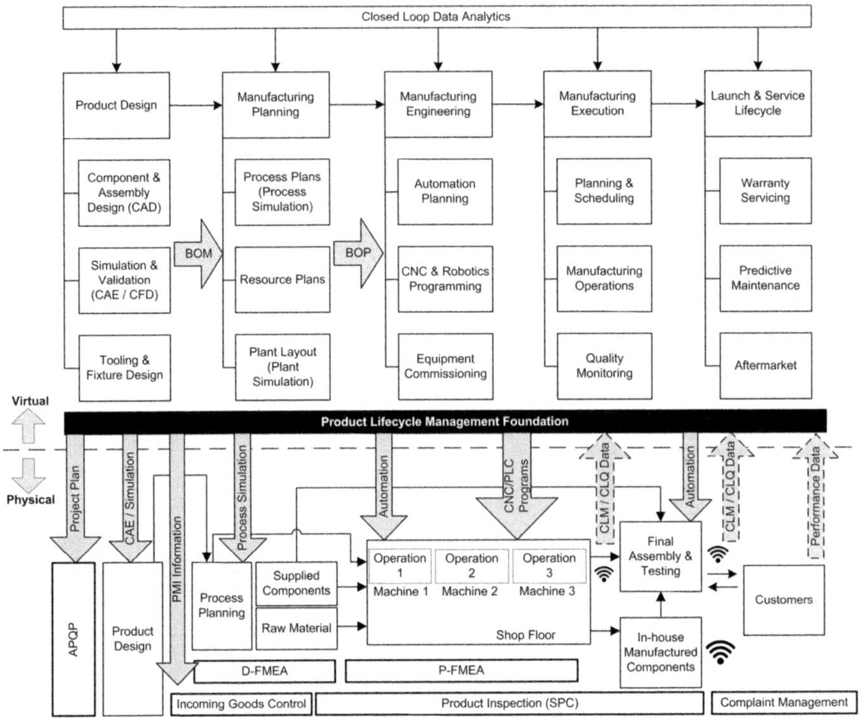

APQP: Advanced Product Quality Planning
D-FMEA: Design Failure Mode and Effects Analysis
P-FMEA: Process Failure Mode and Effects Analysis
CLM: Closed-Loop Manufacturing
CLQ: Closed-Loop Quality

FIGURE 3.11 Quality 4.0: Virtual and physical engineering value chain (modified from Dutta et al., 2021).

Moreover, in Figure 3.11, PLM integrates systems and people (containing cross-functional teams) collaboratively to act as a single source of updated information that can be accessed both in a secure manner and on-demand by the relevant stakeholders. Also, design failure mode and effects analysis (D-FMEA) uses digital tools such as CAE to assess design and performance experiments, while process FMEA (P-FMEA) utilizes digital manufacturing tools for digital validation of processes and plant. Similarly, performance data and real-time inspection is used by industrial-IoT (IIoT) to run real-time SPC, populate dashboards and reports, and perform run-time analytics related to defects and non-conformances thereby improving customer communication and requirements (Dutta et al., 2021).

3.7 CONCEPTUAL FRAMEWORK FOR QUALITY 4.0

Recently, various state-of-the-art have attempted to present the dimensions of Quality 4.0 as a framework that can be implemented (Alzahrani et al., 2021; Armani et al., 2021; Jacob, 2017), but the models are too complex and broad and the systematic application of them will make the transition slower. Therefore, based on the over-arching aim of the chapter and the thorough understanding attained by studying Quality 4.0, its dimensions, and the associated tools, a conceptual framework is proposed, as shown in Figure 3.12. The conceptual framework for Quality 4.0 comprises three elements: 11 research axes of Quality 4.0 (application development, scalability, connectivity, competency, collaboration, culture, leadership, management system, compliance, data, and analytics), real-time quality management (real-time decision-making, data mining, QC/QA, and so on), and Industry 4.0 technologies (big data, artificial intelligence, IoT, autonomous robots, AR/VR, etc.). With this proposed targeted and modern approach, industries/manufacturing setups/organizations can

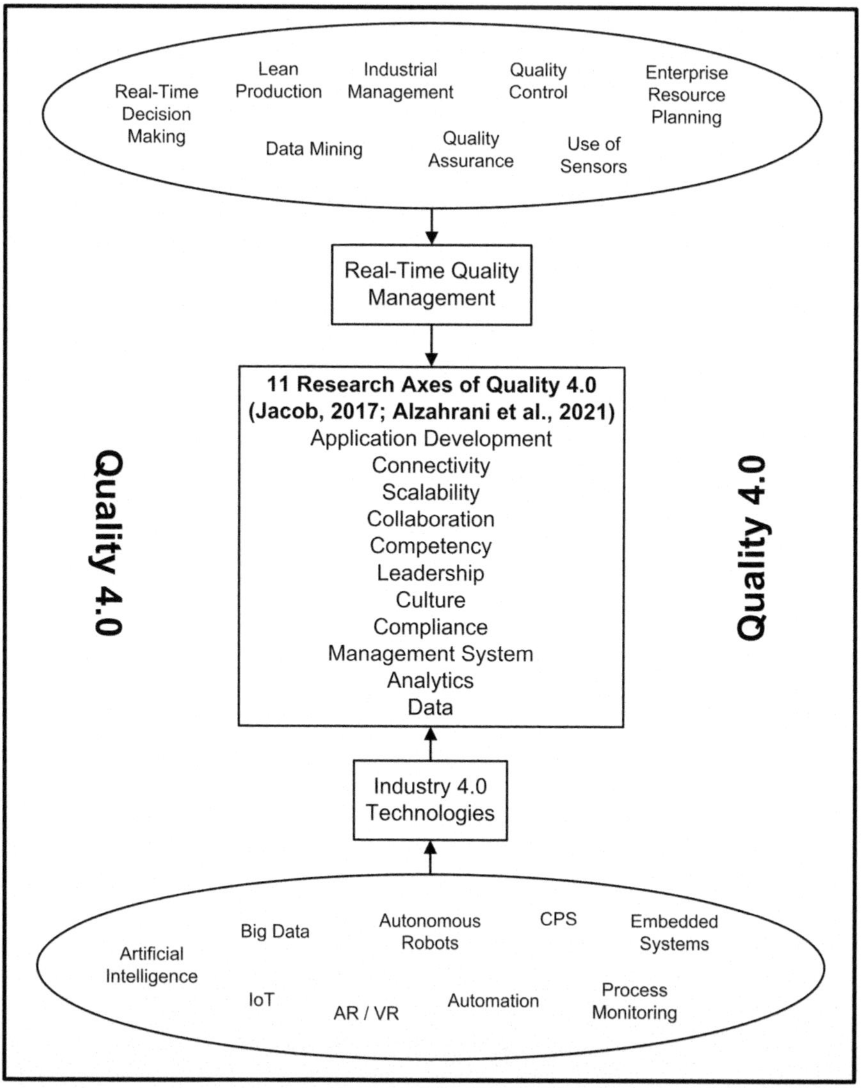

FIGURE 3.12 Conceptual framework to implement Quality 4.0 in industries/organizations.

transition from traditional quality to Quality 4.0 effectively and continuously monitor their quality maturity, thereby allowing diagnostics-based route adjustments.

3.8 DISCUSSION AND CONCLUSION

Quality has evolved naturally from TQM to Quality 4.0, providing competitive advantages to/within organizations such as in manufacturing by developing process and product quality methods in procurement and supply chain, research and development, sales and after-sales, services and logistics, and decision-making, by connecting objects, people, and processes. It is imperative to understand that traditional quality will only be improved and not replaced with the influx of technology, digitalization, and automation-centric environment of Industry 4.0. In other words, Quality 4.0 is the Industry 4.0-resulted advancement wherein traditional quality management practices, such as control charts, are blended with Industry 4.0 technologies like big data, ML, cloud technologies, IoT, connectivity devices, AI, and autonomous robots for real-time quality monitoring to validate products/processes, prevent production errors, rectify the overall system by integrating QC and reliability analysis for predictive maintenance, boost productivity, reduce production costs significantly, and replacement of human labor for effective, efficient, and timely completion of work. Therefore, the dimensions and tools of Quality 4.0 will encourage an environment where the involved human resource has the obligation for all quality actions. Moreover, since success these days is highly correlated with customer satisfaction, the digitalization of processes can help in tracking/transmission of valuable data in real-time, thereby assisting managers/practitioners to provide retrofit measures/remote monitoring on the go with enhanced data visualization.

With all features and benefits associated with Quality 4.0 in the preceding paragraphs, there are many challenges associated with the transition to Quality 4.0 paradigm from the traditional quality practices. These challenges can be broadly categorized into three divisions: human resource-related, organizational, and technological & managerial (Sader et al., 2022). For the first challenge, i.e., human resource-related, future quality personnel must be more competent, sophisticated, and with higher qualifications (having both soft and technical skills) to understand upcoming technologies and effectively apply them.

The second challenge relates to organizational advancement in terms of 'leading the change'. The initiatives of Quality 4.0 are being led more by IT with prime focus on machines rather than by quality teams where the focus should be equally on the labor. Similarly, with cheap sensors, data acquisition via IIoT and big data analytics are readily available, but labor performance is seldom studied. Also, there is a need to assess the impact of Quality 4.0 on lean and Six Sigma, EFQM, ISO 9000 family, and other TQM tools/techniques. Therefore, future quality managers must lead from the front in the transition process by capitalizing on the existing quality practices and embedding them with the Industry 4.0 technologies.

Last, the technological and managerial challenges are associated with the 'readiness' of the organization/manufacturing setup for the transition. Future companies need to assess many factors such as cost-benefit analysis, infrastructure requirements, acquiring credible sources of data and associated data analytical tools, security of data, devices and network reliability, and other issues related to political and social interferences. Consequently, there is no doubt that companies that will successfully transition to Quality 4.0 will attain greater competitiveness by predicting the quality issues of tomorrow, maintaining customer satisfaction through credible performance, and retaining market recognition through more significant levels of stability and excellence.

REFERENCES

Allcock, A. 2018. Nikon talks Quality 4.0. *Machinery*, 176(4276): 49–50. https://www.machinery.co.uk/machinery-features/nikon-talksquality-4-0-industry-4-0 (Accessed on 18 July 2022).

Aldag, M. C., Eker, B. 2018. What is Quality 4.0 in the era of Industry 4.0? 3rd International Conference on Quality of Life. University of Kragujevac.

Alzahrani, B., Bahaitham, H., Andejany, M., Elshennawy, A. 2021. How ready is higher education for Quality 4.0 transformation according to the LNS research framework. *Sustainability*, 13(9): 1–29.

Armani, C. G., Oliveira, K. F., Munhoz, I. P., Akkari, A. C. S. 2021. Proposal and application of a framework to measure the degree of maturity in Quality 4.0: A multiple case study. In: Ram, M. (Ed.) *Advances in mathematics for Industry 4.0*, Academic Press, pp. 131–163.

ASQ. 2018. Industry and Quality 4.0: Bringing them together. https://www.qualitymag.com/articles/95011-industry-and-quality-40-bringing-them-together

ASQ. 2021. Quality 4.0, *American Society for Quality (ASQ)*, North Plankinton Avenue, Milwaukee, WI, USA. available at: https://asq.org/quality-resources/quality-4-0 (accessed 17 July 2022).

Bagozi, A., Bianchini, D., De Antonellis, V., Garda, M., Marini, A. 2019. A relevance-based approach for big data exploration. *Future Generation Computer Systems*, 101(1): 51–69.

Bernal, J. G., Aléson, M. R. 2015. Why and how TQM leads to performance improvements. *Quality Management Journal*, 22(3): 23–37.

Broday, E. E. 2022. The evolution of quality: From inspection to Quality 4.0. *International Journal of Quality and Service Sciences,* 14(3): 368–382.

Bowers, K., Pickerel, T. V. 2019, March.Vox Populi 4.0: Big data tools zoom in on the voice of the customer. *Quality Progress*, 32–39.

Carpinetti, L. C. R. 2012. *Gestão da Qualidade: Conceitos e Técnicas*, Atlas S.A., São Paulo.

Christou, I. T., Kefalakis, N., Soldatos, J. K., Despotopoulou, A-M. 2022. End-to-end iIndustrial IoT platform for Quality 4.0 applications. *Computers in Industry*, 137: 103591.

Chiarini, A. 2020. Industry 4.0, Quality management and TQM world. A systematic literature review and a proposed agenda for further research. *The TQM Journal*, 32(4): 603–616.

Constantinescu, C. L., Francalanza, E., Matarazzo, D., Balkan, O. 2014. Information support and interactive planning in the digital factory: Approach and industry-driven evaluation. *Procedia CIRP*, 25: 269–275.

Drury, C. G. 2000. Human factors and quality: Integration and new directions. *Human Factors and Ergonomics in Manufacturing*, 10: 45–59.

Dutta, G., Kumar, R., Sindwani, R., Singh, R. K. 2021. Digitalization priorities of quality control processes for SMEs: A conceptual study in perspective of Industry 4.0 adoption. *Journal of Intelligent Manufacturing*, 32: 1679–1698.

Elshennawy, A. 2004. Quality in the new age and the body of knowledge for quality engineers. *Total Quality Management & Business Excellence*, 15(5): 603–614.

Emblemsvag, J. 2020. On Quality 4.0 in project-based industries. *TQM Journal*, 32(4): 725–739.

Escobar, C. A., Chakraborty, D., McGovern, M., Macias, D., Morales-Menendez, R. 2021. Quality 4.0 – Green, black and master black belt curricula. *Procedia Manufacturing*, 53: 748–759.

Fonseca, L., Amaral, A., Oliveira, J. 2021. Quality 4.0: The EFQM 2020 model and Industry 4.0 relationships and implications. *Sustainability*, 13, 3107.

Garvin, D. A. 1984. What does 'product quality' really mean? *Sloan Management Review*, 26: 25–43.

Garvin, D. A. 2002. Gerenciando a Qualidade: A Visão Estratégica e Competitive, Qualitymark, Rio de Janeiro.

Hassan, A. 2014. An improved scheme for online recognition of control chart patterns. *International Journal of Computer Aided Engineering and Technology*, 3: 309–321.

Hermann, M., Pentek, T., Otto, B. 2016. Design principles for Industrie 4.0 scenarios. *IEEE 49th Hawaii International Conference on System Sciences (HICSS)*, 3928–3937.

Jacob, D. 2017. Quality 4.0 impact and strategy handbook. *LNS Research*, MaterControl.

Küpper, D., Knizek, C., Ryeson, D., Noecker, J., Quality, E. T. 2019. *Quality 4.0 takes more than technology*. Boston Consulting Group. https://www.bcg.com/publications/2019/quality-4.0takes-more-than-technology.aspx

Li, L. 2018. China's manufacturing locus in 2025: With a comparison of 'made-in-China 2025' and 'Industry 4.0'. *Technological Forecasting and Social Change*, 135: 66–74.

Liao, Y., Deschamps, F., Loures, E. D. F. R., Ramos, L. F. P. 2017. Past, present and future of Industry 4.0 – A systematic literature review and research agenda proposal. *International Journal of Production Research*, 55(12): 3609–3629.

Maganga, D. P., Taifa, I. W. R. 2022. Quality 4.0 conceptualisation: An emerging quality management concept for manufacturing industries. *The TQM Journal*, 35(2): 389–413.

Merce, B. 2014. Integration of management systems as an innovation: A proposal for a new model. *Journal of Cleaner Production*, 82: 132–142.

Müller, J. M., Buliga, O., Voigt, K.-I. 2018. Fortune favors the prepared: How S.M.E.s approach business model innovations in Industry 4.0. *Technological Forecasting and Social Change*, 132: 2–17.

Ngo, Q. H., Schmitt, R. H. 2016. A data-based approach for quality regulation. *Procedia CIRP*, 57: 498–503.

Park, S. H., Shin, W. S., Park, Y. H., Lee, Y. 2017. Building a new culture for quality management in the era of the fourth industrial revolution. *Total Quality Management and Business Excellence*, 28(9/10): 934–945.

Psarommatis, F., May, G., Dreyfus, P-A., Kiritsis, D. 2020. Zero defect manufacturing: State-of-the-art review, shortcomings and future directions in research. *International Journal of Production Research*, 58: 1–17.

Radziwill, N. 2018. Feature Quality 4.0 – Let's get digital: The many ways the fourth industrial revolution is reshaping the way we think about quality. *Quality Progress*, 51(10): 24–29.

Ramezani, J., Jassbi, J. 2020. Quality 4.0 in action: Smart hybrid fault diagnosis system in plaster production. *Processes*, 8(6): 634.

Sader, S., Husti, I., Daroczi, M. 2022. A review of Quality 4.0: Definitions, features, technologies, applications, and challenges. *Total Quality Management & Business Excellence*, 33(9-10): 1164–1182.

Salimova, T., Vatolkina, N., Makolov, V. Anikina, N. 2020. The perspective of quality management system development in the era of Industry 4.0. *Humanities and Social Sciences Reviews*, 8(4): 483–495.

Sebastianelli, R., Tamimi, N. 2002. How product quality dimensions relate to defining quality. *International Journal of Quality & Reliability Management*, 19(4): 442–453.

Singh, J., Ahuja, I. P. S., Singh, H., Singh, A. 2022. Development and implementation of autonomous quality management system (AQMS) in an automotive manufacturing using Quality 4.0 concept – A case study. *Computers & Industrial Engineering*, 168: 108121.

Sony, M., Antony, J., Douglas, J. 2020. Essential ingredients for the implementation of Quality 4.0: A narrative review of literature and future directions for research. *The TQM Journal*, 32(4): 779–793.

Sony, M., Antony, J., Douglas, J. A., McDermott, O. 2021. Motivations, barriers and readiness factors for Quality 4.0 implementation: An exploratory study. *The TQM Journal*, 33(6): 1502–1515.

Souza, F. F. D., Corsi, A., Pagani, R. N., Balbinotti, G., Kovaleski, J. L. 2022. Total quality management 4.0: Adapting quality management to Industry 4.0. *The TQM Journal*, 34(4): 749–769.

Sureshchandar, G. S. 2022. Quality 4.0 – Understanding the criticality of the dimensions using the analytical hierarchy process (AHP). *International Journal of Quality & Reliability Management*, 39(6): 1336–1367.

Taifa, I. W. R., Vhora, T. N. 2019. Cycle time reduction for productivity improvement in the manufacturing industry. *Journal of Industrial Engineering and Management Studies*, 6(2): 147–164.

Zonnenshain, A., Kenett, R. S. 2020. Quality 4.0 – The challenging future of quality engineering. *Quality Engineering*, 32(4): 1–13.

Part II

Manufacturing Systems in Industry 4.0

4 Digital Twins in Smart Manufacturing

Shahid Aziz and Dong Won Jung
Jeju National University, Republic of Korea

Uzair Khaleeq uz Zaman and Anas Bin Aqeel
Department of Mechatronics Engineering, College of Electrical and Mechanical Engineering, National University of Sciences and Technology, Islamabad, Pakistan

4.1 BACKGROUND AND CONCEPT OF DIGITAL TWIN

The Fourth Industrial Revolution, which is also termed Industry 4.0, has arrived in full swing, and it has been influencing and driving all industries, including the high-tech manufacturing industry. It is well known that cyber-physical integration is incumbent for establishing a reliable and highly efficient smart manufacturing (SM) system in Industry 4.0 (Tao et al., 2019). This integration can be achieved by effectively implementing digital twin technology and services (Qi et al., 2018). Digital twin is driving SM after its rapid growth through the evolutional development of new products, services, and technologies, which include smart sensors, artificial intelligence (AI), internet of things (IoT), big data, cloud computing, augmented reality/virtual reality (AR/VR) devices (Židek et al., 2020), modeling and simulation (Shao et al., 2019).

There have been different versions of definitions of digital twin by different individuals and organizations/companies based on the application method in which they utilized digital twins. However, it is also likely to cause misunderstanding due to the several definitions flowing around each technology. A few examples of famous companies' definition of the digital twin is shown in Table 4.1. "A digital twin is a virtual representation of an object or system that spans its lifecycle, is updated from real-time data, and uses simulation, machine learning, and reasoning to help decision making" [IBM]. It can also be defined as a synchronized instance of a digital model or template representing an entity throughout its lifecycle and is sufficient to meet the requirements of a particular use case that the digital twin is meant to address. According to FANUC, the largest maker of industrial robots, "A digital twin is the concept of creating a digital replica of the physical machines, production processes or shop floor layouts in order to generate a number of competitive advantages." Moreover, "a digital twin is an integrated multiphysics, multiscale, probabilistic simulation of an as-built vehicle or system that uses the best available physical models, sensor updates, fleet history, etc., to mirror the life of its corresponding flying twin" (Shafto et al., 2010).

DOI: 10.1201/9781003327523-6

TABLE 4.1
The Definition of Digital Twin According to Seven Different Companies

Company	Definition
IBM	"A Digital twin is a virtual representation of an object or system that spans its lifecycle, is updated from real-time data, and uses simulation, machine learning, and reasoning to help decision making."
Siemens	"Based on the consistent data model across all aspects of the product life cycle, some of the actual operations are accurately and veritably simulated."
General Electric	"Through the virtual models of devices and products, the actual complexities of physical entities are simulated, and insights are projected into applications."
NASA	"The application of interdisciplinary modeling and simulation across the product lifecycle."
ANSYS	"Combined outstanding simulation capabilities with powerful data analysis capabilities, it is to help enterprises gain strategic insights."
PTC	"PLM process is extended into the next design cycle to create a closed-loop product design process and help achieve predictive maintenance of the product."
FANUC	"A digital twin is the concept of creating a digital replica of the physical machines, production processes or shop floor layouts in order to generate a number of competitive advantages."

The digital twin can predict virtually everything that will happen in the physical world, thus providing valuable insights for future forecasting and development. This also allows testing and a better understanding of the product in the early stage, thus minimizing downtime and reducing cost. Digital twin technology is the future of designing and manufacturing a product, process, or service.

The digital twin is being used in the development of robots (Girletti et al., 2020) and autonomous vehicles and their sensor suites to enable testing in traffic and environment simulations. The digital twin implementation has a huge part to play in the testing, development, and validation of autonomous vehicles. The digital twin is also helping the healthcare industry through data by analyzing different circumstances of individual patients for their performance and by comparing them to the population and finding patterns to see trends. The digital twin also helps regulate and monitor the energy generation and capacity, especially wind turbines that can utilize the digital twin to integrate energy data and analyze energy growth avenues.

Supercomputing is the driving force of discovery in every field, from scientific to industrial, allowing researchers to understand the behavior of the smallest particles and visit the furthest expanses of the universe to unlock the meaning of life with digital twins; it is giving industries superpowers to time travel, letting them explore an infinite number of futures and decipher the past through different lenses. With million-X higher performance powered by accelerated computing, data center scalability, and AI, supercomputing will unlock new opportunities for us all.

This chapter also reviews the recent development of digital twin technologies in SM with special emphasis on smart product design, smart biomanufacturing, and IoT for SM from the perspective of Industry 4.0. This research work is expected to provide an effective guideline and broader view to the manufacturers and industry players of the key applications of the digital twin in SM. Successful stories of the world's reputed companies for utilization of digital twin for SM have been added to motivate individuals and companies who are planning to adopt digital twin technology for implementation.

4.2 DIGITAL TWIN-DRIVEN SMART PRODUCT DESIGN

The innovations being done in almost every technology depends heavily on digital twins, which delivers virtual representation of real-world products, systems, and urban infrastructures. For example, the product design of an electric motor can benefit from its digital twin, which can unveil its physical form, and also analyze its mechanical (rotation of the shaft, thermal conductivity) as well as electrical functions (current, voltage, sensors data). The digital twin greatly influences development, production, and operation by evolving through data flow, feedback via user experience, and incoming new data. A product's behavior can be simulated and analyzed well before its physical replica has been manufactured during product development. Three-dimensional printing for product design also relies almost entirely on digital twins. In a recent study performed by Siemens for the mixing of gases in micro-mixers, insights from the simulation of form and flow behavior were combined with generative algorithms. This helped Siemens to develop a unique micro-channel shape and configuration, which increased the mixing efficiency significantly. Digital twins can even help to simulate entire factories, including individual machines and their processes. As an example, let's consider the case of milling robots which experience large forces during the milling operation, leading to inaccurate movements. This problem can be solved by estimating these forces that push the robot away and compensating them in real time, keeping the robot in its path. With regards to operations, sudden disruptions caused due to sensor data of any real point in real-time can be compared to the simulation of that point and reliably predict the point parallel to operations using a digital twin. Digital twin opens new ways for development, production, and operations.

4.2.1 Case Study: Siemens

In the current car manufacturing industries, the development of cars is mostly done in a virtual environment. Siemens NX CAD is being used for successful product designing of vehicles. Automotive designers make their first model with clay to start with the design process, which is then converted into actual products through NX by automotive engineers. The digital twin of the car is created in the digital Enterprise solution portfolio. This enables the optimization of the product design before it's finally built.

Similarly, with the growth of the electric vehicle and energy storage industry, the demand for lithium-ion batteries is still proliferating, so battery manufacturers are focusing on optimizing and improving their processes to maintain the continuity of business in the market. Siemens provides proven automation and digitization solutions across multiple industries to help grow and sustain these businesses. Siemens digital twin technology facilitates manufacturing from automation and

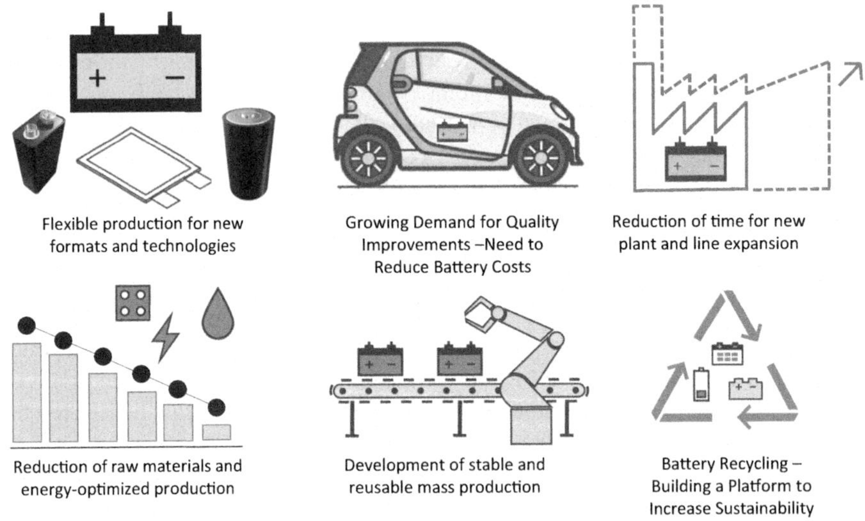

FIGURE 4.1 Battery production process (adapted from Siemens Korea Digital Industries (siemens.com Global website)).

drives technology to production planning and design software; Siemens can help to optimize every step of the battery manufacturing operation.

By establishing more flexible, transparent, and efficient processes in all areas of the cell, module, and pack production, one can ensure continued success in battery production and accelerate time to market, as shown in Figure 4.1.

4.2.1.1 Digital Twin in the Battery Industry

Siemens is the best partner to support industry-specific solutions for companies along the entire value chain, providing digital enterprise solutions optimized for battery manufacturing workflows using digital twins as described in its industry-specific solutions on (https://www.industry.siemens.co.kr/product/list.php?code=9&cat_flag=I).

Digital twins show an optimal virtual model of a product or production plant, intuitively show development throughout the entire lifecycle, and easily show operators to predict behavior, optimize performance, and gain insights from previous design and production experiences.

Siemens's comprehensive digital twin concept consists of three components, as shown in Figure 4.2: a digital twin of products, a digital twin of production, and a digital twin of performance of products/production. Siemens is the only company that can make an offer from a holistic point of view, as it provides sufficient industry-specific expertise and optimized tools.

4.2.1.2 Real-World Battery Performance Improvement Through Virtualization

A product's digital twin integrates all technology domain information into one data model. This allows simulation, testing, and battery performance optimization within the virtual environment to identify and correct possible problems or defects before

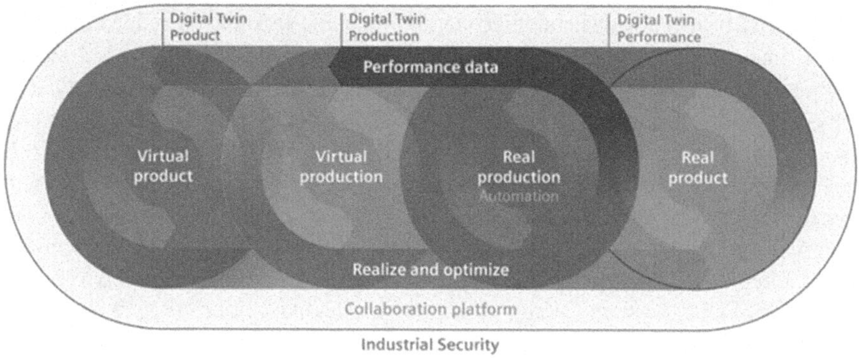

FIGURE 4.2 Components of Siemens' Digital twin concept (adapted from Siemens Korea Digital Industries (siemens.com Global website)).

actual battery mass production. These continuous data can be used as basic data in other fields of work, such as preliminary research, design, and instrumentation.

4.2.1.3 Optimization of the Entire Plant

As battery products are virtually designed and tested before production, production lines can also be planned, simulated, and optimized in a virtual environment with a digital twin implementation.

4.2.1.4 Data-Driven Optimization of Production and Product Performance

Currently, batteries and production processes generate vast amounts of data for product versatility and efficiency. Digital twins allow for continuous improvement by embracing and analyzing this production data in a virtual environment which gives enough data to make decisions in the real world.

In the battery production environment, these strengths enable operators to improve their products and create new business opportunities through accurate analysis of production data. Siemens' software cover the entire production process, from graphical modeling to virtual commissioning and line monitoring. Siemens utilized the digital twin to develop a world record-setting of an electric aircraft motor that not only weighs 50 kilograms but is also five times more powerful than comparable electric motors.

4.3 DIGITAL TWIN FOR SMART BIOMANUFACTURING

Due to the ineffective methods being adopted, developing a drug may cost a billion dollars and an average of ten years. Cell cultures in Petri dishes do not resemble organs or human diseases, and animal experiments are often unsuitable because animals are not human beings. Therefore, drug development is costly and takes much time. That is also true for toxicity tests for cosmetics, chemicals, and food products. Organ-on-a-chip (OOAC) technology has recently been introduced into the healthcare and personalized medicine industry to revolutionize biopharma research, development, and manufacturing (Van Den Berg et al., 2019). OOAC technology has the potential to use artificial intelligence and machine learning to reduce drug discovery times and costs and the potential to replace animal testing (Li et al., 2022) by incorporating bioreactors, and tissue culture (Tay et al., 2016) technologies. OOAC is a physical model of an organ or organs replicated on a device called a "chip." The chip involves growing tiny versions of organs from living cells in cavities and channels. In these devices, we can make three-dimensional cell cultures of human cells in an environment that closely mimics what is actually present in the human body. The living cells are provided nutrients and oxygen via cellular media and extracellular matrix circulated through microchannels. The whole system is known as the Microphysiological system (Wang et al., 2018; Ahadian et al., 2018; Wang et al., 2016; Skardal et al., 2016; Zhang et al., 2017), which consists of the OOAC device, pumping system (Chen et al., 2019; Yang et al., 2019; Li et al., 2019; Lohasz et al., 2019; Edington et al., 2018; Satoh et al., 2018), tubing, nutrients' reservoirs, oxygen cylinder, sensors, and process monitoring equipment (Zhang et al., 2018). Some researchers have tried to integrate multiple OOAC devices to form a micro-multiphysiological system termed as body-on-a-Chip (Sung et al., 2013) that replicates the dynamics of the whole-body response. Medicine/drugs can be tested on individual OOAC devices to understand the response of medicine to the cells, and the same phenomenon can be translated to have a closer look at the whole body's response to the introduction of a particular medicine. Through this technology, drug testing is being done for the most common diseases, i.e., diabetes, kidney (Pietilä et al., 2014; Jang et al., 2013), lung (Huh et al., 2012; Geraili et al., 2018; Baker et al., 2011; Huh et al., 2013), liver (Gröger et al., 2018; Lee et al., 2018), and cardiovascular (Ahn et al., 2018; Parker et al., 2019) diseases. Digital twins are digital replicas of an object, process (Silfvergren et al., 2021), or system (Sundqvist et al., 2022) – and in this case: a human patient. In the future, a digital copy of every human body – a digital twin – may be used to help humans live healthy life. To achieve this goal, behavioral scientists, psychologists, doctors, and software developers, are working together to develop mathematical models (Herrgårdh et al., 2021a, 2021b, Herrgårdh et al., 2022) as tools for better health (Sundqvist 2022). In recent studies, various biosensors have been implemented in the OOAC systems for online monitoring of vital parameters like pH (Mousavi et al., 2016), oxygen (Brennan et al., 2014), CO_2, Virtual reality (VR) (TEER) (Maoz et al., 2017; van der Helm et al., 2019), and various other biomarkers. A recent research work demonstrates the use of albumin immunosensors to monitor the microphysiological system of liver-on-a-chip (Asif et al., 2021), as shown in Figure 4.3. In another work, real-time monitoring of liver fibrosis was done in a microphysiological system via embedded sensors (Farooqi et al., 2021). The actual image of the liver fibrosis-on-chip and the components associated with it have been shown in Figure 4.4. The same group developed a microfluidic chip platform enclosure with a microfluidic chip connected to a micropump for media circulation and to an optical pH sensor for pH measurements (Ali et al., 2020). A portable fluorescence microscope is also installed right above the microfluidic organ chip for real-time visual monitoring. Other components include bubble remover, heater, fan, and control stages, as shown in Figure 4.5.

Using digital twins technology, the OOAC manufacturing industry can save a lot of time and resources by making digital twins of individual organs and

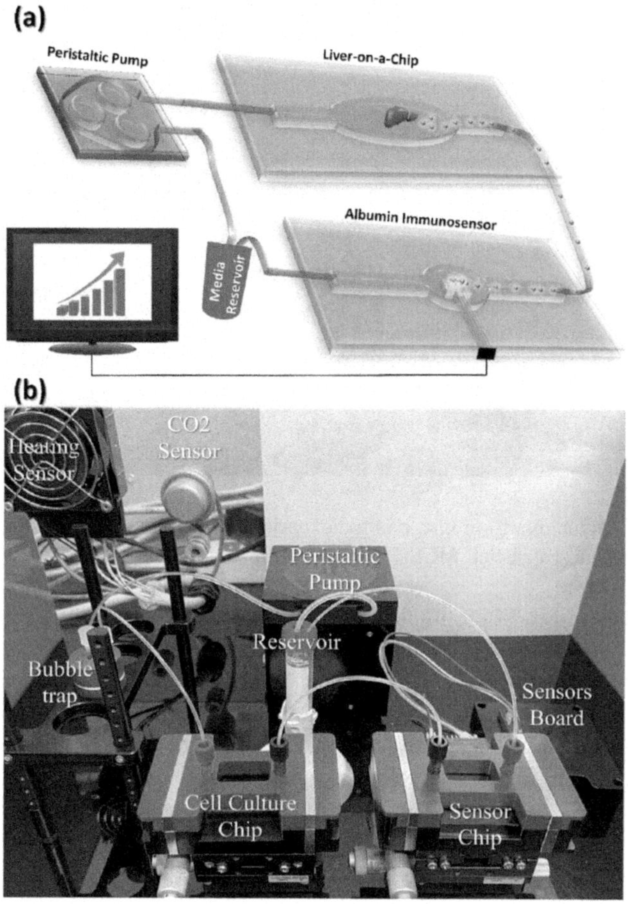

FIGURE 4.3 (a) Schematic of microphysiological system of liver-on-a-chip with albumin immunosensor; (b) actual image of the microphysiological system (adapted from Asif et al., 2021).

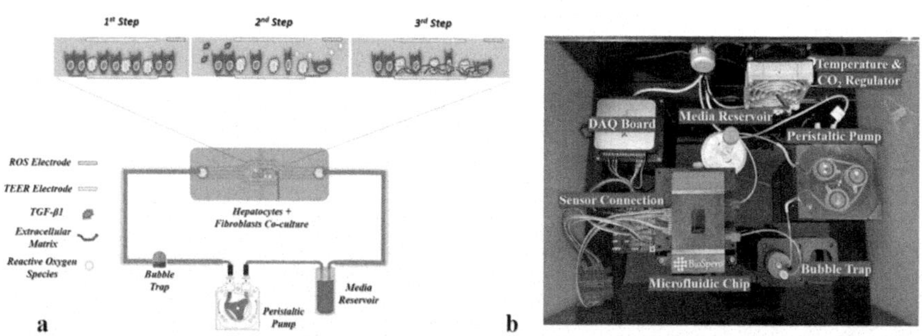

FIGURE 4.4 (a) The liver fibrosis-on-chip schematic; (b) The actual image of the liver fibrosis-on-chip device and associated components (adapted from Farooqi et al., 2021).

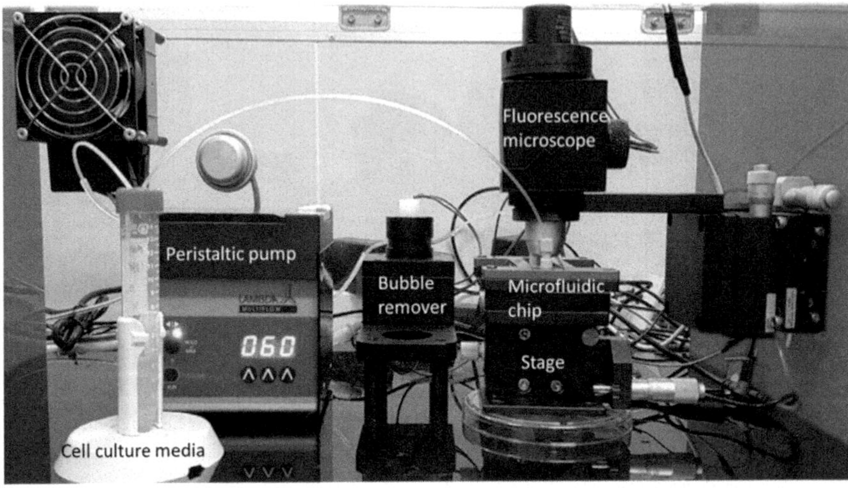

FIGURE 4.5 Actual platform image of the experimental setup for the online monitoring of ROS in breast cancer cell line MCF-7 (adapted from Ali et al., 2020).

multi-organ platforms. Microfluidic simulations of the tubing, pumps, reservoirs, and cavities can help in the SM of the physical structure of the OOAC platform. Cell growth-rate simulations, as well as soft sensors for monitoring vital parameters, can help in the SM of the sensors and cell-culture layers within the channels and cavities.

Biomicrofluidic researchers at the Delft University of Technology have proposed a flow control microfluidic device for next-generation OOAC experiments, as shown in Figure 4.6 (Özkayar et al., 2022). This configuration is suitable for kidney-on-a-chip

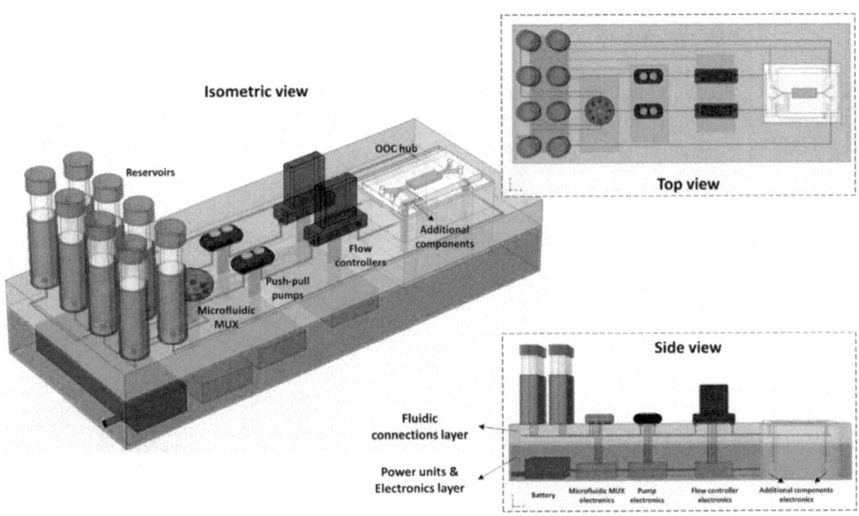

FIGURE 4.6 A microfluidic platform architecture of push-pull mechanism designed for double channel and double chamber OOAC device (adapted from Özkayar et al., 2022).

and uses a modular configuration of a push-pull mechanism. In addition, the flow of fluids from different reservoirs is controlled by the microfluidic multiplexer added to the second channel. Two layers of fluidic and electronics are connected to the microfluidic components via interconnections. The OOAC hub window is used for inverse and upright microscopy. The desired configurations can be achieved when needed by permitting the exchange of fluidic components in the top layer (out of plane).

4.4 DIGITAL TWINS AND THE INTERNET OF THINGS FOR SM

Digital twin for IoT is the way of virtually representing the elements and the dynamics of IoT operation and its working throughout its lifecycle. The ways in which the design, build, and operations of an IoT device are constructed are heavily influenced by the digital twin. In the design phase, different aspects of engineering work together synergistically to collaborate into a single facility of operational-oriented design. Particularly, the physical components, along with the physical bill of materials (BOM), collaborate with the virtual components such as software, sensors, chips, etc. This collaboration brings out the highest quality product by virtue of the digital twin. In the build phase, the digital twin helps in better understanding the influence of the product's tolerances on the devices that make the product. Moreover, its also about the improvement of the manufacturing process through the correction of tolerances and outcomes that are desired in the product.

Lastly, the actual operation of the product is facilitated by the digital twin. During the product's life cycle, the products are significantly influenced by their environmental changes, and they go through various physical and virtual changes over time, so the digital twins need to adjust to those changes with the products as they age. This feedback through the digital twin helps facilitate operations. It also helps to improve the design and manufacturing through the lessons that are learned and the recalibration that takes place along the way.

4.4.1 Case Study: Covestro

As project manager of Covestro's digital ChemLab, Lennart oversees supporting R&D labs with digital methods and new technologies. He aims to boost R&D innovative potential by allowing researchers to collect more results and insights more quickly from fewer experiments. Ultimately this approach is expected to shorten product development timelines.

To make a significant impact, Lennart knew that he had to focus on reducing the time researchers spend on data collection while simultaneously increasing data quality at the point of experimentation. He decided to leverage recent advances in speech recognition technologies to enable hands-free documentation at the bench.

LabTwin (https://www.labtwin.com/) offers the opportunity to tackle several challenges with a single digital solution. By providing hands-free documentation to lab workers, LabTwin enabled both an increase in the quantity of data recorded without slowing down experiments and a decrease in the risk of contamination by avoiding the need to go back and forth between lab benches and offices. By offering researchers a method of contemporaneous data capture and data access, LabTwin could also reduce the number of manual data entry errors and prevent data loss, thus improving data quality. Finally, as the collected data is automatically digitized, scientists no longer need to retype their notes at the end of each experiment, saving significant time and ensuring original data records are truly contemporaneous.

As a first step, Covestro decided to roll out LabTwin for two very different use cases leveraging the various capabilities of the digital lab assistant. The first use case was to guide the preparation of polymer mixtures with the IoT integration. The second use case was to support documentation of time-sensitive foaming processes with hands-free data capture.

4.4.1.1 Use Case 1: Polymer Formulation

Scientists prepare the polymer foam mixture by following a recipe that lists the different components to be weighed and mixed.

4.4.1.1.1 Challenges

Data Quality and Integrity: A tolerance margin allows the actual reagent weight to differ from the recipe weight. These variations need to be recorded as they can significantly impact the result.

4.4.1.1.2 Contamination Risk

Using printed protocols for polymer recipes or going back and forth to the office from the lab to access data is a potential risk of contamination with hazardous chemicals.

4.4.1.1.3 LabTwin Solution

As a digital lab assistant, LabTwin can verbally guide Covestro scientists through formulation steps and streamline their documentation by harnessing IoT integration. Captured data is automatically structured and enriched with metadata, increasing data quality and integrity. In addition, Covestro scientists leverage this human-machine interaction to maintain the recipe ratio with integrated recalculations, saving time and reducing potential waste of components. Enabling digital data access and capture can strongly reduce the need for paper in the lab, and at the end of the day, data can be exported or systematically uploaded to the data repository, cutting down extra data processing. The workflow of Covestro's digitalization with LabTwin for polymer formulation is shown in Figure 4.7.

4.4.1.2 Use Case 2: Foam Characterization

Once the polymer reagents are mixed, the foam starts forming. Scientists must capture the different stages of foam formation, as this information will be used for product development. Researchers must also record observations and measurements from various characterization tests.

4.4.1.2.1 Challenges

Data Quality and Integrity: The foaming process is very fast; therefore, scientists must handle the experiment while mentally storing timings and observations,

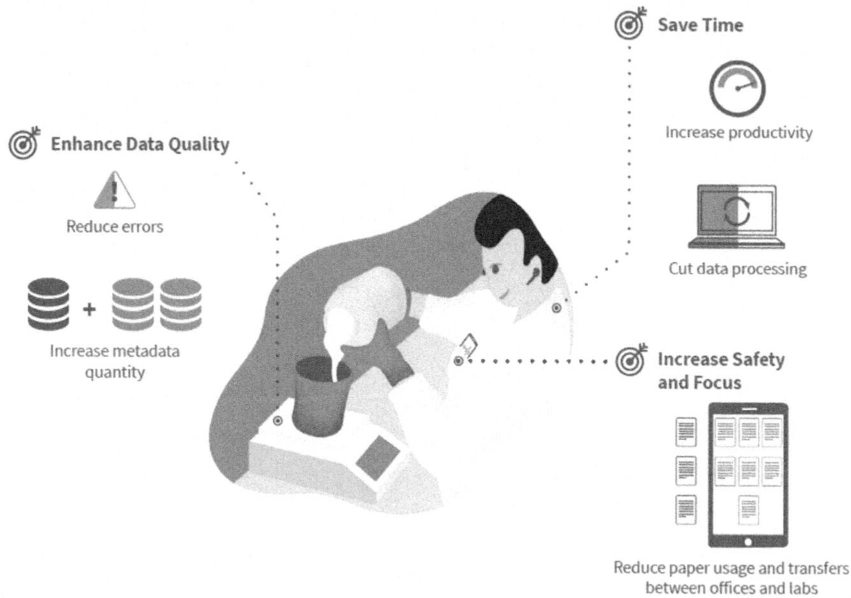

FIGURE 4.7 Covestro aims to increase R&D data quantity and quality, reduce paper as well as gain time using LabTwin's solution.

which would be written down later. Such multitasking increases the risk of errors and data loss.

4.4.1.2.2 Double Documentation

Scientists record their observations and measurements on paper at the point of experimentation and then type the data into computers after the experiment.

4.4.1.2.3 LabTwin Solution

With hands-free documentation, Covestro scientists can keep their eyes on the foaming process while voicing their observations, which LabTwin automatically digitizes and tags with timestamps. Researchers can easily augment the digital data with pictures.

Using LabTwin to support documentation of foaming workflow can significantly improve data quality and integrity. Covestro scientists capture more parameters with this feature, reduce the need to use paper in the lab, and will significantly decrease time wasted on retyping data. The workflow of how Covestro streamlines foaming documentation with LabTwin is shown in Figure 4.8.

4.4.2 Case Study: Digital Twin Solution in Machining Operation by FANUC

FANUC, a world leader in robotics and automation solutions, has developed technology for implementing digital twin technology in machining operations. This

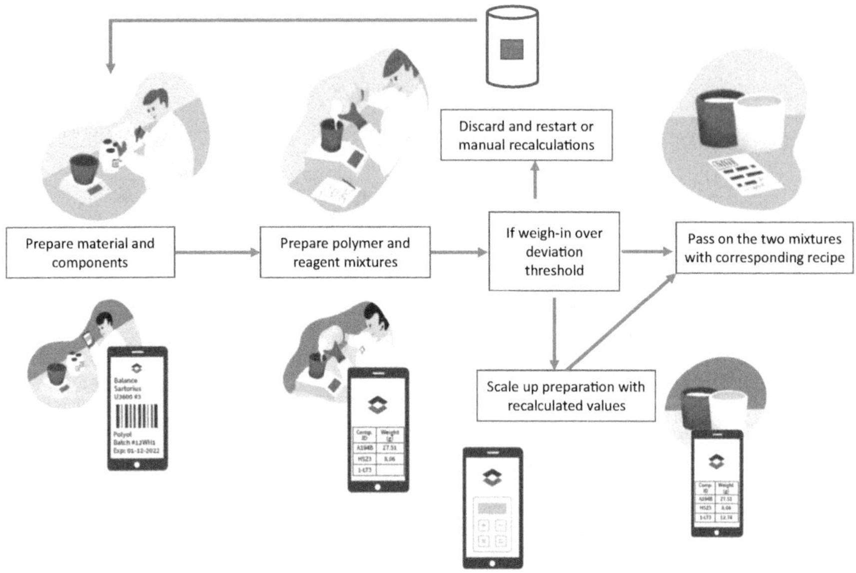

FIGURE 4.8 How Covestro digitalizes its polymer formulation workflow with LabTwin.

has been achieved using the newly released FANUC software, SURFACE ESTIMATION, on Victor Taichung's 5-axis machining center Vcenter-AX630. The objective of the application was to foresee the actual cutting texture on part surface by digital twin. The operation was affected by several factors such as (a) the mismatch between the tool paths generated by CAM and the actual cutting path; (b) unsatisfactory parameter settings of CNC control; (c) inefficient servo system response; and (d) thermal displacement, to name a few. The new software from FANUC has features to improve the shortcomings of the existing system. The software includes the machine servo parameters into the system to match the actual cutting operation. This gives the user the option to run the program on the real part only if the simulated result is satisfactory.

4.5 APPLICATION FRAMEWORK FOR PRODUCT/PROCESS DESIGN USING DIGITAL TWINS

4.5.1 Future Trends

The digital twin market was valued at USD 10.27 billion in 2021, and it is expected to reach a value of USD 61.45 billion by 2027, registering a CAGR of 34.48% over the forecast period, 2022–2027. The increase in adoption of 3D printing, sensors, and AI in the SM industries, i.e., product designing, healthcare, IoT, etc., is going to cause an increase in demand for digital twins technology and services. It is expected that simulation technologies for digital twins in smart manufacturing could grow at a rate of 7.1% to USD 2.6 billion by 2030 (*Global Digital Twin Market, By Type, By Technology, By Application, By End User, By Region, Competition, Forecast & Opportunities, 2017–2027F*).

4.5.2 Application Framework

If one has the capability to utilize digital twin technology, it certainly has enormous benefits for the manufacturer because it can provide value in multiple dimensions. It can improve the safety of operations because the digital twin manufacturer can first simulate a lot of scenarios, even before putting the physical product in, to make sure that the product actually works as it is designed. Even after the product is being used, the physical product is being manufactured and delivered to the customer, and they can constantly get real-time information, and a lot of times, can predict the process on certain days. Any deviation in the product performance can be predicted, and an early warning system can be developed, which can reduce uncertainty and help improve safety.

Considering the field of study, i.e., digital twin in SM in full development, more research studies are required to contribute. To have a broader view of the digital twin in SM, researchers and industrialists from different countries should discuss collaborations on international forums. One such event was organized by the University of Applied Sciences and Arts of Switzerland, i.e., CMS 2022 55th CIRP International Conference on Manufacturing Systems.

In smart biomanufacturing, Multi OOACs will be developed through digital twins, which relate to channels through which fluids can be pumped, and therefore medicines can be tested. Initially, only vital organ chips will be developed for the lung, liver, kidney, and heart. However, in the future, it will be possible to make these chips for all kinds of organs of the human body. This will be an important development in smart biomanufacturing because this standardized and modular approach is ideal for workflows in the pharmaceutical, food, and cosmetics industries. It requires a dedicated and collaborative team consisting of technical people all the way up to biomedical researchers. There is also a great need for the support of industrial partners, from makers to applicators. The digital twin-enabled smart manufacturing of multi-OOAC will lead to better, cheaper, and more effective drug development, and this also applies

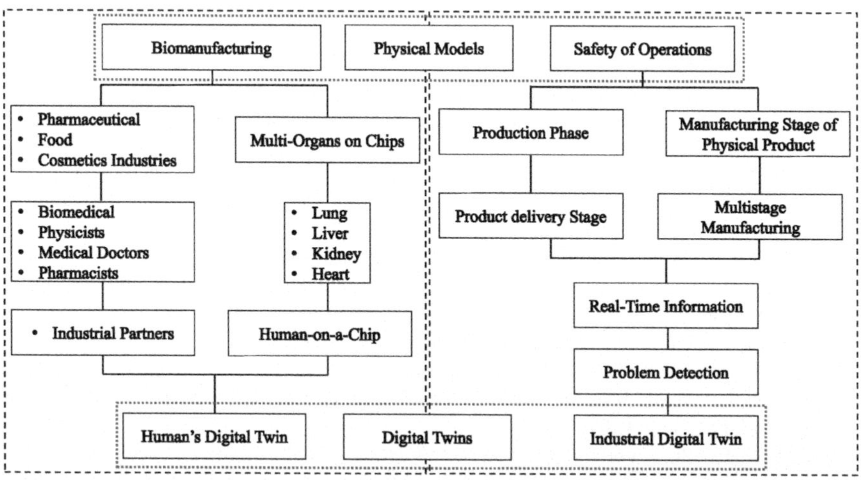

FIGURE 4.9 Overview of the application framework for Digital Twin for Smart Manufacturing.

to toxicity testing in the cosmetics and food industries. It is also expected to boost the world economy and open up a new market in OOAC. The overall application framework for digital twins in smart manufacturing is given in Figure 4.9.

ACKNOWLEDGMENTS

This work was supported by the Korea Institute of Energy Technology Evaluation and Planning (KETEP) grant funded by the Korea Government (MOTIE) (20206310100050, A development of remanufacturing technology on the grinding system of complex-function for high-precision).

REFERENCES

Ahadian, S., Civitarese, R., Bannerman, D., Mohammadi, M. H., Lu, R., Wang, E., Davenport-Huyer, L., Lai, B., Zhang, B., Zhao, Y., Mandla, S. 2018. Organ-on-a-chip platforms: A convergence of advanced materials, cells, and microscale technologies. *Advanced Healthcare Materials*, 7(2): 1700506.

Ahn, S., Ardoña, H. A. M., Lind, J. U., Eweje, F., Kim, S. L., Gonzalez, G. M., Liu, Q., Zimmerman, J. F., Pyrgiotakis, G., Zhang, Z., Beltran-Huarac, J. 2018. Mussel-inspired 3D fiber scaffolds for heart-on-a-chip toxicity studies of engineered nanomaterials. *Analytical and Bioanalytical Chemistry*, 410(24): 6141–6154.

Ali, M., Kim, Y. S., Khalid, M. A. U., Soomro, A. M., Lee, J. W., Lim, J. H., Choi, K. H., Ho, L. S. 2020. On-chip real-time detection and quantification of reactive oxygen species in MCF-7 cells through an in-house built fluorescence microscope. *Microelectronic Engineering*, 233: 111432.

Asif, A., Park, S. H., Soomro, A. M., Khalid, M. A. U., Salih, A. R. C., Kang, B., Ahmed, F., Kim, K. H., Choi, K. H. 2021. Microphysiological system with continuous analysis of albumin for hepatotoxicity modeling and drug screening. *Journal of Industrial and Engineering Chemistry*, 98: 318–326.

Baker, M. 2011. Technology feature: A living system on a chip. *Nature*, 471(7340).

Brennan, M. D., Rexius-Hall, M. L., Elgass, L. J., Eddington, D. T. 2014. Oxygen control with microfluidics. *Lab on a Chip*, 14(22): 4305–4318.

Chen, Z., He, S., Zilberberg, J., Lee, W. 2019. Pumpless platform for high-throughput dynamic multicellular culture and chemosensitivity evaluation. *Lab on a Chip*, 19(2): 254–261.

Edington, C. D., Chen, W. L. K., Geishecker, E., Kassis, T., Soenksen, L. R., Bhushan, B. M., Freake, D., Kirschner, J., Maass, C., Tsamandouras, N., Valdez, J. 2018. Interconnected microphysiological systems for quantitative biology and pharmacology studies. *Scientific Reports*, 8(1): 1–18.

Farooqi, H. M. U., Kang, B., Khalid, M. A. U., Salih, A. R. C., Hyun, K., Park, S. H., Huh, D., Choi, K. H. 2021. Real-time monitoring of liver fibrosis through embedded sensors in a microphysiological system. *Nano Convergence*, 8(1): 1–12.

Geraili, A., Jafari, P., Hassani, M. S., Araghi, B. H., Mohammadi, M. H., Ghafari, A. M., Tamrin, S. H., Modarres, H. P., Kolahchi, A. R., Ahadian, S., Sanati-Nezhad, A. 2018. Controlling differentiation of stem cells for developing personalized organ-on-chip platforms. *Advanced Healthcare Materials*, 7(2): 1700426.

Girletti, L., Groshev, M., Guimarães, C., Bernardos, C. J., de la Oliva, A. 2020, December. An intelligent edge-based digital twin for robotics. In *2020 IEEE Globecom Workshops (GC Wkshps*, 1–6. IEEE.

Gröger, M., Dinger, J., Kiehntopf, M., Peters, F. T., Rauen, U., Mosig, A. S. 2018. Preservation of cell structure, metabolism, and biotransformation activity of liver-on-chip organ models by hypothermic storage. *Advanced Healthcare Materials*, 7(2): 1700616.

Herrgårdh, T., Madai, V. I., Kelleher, J. D., Magnusson, R., Gustafsson, M., Milani, L., Gennemark, P., Cedersund, G. 2021a. Hybrid modelling for stroke care: Review and suggestions of new approaches for risk assessment and simulation of scenarios. *NeuroImage: Clinical*, 31: 102694.

Herrgårdh, T., Li, H., Nyman, E., Cedersund, G. 2021b. An updated organ-based multi-level model for glucose homeostasis: Organ distributions, timing, and impact of blood flow. *Frontiers in Physiology*, 12: 619254.

Herrgårdh, T., Hunter, E., Tunedal, K., Örman, H., Amann, J., Navarro, F. A., Martinez-Costa, C., Kelleher, J. D., Cedersund, G. 2022. Digital twins and hybrid modelling for simulation of physiological variables and stroke risk. *bioRxiv*.

Huh, D., Leslie, D. C., Matthews, B. D., Fraser, J. P., Jurek, S., Hamilton, G. A., Thorneloe, K. S., McAlexander, M. A., Ingber, D. E. 2012. A human disease model of drug toxicity-induced pulmonary edema in a lung-on-a-chip microdevice. *Science Translational Medicine*, 4(159): 159ra147–159ra147.

Huh, D., Kim, H. J., Fraser, J. P., Shea, D. E., Khan, M., Bahinski, A., Hamilton, G. A., Ingber, D. E. 2013. Microfabrication of human organs-on-chips. *Nature Protocols*, 8(11): 2135–2157.

Jang, K. J., Mehr, A. P., Hamilton, G. A., McPartlin, L. A., Chung, S., Suh, K. Y., Ingber, D. E. 2013. Human kidney proximal tubule-on-a-chip for drug transport and nephrotoxicity assessment. *Integrative Biology*, 5(9): 1119–1129.

Lee, S. Y., Sung, J. H. 2018. Gut–liver on a chip toward an in vitro model of hepatic steatosis. *Biotechnology and Bioengineering*, 115(11): 2817–2827.

Li, J., Chen, J., Bai, H., Wang, H., Hao, S., Ding, Y., Peng, B., Zhang, J., Li, L., Huang, W. 2022. An overview of organs-on-chips based on deep learning. *Research*, 2022:20.

Li, Z., Seo, Y., Aydin, O., Elhebeary, M., Kamm, R. D., Kong, H., Saif, M. T. A. 2019. Biohybrid valveless pump-bot powered by engineered skeletal muscle. *Proceedings of the National Academy of Sciences*, 116(5): 1543–1548.

Lohasz, C., Frey, O., Bonanini, F., Renggli, K., Hierlemann, A. 2019. Tubing-free microfluidic microtissue culture system featuring gradual, in vivo-like substance exposure profiles. *Frontiers in Bioengineering and Biotechnology*, 7: 72.

Maoz, B. M., Herland, A., Henry, O. Y., Leineweber, W. D., Yadid, M., Doyle, J., Mannix, R., Kujala, V. J., FitzGerald, E. A., Parker, K. K., Ingber, D. E. 2017. Organs-on-chips with combined multi-electrode array and transepithelial electrical resistance measurement capabilities. *Lab on a Chip*, 17(13): 2294–2302.

Mousavi Shaegh, S. A., De Ferrari, F., Zhang, Y. S., Nabavinia, M., Binth Mohammad, N., Ryan, J., Pourmand, A., Laukaitis, E., Banan Sadeghian, R., Nadhman, A., Shin, S. R. 2016. A microfluidic optical platform for real-time monitoring of pH and oxygen in microfluidic bioreactors and organ-on-chip devices. *Biomicrofluidics*, 10(4): 044111.

Özkayar, G., Lötters, J. C., Tichem, M., Ghatkesar, M. K. 2022. Toward a modular, integrated, miniaturized, and portable microfluidic flow control architecture for organs-on-chips applications. *Biomicrofluidics*, 16(2): 021302.

Parker, K. K. 2019. Designer assays for your sick, subdivided heart. *Cell*, 176(4): 684–685.

Pietilä, I., Vainio, S. J. 2014. Kidney development: An overview. *Nephron Experimental Nephrology*, 126(2): 40–44.

Qi, Q., Tao, F. 2018. Digital twin and big data towards smart manufacturing and industry 4.0: 360 degree comparison. *IEEE Access*, 6: 3585–3593.

Satoh, T., Sugiura, S., Shin, K., Onuki-Nagasaki, R., Ishida, S., Kikuchi, K., Kakiki, M., Kanamori, T. 2018. A multi-throughput multi-organ-on-a-chip system on a plate formatted pneumatic pressure-driven medium circulation platform. *Lab on a Chip*, 18(1): 115–125.

Shao, G., Jain, S., Laroque, C., Lee, L. H., Lendermann, P., Rose, O. 2019 December. Digital twin for smart manufacturing: The simulation aspect. In *2019 Winter Simulation Conference (WSC)*, 2085–2098. IEEE.

Shafto, M. Conroy, M. Doyle, R. Glaessgen, E. Kemp, C. LeMoigne, J. Wang, L. 2010. *Draft modeling, simulation, information technology and processing roadmap*. National Aeronautics and Space Administration, https://www.nasa.gov/pdf/501321main_TA11-MSITP-DRAFT-Nov2010-A1.pdf (accessed on 7 June 2023).

Silfvergren, O., Simonsson, C., Ekstedt, M., Lundberg, P., Gennemark, P., Cedersund, G. 2021. Digital twin predicting diet response before and after long-term fasting. *PLOS Computational Biology*, 18(9):e1010469

Skardal, A., Shupe, T., Atala, A. 2016. Organoid-on-a-chip and body-on-a-chip systems for drug screening and disease modeling. *Drug Discovery Today*, 21(9): 1399–1411.

Sundqvist, N., Grankvist, N., Watrous, J., Mohit, J., Nilsson, R., Cedersund, G. 2022. Validation-based model selection for 13C metabolic flux analysis with uncertain measurement errors. *PLOS Computational Biology*, 18(4): e1009999.

Sundqvist, N., Sten, S., Engström, M., Cedersund, G. 2022. Mechanistic model for human brain metabolism and the neurovascular coupling, *PLOS Computational Biology*, 18(12):e1010798

Sung, J. H., Esch, M. B., Prot, J. M., Long, C. J., Smith, A., Hickman, J. J., Shuler, M. L. 2013. Microfabricated mammalian organ systems and their integration into models of whole animals and humans. *Lab on a Chip*, 13(7): 1201–1212.

Tao, F., Zhang, M., Nee, A. Y. C. 2019. *Digital twin driven smart manufacturing*. Academic Press.

Tay, A., Pavesi, A., Yazdi, S. R., Lim, C. T., Warkiani, M. E. 2016. Advances in microfluidics in combating infectious diseases. *Biotechnology Advances*, 34(4): 404–421.

Van Den Berg, A., Mummery, C. L., Passier, R., Van der Meer, A. D. 2019. Personalised organs-on-chips: Functional testing for precision medicine. *Lab on a Chip*, 19(2): 198–205.

van der Helm, M. W., Henry, O. Y., Bein, A., Hamkins-Indik, T., Cronce, M. J., Leineweber, W. D., Odijk, M., van der Meer, A. D., Eijkel, J. C., Ingber, D. E., van den Berg, A. 2019. Non-invasive sensing of transepithelial barrier function and tissue differentiation in organs-on-chips using impedance spectroscopy. *Lab on a Chip*, 19(3): 452–463.

Wang, Y. I., Carmona, C., Hickman, J. J., Shuler, M. L. 2018. Multiorgan microphysiological systems for drug development: Strategies, advances, and challenges. *Advanced Healthcare Materials*, 7(2): 1701000.

Wang, Z., Samanipour, R., Kim, K. 2016. Organ-on-a-chip platforms for drug screening and tissue engineering. *Biomedical Engineering: Frontier Research and Converging Technologies*, 209–233.

Yang, Y., Fathi, P., Holland, G., Pan, D., Wang, N. S., Esch, M. B. 2019. Pumpless microfluidic devices for generating healthy and diseased endothelia. *Lab on a Chip*, 19(19): 3212–3219.

Zhang, B., Korolj, A., Lai, B. F. L., Radisic, M. 2018. Advances in organ-on-a-chip engineering. *Nature Reviews Materials*, 3(8): 257–278.

Zhang, Y. S., Aleman, J., Shin, S. R., Kilic, T., Kim, D., Mousavi Shaegh, S. A., Massa, S., Riahi, R., Chae, S., Hu, N., Avci, H. 2017. Multisensor-integrated organs-on-chips platform for automated and continual in situ monitoring of organoid behaviors. *Proceedings of the National Academy of Sciences*, 114(12): E2293–E2302.

Židek, K., Piteľ, J., Adámek, M., Lazorík, P., Hošovský, A. 2020. Digital twin of experimental smart manufacturing assembly system for industry 4.0 concept. *Sustainability*, 12(9): 3658.

5 Robotic Systems

Kanwal Naveed and Uzair Khaleeq uz Zaman
Department of Mechatronics Engineering, College of Electrical and Mechanical Engineering, National University of Sciences and Technology, Islamabad, Pakistan

5.1 INTRODUCTION: BACKGROUND AND DRIVING FORCES

The world has seen dramatic advancements over the past couple of decades due to the introduction of robotics in almost every line of work, whether home utilities, high-end manufacturing industries, or military applications. Robotic systems were first introduced in manufacturing industries in the early 1900s, and since then, their diverse applications have only been on the rise in manufacturing domains. George Devol introduced the very first robotic arm in the mid-1950s, which proved to be a big revolution in the field of industry. Modern science is researching making these robotic arms collaborative and work under or with a human, thus called collaborative robots.

Depending on the application preference, industries can deploy mobile or posted/fixed robots. Mobile robots are mostly used where the robot is required to pick and place objects in various locations in the presence of non-stationary humans/laborers. Fixed robots are more prevalent in the manufacturing industry, where the parts are usually ready and already in place. Both types of robots, when deployed in the industry, can work together to achieve complete automation from part assembly to packaging and finally transporting them to various locations in the industry, thereby reducing the manufacturing downtime to a minimum, caused mainly by human errors. Although human interference is undeniable, it can be minimized and assisted by collaborative robots in the manufacturing domain. Similarly, incorporating some heavy-duty autonomous mobile robots can avoid accidents and hazardous situations since the robots will perform tedious tasks in place of humans.

In addition to improving manual labor, robotic systems have self-improving capabilities. This is made possible with the help of the Internet of Things (IoT), where operational patterns are stored over cloud connectivity. This data is then analyzed to improve the system's reliability and efficiency over time. A self-predicting robot that can inform of its maintenance requirements and schedule can bring a revolution in the manufacturing industry on an unmatched level.

5.2 ROLE OF ROBOTIC SYSTEMS IN A MANUFACTURING SYSTEM

The role of robotics in manufacturing systems is growing as quickly as ever. According to a survey (World Robotics Report, 2020), over 2.7 million robots

DOI: 10.1201/9781003327523-7

currently operate in various industrial and manufacturing systems. These applications range from the highly mobile environment (mobile robotics) to fixed isolated automated systems (human-collaborative robots).

5.2.1 Mobile Robotics in Manufacturing Systems

For manufacturing systems, mobile robots are usually deployed in applications concerning the loading/unloading of docks and pallets. These mobile robots can perform a task on order and collaborate with the humans around them with the help of manipulators. It is crucial to understand the abilities and shortcomings of various available types of mobile robots to incorporate them in any manufacturing system having varying industrial requirements. Table 5.1 provides a detailed classification of mobile robots for their level of autonomy.

These robots can be equipped for human collaboration irrespective of their autonomy; however, it requires much planning and coordination. Making sure that the robot's path is cleared and it will not collide with any walking human that might interfere in the designated area of the concerned robot, the correct and timely delivery of the package, and timely recharging of the robot itself, are a few of the salients. An offline coordination training or an on-the-go learning behavior strategy can handle these challenges. Moreover, researchers have introduced various approaches to enable mobile robots to collaborate with humans. Sisbot et al. (2007) worked on a mobile robot with the ability to move around in the presence of people while considering their comfort. Pedersen et al. (2012) introduced a guidance-based method that can teach a robot to perform 'pick and place' tasks with the help of advanced guided gestures from a human expert. Similarly, Rosenthal and Veloso (2012) worked on a navigating robot that could locate humans and communicate with them to carry out various tasks.

5.2.2 Fixed-Collaborative Robots in Manufacturing Systems

Today's fast-changing manufacturing capabilities demand a more sophisticated system than a regular industrial robot and human collaboration that calls for the

TABLE 5.1
Mobile Robot Classification

Level of Autonomy (State of the Art)		
No Autonomy	**Semi-Autonomous**	**Full Autonomy**
• Requires a magnetic makers or wires to move around	• May require an initial map of environment	• Requires no external navigation support
• No integrated sensor systems	• Use a comprehensive sensor system	• Requires an integrated sensor system
• Less flexibility with respect to navigation	• Little or no flexibility	• Highly flexible systems wrt to navigation abilities

TABLE 5.2
Application of Cobots

Types of Cobots			
Force Limiting	**Speed and Separation**	**Safety Monitored**	**Hand Guided**
• Corners are rounded up to avoid edges • Intelligent collision sensors	• Requires minimal human assistance • Uses a vision monitoring system to keep an eye on safe proximity area	• Similar characteristics to Speed and Separation • Can sense and stop when a human is in proximity	• Easier to program and allows a human to teach • Can achieve the minimum downtime

introduction of collaborative robots, thereby producing more effective and flexible manufacturing. While traditional robots can transport heavy payloads, fixed collaborative robots are not only able to carry out assembly/de-assembly tasks but are also able to save workspace (Audi, 2015). Also, depending on the cobot features, their potential application might change in various systems, as seen in Table 5.2.

5.3 ROBUST MEASUREMENTS IN A MANUFACTURING ENVIRONMENT

One key feature that differentiates a smart manufacturing system from a traditional one is the presence of sensors capable of sensing, learning, organizing, and maintaining the information to incorporate the system's abilities and behavior analysis processes. In other words, the inclusion of sensors leads a manufacturing system toward smartness by enabling it to adjust to environmental needs and changes. Such an inclusion leads to minimal or no human input toward the manufacturing workstation. Instead, the human and the machine can co-exist using advanced sensors and big data analytics.

5.3.1 Modern Sensing Techniques

A sensor is an electronic device consisting of sensitive materials that can sense the changes in their surroundings, for example, temperature variation or a high pulse in current through a wire, etc. Sensors can be broadly categorized as active or passive sensors. Smart manufacturing processes employ various types of sensors to help in factory operations' robot control, product movement, assembly, and milling, thereby enhancing manufacturing efficiency. This chapter will subsequently discuss some essential types of sensors commonly used in smart factories. The summary of such sensors is explained in Table 5.3.

5.3.2 Sensor Fusion

The term 'sensor fusion' is self-explanatory in that multiple sensors are fused to generate various sensory signals rather than applying only one sensor and relying on

TABLE 5.3
Summary of Few Important Sensors Employed in Robotics (Adapted from Kalsoom et al., 2020)

	Modern Sensors
	Temperature Sensors
Thermistor	These sensors are made up of ceramic and comes with a temperature range of –50 to 250°C. These types of sensors are used mostly in the automobile industry due to their fast thermal response.
Thermocouple	Variation is temperature produces a voltage variation for temperature reading. These sensors are inexpensive as compared to the other two types thus mostly employed in industry-based measurements. The temperature range varies from –250 to 3000°C
Thermometer	Mainly constructed of platinum, these sensors offer a temperature range of –200 to 850°C on average. Mostly, such sensors find their application in overload protections and automotives. These sensors are highly stable with high precision output.
	Pressure Sensors
Resonance	These sensors consist of a diaphragm and a magnetic coil connected to one another which vibrates depending upon the amount of pressure. Such sensors are expensive, silicon is an example of a resonant sensor.
Capacitance	These sensors are highly sensitive to pressure change as their capacitance changes with change in pressure. Such sensors are most used.
Piezoelectric	When a pressure is applied on a piezoelectric material, a charge is applied which is directly proportional to the amount of applied force. Applied in high impedance circuits but can measure dynamic pressure changes.
	Force Sensors
Load Cells	These sensors consist of a variety of different material cells including capacitive, inductive, hydraulic, etc. such sensors come with a good accuracy and are less expensive
Strain gauge	The resistance varies with applied force. These sensors come in small sizes, high resolution but low accuracy.
Force Sensing resistor (FSR)	These sensors utilize a piezo-sensitive resistor to sense the applied force. These sensors are small, low cost, and low precision.

only one type of information from the environment. Sensor fusion mostly finds applications in mobile automation in manufacturing and collaborative robots. The significant advantages of sensor fusion include the following:

- Increase in reliability
- Enhanced coverage of required parameters
- Precision improvement

Various types of algorithms have been developed to apply sensor fusion. Some of these are discussed in the following sections.

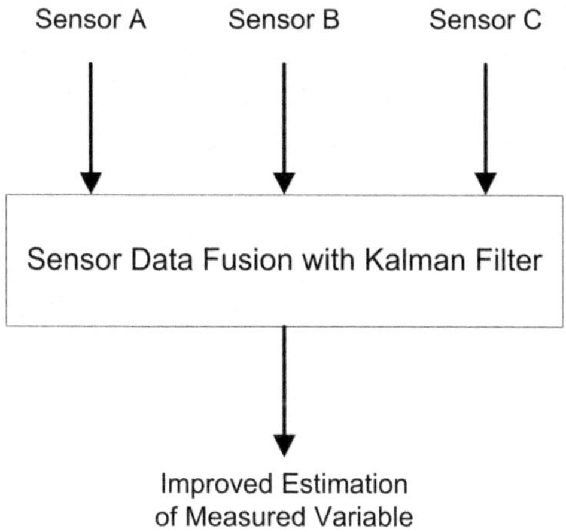

FIGURE 5.1 Sensor fusion concept.

5.3.2.1 Kalman Filters for Sensor Fusion

The primary aim of Kalman filters is to combine/fuse data from multiple (noisy) sensors and use Kalman filters to attain an improved estimate of the variable under observation. Such a variable can be temperature, pressure, sound, etc., as shown in Figure 5.1. One of the essential advantages of Kalman filters is the reduction in noisy observation of sensors by using an improved online estimate of the variable. Moreover, this type of sensor fusion proves to have much better performance in probabilistic processes.

5.3.2.2 Machine Learning-Based Sensor Fusion

Neural networks (NN) are complex replicas of neurons in the form of mathematical models which can classify various characteristics by operating in parallel. NNs have proven robust and more stable than the traditional Kalman filters approach. Initially initialized with random weights, the NN can easily be trained to learn how a system or a particular observation variable behaves to estimate/predict its future behavior. The weights of the system keep on updating as the system behavior varies, increasing the system's robustness (see Figure 5.2). NN-based sensor fusion is best applicable for the navigation-based design.

5.4 PLANNING

When deployed in a real-time industrial/manufacturing environment, any robotic system requires some planning to interact with its surroundings and humans effectively and safely. Not only is it essential that a helper robot completes its assigned task, but it should be able to do it without disrupting nearby objects or harming any human coming in or interacting with the robot. For this purpose, the scientific society has been focusing on developing effective planning methods for robotic systems to generate a path or trajectory for successful operation completion. Whether it is a mobile robot, a drone, a static robotic arm, or a combination of the

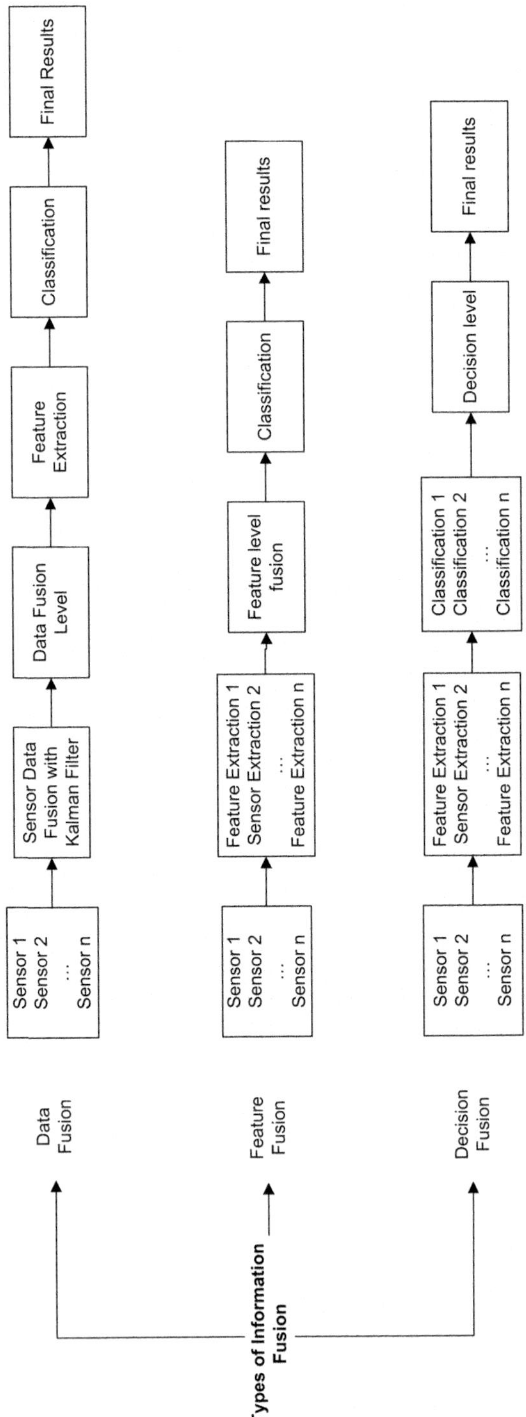

FIGURE 5.2 Neural network-based sensor fusion.

three, it will use the available sensors to understand its surroundings and come up with the best possible plan to achieve its task.

5.4.1 Path Planning for Wheeled Mobile Robots

It is undeniable that position mobility is the most crucial factor in most task executions. Mobile robots are everywhere, from households to industrial settings and war zones. One can find mobile robots carrying out various tasks which are otherwise tedious or too dangerous for humans. Moreover, path planning in any industrial setting depends on prior environmental information. Based on that, the path planning for mobile robots can be divided into local and global planning. The local path planning approach assumes very little or no knowledge of the environment (Li et al., 2011) as opposed to global planning, which assumes that the environment is known beforehand. Both approaches are currently being applied to the navigational planning of mobile robots in an industrial setting. Figure 5.3 shows the navigational approaches which can be applied. Among all these approaches, the five most researched approaches are genetic algorithm (GA), particle swarm optimization (PSO), artificial potential field (APF), ant colony optimization (ACO), and A* algorithm (Zhang et al., 2018).

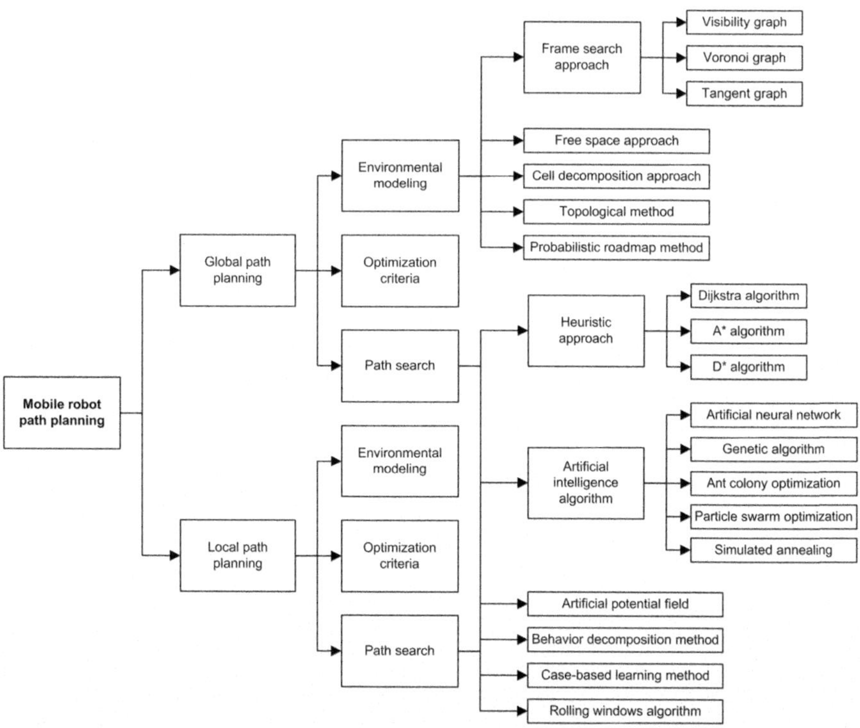

FIGURE 5.3 Technique classification for path planning (adapted from Zhang et al., 2018).

5.4.1.1 A* Planning Approach

A* algorithm was first introduced in the early 1970s (Hart et al., 1968). It works on finding the shortest path from one point to another, just like the Dijkstra algorithm. It starts from an initial node and works to the final node by minimizing a cost function. The cost function value is dependent upon the optimal path from the initial to the final node.

5.4.1.2 The Genetic Algorithm Approach

The genetic algorithm is a simple algorithm with an efficient optimal path search capability. GA works on the principle of creating a cluster of chromosomes. The chromosomes represent all the possible path solutions to any given scenario. As a second step, a fitness value is assigned to each entity depending on the objective at hand. The basic flowchart to implement GA for a robotic manipulator (Zhao et al., 1994) is shown in Figure 5.4.

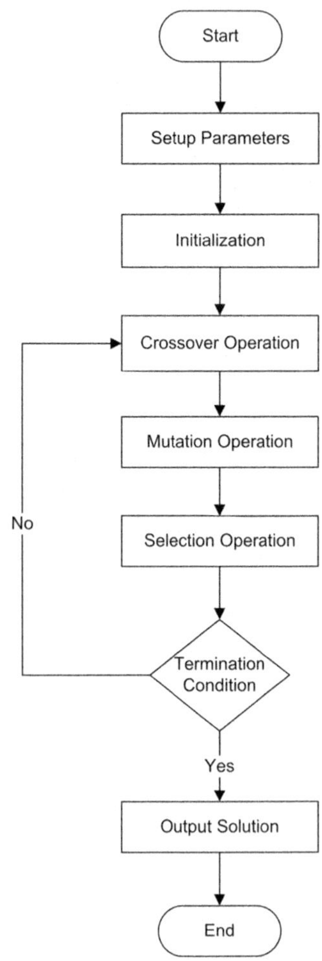

FIGURE 5.4 Basic flow diagram of GA algorithm.

5.4.1.3 Ant Colony Optimization Algorithm

ACO algorithm works on the principle of positive feedback. This approach is applied chiefly to global path planning scenarios (Wang et al., 2009) but can also work for local planning. ACO works on the principle through which ants search for food, mainly called ant foraging. This process is comparable to the path planning principles. However, the technique suffers from problems of high computational efforts and premature convergence.

5.4.2 Trajectory Planning for Manipulators

In the modern manufacturing industry, a robot manipulator is required to perform different high-level tasks, including assembly, pick and place, and maintenance (Ata, 2006) while avoiding any obstacle collision, or harming a collaborative human. Such a scenario requires sophisticated end-to-end trajectory planning for a manipulator called 'motion planning'. The overall task of planning for industrial manipulators can be broken down into steps, as shown in Figure 5.5. The manipulator job may be a simple pick and place. However, to carry that out, the complete

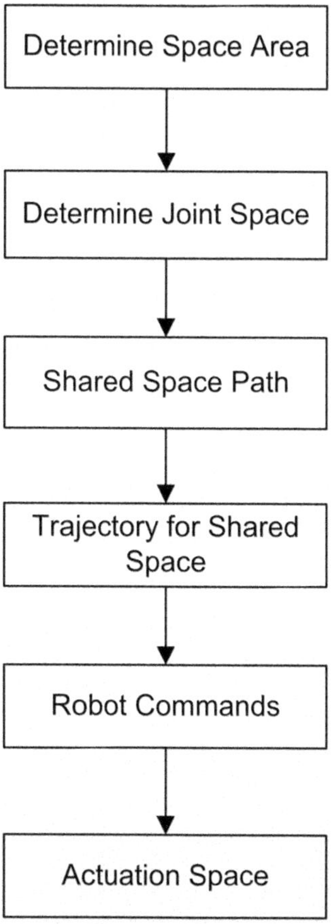

FIGURE 5.5 Planning technique for industrial manipulators.

workspace is to be defined, which includes picking the object with an initial position and orientation and moving the robot's end effector toward the final position and orientation where the object is to be placed. Here, trajectory planning occurs where the manipulator must find an optimal path toward the goal position while performing collision avoidance. Trajectory planning has been an area of focus for researchers for many years. Various methods exist and are proven to be useful.

5.4.2.1 Probabilistic Roadmap

Probabilistic roadmap (PRM) method builds a roadmap between the initial and final position by focusing on creating connectivity of free space. The road map is then defined as a graph 'G' whose vertices are the connectivity points of free space. The path planning between an initial and goal position is answered by creating a road map and then locating these two positions as connected points of the roadmap. The probabilistic nature of this method makes it harder to find out the convergence time of this method despite the ease of implementation (LaValle et al., 2004).

5.4.2.2 Rapidly Exploring Random Trees

The basic idea behind rapidly exploring random trees (RRTs) goes back to the 1950s (Bellman et al., 1957), depending on optimal control theory and non-holonomic randomized path planning. This approach builds on an initial state by making a search tree whose vertex represents new states, and the edges are represented by an applied input, creating a shift from the initial to the next state. This method is mostly applied to problems with smaller degrees of freedom because of their dependency on dynamic programming.

5.4.3 Drone Flight Planning

In industrial areas, drones find their applications in inventory scanning and part delivery services. These applications require drones to be equipped with a camera for target recognition. Some drones are also equipped with light detection and ranging (LiDAR) and utilize sensor fusion to achieve better precision. The path planning for drones is mainly divided into two broad categories: optimal planning and heuristic planning. The optimal planning primarily includes optimization of parameters and exhaustive methods (Fu et al., 2018), while the heuristic methods are A*algorithm, ACO, etc., which are also explained in section 5.4.1.

5.5 CONTROL

Control strategy for a robotic system is a systematic approach to ensure the desired results and behavior from a robot, whether it is a mobile robot, a cobot, or a drone. In this era of Industry 4.0, it has become imperative to include autonomous robots in manufacturing applications, thereby proposing new challenges to the control community.

5.5.1 Classical Control Techniques for Robotic Systems

The classical control systems are suitable for industrial applications where the task is pre-defined with fixed parameters and not prone to uncertainties or changes.

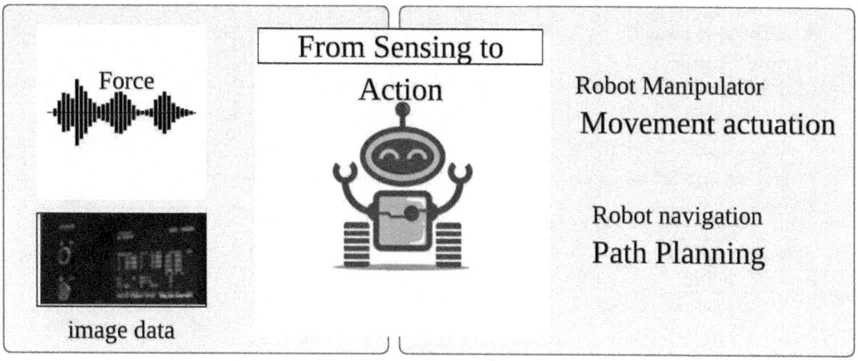

FIGURE 5.6 Interaction planning for a humanoid robot.

5.5.1.1 Position Control

This control is applied to any continuous and repetitive task for example a cleaning robot in an industrial or household setup. Position control takes advantage of inverse robot kinematics to achieve desired robot motion. For example, the inverse Jacobian was utilized by Yamazaki et al. (2010) to plan the robot's upper body motion. A generic representation of Position control for a humanoid robot is shown in Figure 5.6.

5.5.1.2 Force Control

Future control (FC) utilizes available sensor measurements to assess the robot's interaction with the environment. Sato et al. (2011) presented a controller capable of tracking force to control the resulting torque. Ortenzi et al. (2014) developed a torque minimization control to increase the manipulator stability while performing board wiping tasks. Being a classical control approach, FC is limited by a controlled environment and finds it extremely difficult to adapt where the environment is constantly changing, for example, varying forces on robot tires due to uneven terrain.

5.5.2 Practical Robotic Control Systems with Adaptive Techniques

The real-time industrial environment is constantly changing, limiting the scope of the classical control approach and calling for control methods with the ability to adjust as per environment changes continually or if some parameters of the robotic system are uncertain or unknown. For adaptive control, the designed system doesn't depend on any prior data or system behavior information. Adaptive control has three major types: model reference adaptive controller (MRAC), self-tuning, and gain schedule. Among these, the MRAC is the most used algorithm. One such application of MRAC was proposed by Naveed et al. (2014) in the presence of unknown and uncertain robotic parameters.

5.5.3 Noisy Manufacturing Facilities and Robust Control

Real-time manufacturing demands a more sophisticated and robust control algorithm design that can adjust itself according to the changing environment and learn

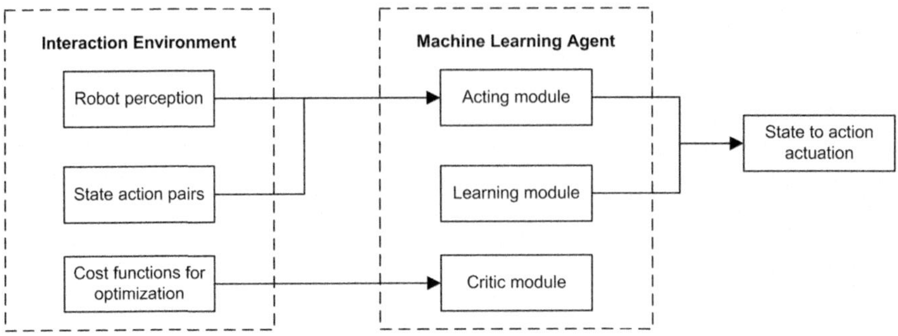

FIGURE 5.7 Machine learning control algorithm.

from past behaviors to improve its current state. Robust control is also required in the case of manipulators interacting with a dynamic environment, thereby necessary for an effective and safe environment interaction to carry out a task at hand.

The manufacturing process in its current state is highly dynamic, making it almost impossible for the classical or adaptive control techniques to predict and analyze the system behavior as one complete picture in such a short time (Qi et al., 2019). NN-based machine learning (ML) algorithms are becoming increasingly popular with time due to their ability to learn quickly about their surroundings (LeCun et al., 2015). The ML process can be categorized into three major types: supervised learning, reinforcement learning, and unsupervised learning. A basic flowchart to understand the basic idea behind a general ML algorithm is shown in Figure 5.7.

5.5.4 End-to-End Control Design with Optimal Control

Optimal control methods are used when a manufacturing system is modeled as a piecewise deterministic system that can be controlled via a stochastic control approach. Such systems are explained by Davis et al. (1984) and Boukas et al. (1994). Also, an optimal control system can be designed to achieve manufacturing tasks in a dynamic environment where the system is prone to failures due to noise, disturbances, or machine ageing (Boukas & Yang 1996). Their research (1996) designed an optimal control method for smooth production planning and system maintenance. Along similar lines, Akella and Kumar (1986) presented an optimal control approach for the production rate of a failure-prone manufacturing system.

5.6 MODERN ROBOTICS FOR MANUFACTURING SYSTEMS

With the advent of the Industry 4.0 concept, robotics is also evolving to meet modern industrial requirements and needs. Many modern and high-end robotic systems, including collaborative robots, mobile robots with manipulators, and drones, are equipped with sophisticated learning behavior modules.

5.6.1 Collaborative Robots

Collaborative robots, also called cobots, are explicitly designed to work alongside humans and share the same workspace to achieve a task. However, human safety while working with a cobot has been an area of active research (Hägele et al., 2016). In a study presented by Landi et al. (2016), the cobot is equipped with a sensor and movement guide to ensure human safety in an industrial setup. The concept of force limitation to avoid harming a human was presented by Ferraguti et al. (2017), where a robot can reduce its force when in close proximity to a human. Some of the significant robotic systems including cobot producers in the world are ABB, KUKA, and Omron robotics. A brief review and analysis of cobot control and planning strategies were presented by Sherwani et al. (2020). The authors discussed in detail the applications and advantages of cobots in the manufacturing industry.

5.6.2 Adaptive Learning Techniques

Classic supervised ML approaches require a lot of data collection to attain a well-trained system that can adapt to a dynamic environment. However, the involvement of humans and smart manufacturing systems offers a great deal of dynamic variation in the robot's surroundings. Under these circumstances, the classical ML approaches cannot adapt to unforeseen events, leading to the failure of the manufacturing system, which gives rise to adaptive learning systems that can evolve by learning from the online data in a dynamic system (Zliobaite et al., 2012). To achieve this target, a robotic system is usually equipped with automated ML detectors (Kedziora et al., 2020), which can collect data invariably without any human intervention. However, model generalization for successfully deploying an adaptive ML-based manufacturing system poses an issue due to constantly changing worker behavior. Therefore, careful research is needed to be carried out in the field of Adaptive ML model generalization, automation of the model process, and generic solution development for a variety of manufacturing environments.

5.7 ROBOTIC SYSTEMS IN MANUFACTURING INDUSTRY – CASE STUDIES

5.7.1 Case Study 1 – Design and Development of an Automated Guided Vehicle for Warehouse Automation of a Textile Factory in Pakistan

The transport/movement of material/resources within shop floor limits or to/from warehouses is referred to as 'material handling' in manufacturing. Automated guided vehicles (AGVs) follow a pre-defined path for navigation and are becoming the recent trend in material handling. Globally, various applications are associated with AGVs in automotive, food and beverages, pharmaceutical, textile, electronics, and other industries. However, to remain an attractive destination for international buyers, many developing nations (such as Vietnam and Bangladesh) are taking steps to modernize industry, train their workforce, and meet global standards. Pakistan, in a similar fashion and considering the rapidly changing international business scenarios, has been looking forward to its primary industry, i.e., textiles, for automation and digitalization. The textile industry has remained the backbone of Pakistan's economy. It contributes one-fourth to the total value added by the industrial sector, with an average share of 61.24% in national exports (Pakistan Economic Survey, 2022).

Moreover, warehouse, inventory, and shipping are the three essential areas of the textile industry. A well-managed warehouse can streamline and optimize many aspects, such as repetitive tasks and unnecessary manual/labor-intensive work, which otherwise result in wastage of useful manhours, losses in terms of efficiency, and less visibility across the departments and processes. Such issues are commonly found in many textile units in Pakistan, and there was a need for a local solution that considers the individual requirements of Pakistan's textile industry.

Considering the problem, an AGV was designed and developed by the National University of Sciences and Technology, Pakistan, in collaboration with FAST National University of Computer & Emerging Sciences, Pakistan. The target industry was Crescent Textile Mills, Pakistan. The developed AGV is shown in Figure 5.8. The overall scope of work was divided into two heads: electronic and mechanical. The electronics head contained work and equipment inspired by the 'open-source automatic robot (AMR) design' by Robotec Solutions (Robotec, 2021). Moreover, two permanent magnet brushless motors were connected with the system's core, i.e., the FBL2360 dual channel motor controller. The same controller also acted as the robot's supervisory and navigation computer. The battery management system was connected to a lithium-ion battery pack that served as the power source. Also, at the docking stations, the charge contact allowed the robot to connect to the battery charger. The robot followed magnetic tapes affixed to the floor with the help of magnetic track sensors to follow a pre-defined path. It also included a Wi-Fi adapter for communication with a host computer.

FIGURE 5.8 Developed AGV for the textile industry.

FIGURE 5.9 First view of the mechanical CAD design of developed AGV.

The proposed AGV used the 'differential steering' method, which is popular for all movements (forward and backward), including sharp turns. The fully symmetrical design further facilitated the operation of AGV in both directions. In addition, there is often a requirement from the industry for a manual driving option for AGVs. For the said purpose, remote control (RC) radio was added to connect directly to the motor controller's free inputs that could easily switch from automatic to manual command. Finally, ultrasonic sensors were used for obstacle avoidance and safety. A script was also written to react to any impact conditions, and a small, inexpensive piezo buzzer was made active during movement to alert people of the robot's presence.

For the mechanical head, the specifications were shared by the industry, i.e., Crescent Textile Mills, Pakistan. The AGV was required to carry a load of up to 200 kg, which required rigorous material selection. Moreover, the purpose of the AGV was concentrated not only on lifting the desired load but also on moving smoothly with consistency. Structural, modal, and harmonic analyses were performed in SolidWorks while designing components such as beams, chassis, motor brackets, tire rim, bearing brackets, shafts, shaft flanges, and battery holding frame. The complete mechanical CAD design for the AGV is shown in Figures 5.9 and 5.10. The AGV was deployed in Crescent Textile Mills, Pakistan, in June 2022 and is currently under operation.

5.7.2 Case Study 2 – Automated Deliverance of Goods by an AGV

The need for AGV for transportation and general mobility applications in the industry gives rise to the need for sophisticated planning and control algorithms to carry out any given task in the presence of external environmental disturbances. Thus, it can also be inferred that the robot's operating environment is the crucial factor in deciding and selecting the required AGV type and its planning and control algorithm (Nguyen et al., 2020). A study by Bajestani and Vosoughinia (2010) presented a simple and cheap ATMEGA-based landmark navigation system for an AGV. Though the AGV gave excellent results, the approach needed to be more

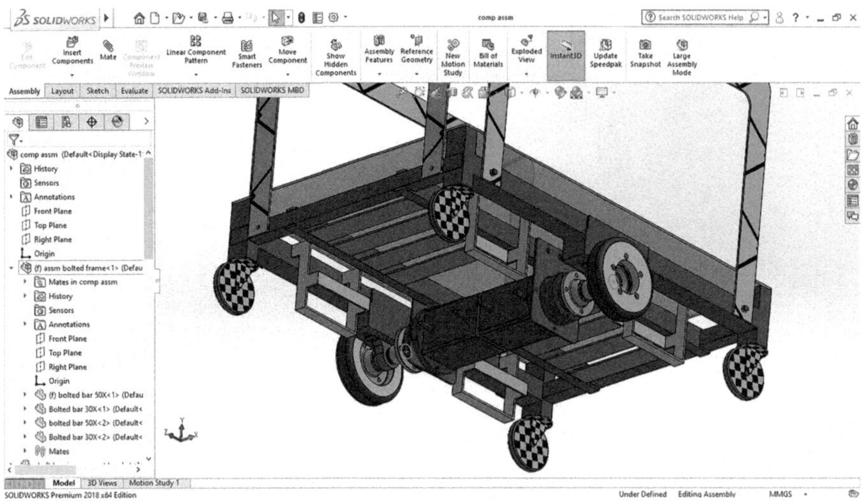

FIGURE 5.10 Second view of the mechanical CAD design of developed AGV.

complex and able to adapt to any uncertainties. Kishor (2012) presented a better approach using an NN-based AGV, which utilized vision-based scanners to navigate and sense the surroundings. However, with no landmark information, more than a single camera input is needed for the smooth operation of an AGV in a dynamic industrial setup, thereby raising the need for sensor fusion. The AGV system Rex and Klemets (2019) developed uses a laser scanner and camera input. The case study was performed using a MiR200 in Volvo Trucks' factory in Tuve, Sweden. The main task of the AGV was to deliver materials from one assembly place to another in an efficient and safe method in the presence of human worker interference. The study used four steps, as illustrated in Figure 5.11, to carry out the

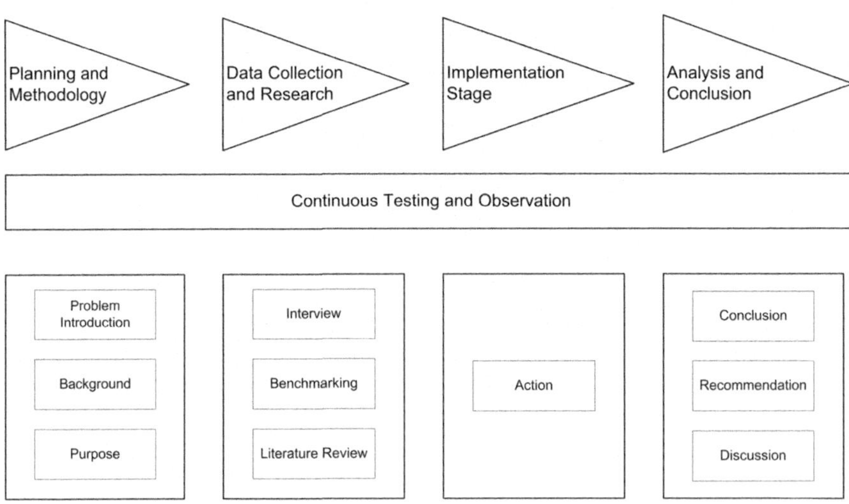

FIGURE 5.11 Steps for carrying out delivery task.

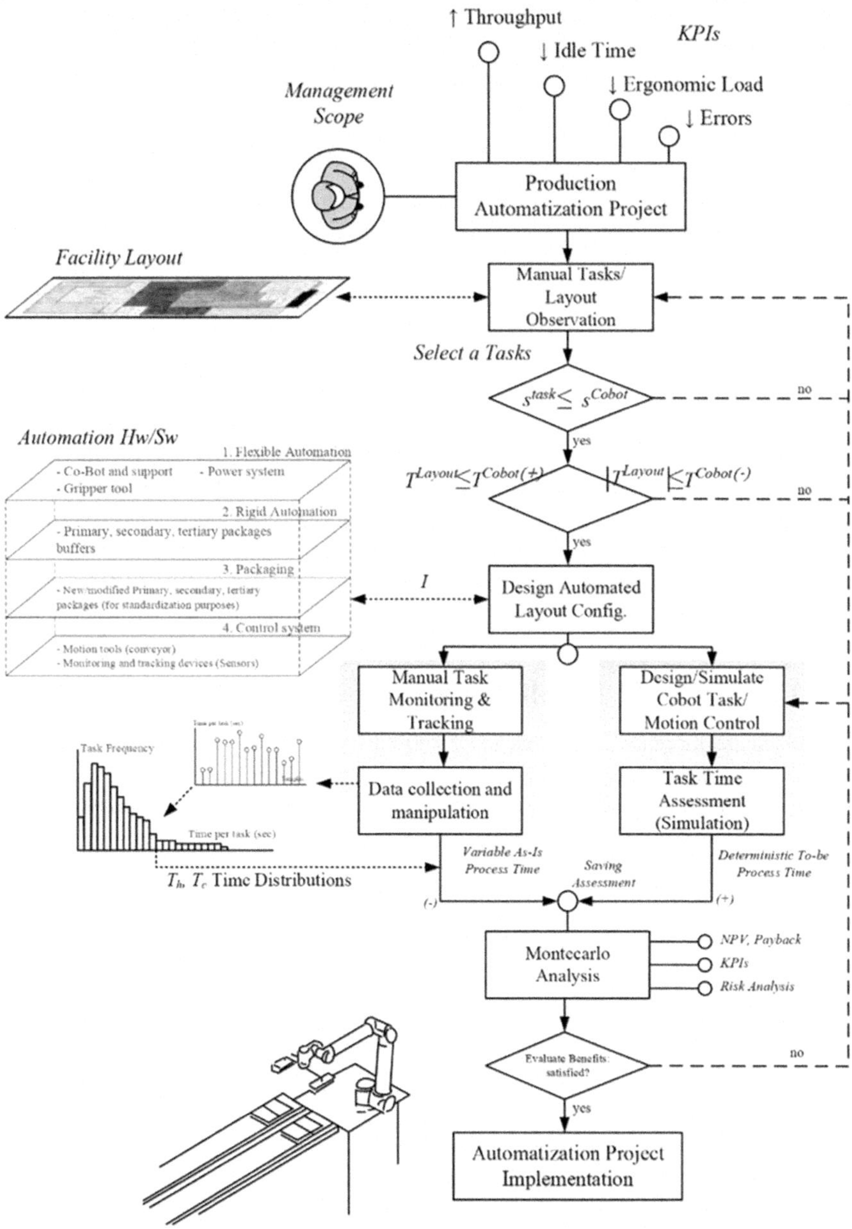

FIGURE 5.12 Food packaging flow adopted for manipulator (adapted by Accorsi et al., 2019).

delivery task. However, the method suffered from the absence of an in-depth control algorithm and a lack of compatibility tests which failed AGV to handle various unforeseen scenarios despite sensor fusion.

5.7.3 Case Study 3 – An Application of Collaborative Robots in the Food Production Facility

Although the food industry is one of the most significant leading sectors in any country's economy, the real-time deployment of robotics is slow for them mainly due to the sensitivity of food items. Therefore, a case study was presented by Accorsi et al. (2019) to understand the effectiveness of collaborative robots in the food catering industry. While performing the analysis, it was kept in mind that the catering industry's primary goal is to produce cheap food customized to the customer's needs and tightly packaged and stored to avoid quality loss and contamination. The case study (Accorsi et al., 2019) was performed in a catering industry that produced over ten thousand packages per day (Penazzi et al., 2017). The scope of the cobot was to reduce the operator's load, decrease the idle time of the active labor, and improve the overall packaging rate. The methodology adopted by this approach is described in Figure 5.12. The analysis of the cobot performance concluded that the robotic system could automate the promising tasks for the decision-makers.

5.8 CONCEPTUAL FRAMEWORK AND PROSPECTS

With a drastic increase in the usage of robotic systems in industrial setups, it has become imperative that next-generation robots can meet the industry's dynamic demands. From zero perception of surroundings to custom and cloud manufacturing,

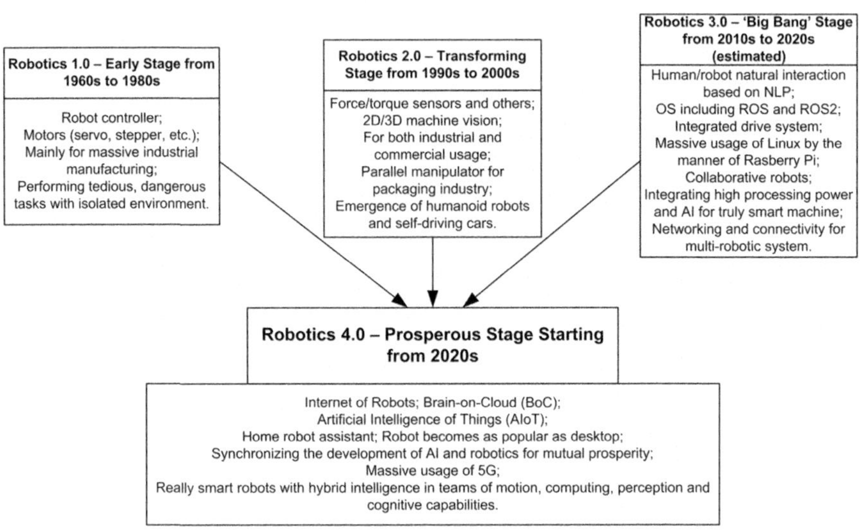

FIGURE 5.13 Robotics 1.0 to Robotics 4.0 – a journey explained (adapted from Gao et al., 2020).

robotics has come a long way. The current era has seen numerous advances, including NN interactions (Xu et al., 2019), digital twinning (He et al., 2018), and human-robot collaboration (Nikolakis et al., 2019). However, with all these advancements, robotics has yet to undergo updates to cope successfully with the emerging challenges of the ever-growing industry. This further demands the era of 'Robotics 4.0' in addition to Industry 4.0 to bring out the true revolution of industrial setups. A conceptual framework to achieve robotics 4.0 is presented in Figure 5.13 (Gao et al., 2020), where the focus was laid on the synchronization of artificial intelligence (AI) and robotics. Further, 5G and a new concept based on AI-based IoT were used. Including these new methods will guarantee a drastic advancement in the world of robotics, thereby taking Industry 4.0 to a new era of prosperity.

REFERENCES

Accorsi, R., Tufano, A., Gallo, A., Galizia, F. G., Cocchi, G., Ronzoni, M., Abbate, A., Manzini, R. 2019. An application of collaborative robots in a food production facility. *Procedia Manufacturing*, 38: 341–348.

Akella, R., Kumar, P. 1986. Optimal control of production rate in a failure prone manufacturing system. *IEEE Transactions on Automatic control*, 31(2): 116–126.

Ata, A. A., Myo, T. R. 2006. Collision-free trajectory planning for manipulators using generalized pattern search. *International Journal of Simulation Modelling*, 5(4):145–154.

Audi, A. G. 2015. New human-robot cooperation in Audi's production processes [online], Accessed on 27 November 2022.

Bajestani, S. E. M., Vosoughinia, A. 2010. Technical report of building a line follower robot. In *2010 International Conference on Electronics and Information Engineering*, 1: V1–V1.

Bellman, R., Kalaba, R. 1957. On the role of dynamic programming in statistical communication theory. *IRE Transactions on Information Theory*, 3(3): 197–203.

Boukas, E. K., Yang, H. 1996. Optimal control of manufacturing flow and preventive maintenance. *IEEE Transactions on Automatic Control*, 41(6): 881–885.

Boukas, E. K., Zhu, Q., Zhang, Q. 1994. Piecewise deterministic Markov process model for flexible manufacturing systems with preventive maintenance. *Journal of Optimization Theory and Applications*, 81(2): 259–275.

Davis, M. H. 1984. Piecewise-deterministic Markov processes: A general class of non-diffusion stochastic models. *Journal of the Royal Statistical Society: Series B (Methodological)*, 46(3): 353–376.

Ferraguti, F., Landi, C. T., Secchi, C., Fantuzzi, C., Nolli, M., Pesamosca, M. 2017. Walk-through programming for industrial applications. *Procedia Manufacturing*, 11: 31–38.

Fu, Z., Yu, J., Xie, G., Chen, Y., Mao, Y. 2018. A heuristic evolutionary algorithm of UAV path planning. *Wireless Communications and Mobile Computing*, 2018, doi: 10.1155/2018/2851964.

Gao, Z., Wanyama, T., Singh, I., Gadhrri, A., Schmidt, R. 2020. From industry 4.0 to robotics 4.0—A conceptual framework for collaborative and intelligent robotic systems. *Procedia Manufacturing*, 46: 591–599.

Hägele, M., Nilsson, K., Pires, J. N., Bischoff, R. 2016. Industrial robotics. In *Springer Handbook of Robotics*, 1385–1422.

Hart, P. E., Nilsson, N. J., Raphael, B. 1968. A formal basis for the heuristic determination of minimum cost paths. *IEEE Transactions on Systems Science and Cybernetics*, 4(2): 100–107.

He, Y., Guo, J., Zheng, X. 2018. From surveillance to digital twin: Challenges and recent advances of signal processing for industrial internet of things. *IEEE Signal Processing Magazine*, 35(5): 120–129.

Kalsoom, T., Ramzan, N., Ahmed, S., Ur-Rehman, M. 2020. Advances in sensor technologies in the era of smart factory and industry 4.0. *Sensors*, 20(23): 6783.

Kedziora, D. J., Musial, K., Gabrys, B. 2020. Autonoml: Towards an integrated framework for autonomous machine learning. *ArXiv Preprint ArXiv*, 2012: 12600.

Kishor K. 2012. Design of automated guided vehicle. *IJARC*, 3(1).

Landi, C. T., Ferraguti, F., Secchi, C., Fantuzzi, C. 2016. Tool compensation in walk-through programming for admittance-controlled robots. In *IECON 2016-42nd Annual Conference of the IEEE Industrial Electronics Society*, 5335–5340.

LaValle, S. M., Branicky, M. S., Lindemann, S. R. 2004. On the relationship between classical grid search and probabilistic roadmaps. *The International Journal of Robotics Research*, 23(7–8): 673–692.

LeCun, Y., Bengio, Y., Hinton, G. 2015. Deep learning. *Nature*, 521(7553): 436–444.

Li, P., Huang, X., Wang, M. 2011. A novel hybrid method for mobile robot path planning in unknown dynamic environment based on hybrid DSm model grid map. *Journal of Experimental & Theoretical Artificial Intelligence*, 23(1): 5–22.

Naveed, K., Khan, Z. H., Hussain, A. 2014. Adaptive trajectory tracking of wheeled mobile robot with uncertain parameters.*Computational Intelligence for Decision Support in Cyber-physical Systems*, 237–262.

Nguyen D. D., Tran, H. T., Nananukul, N. 2020. A dynamic route-planning system based on industry 4.0 technology. *Algorithms*, 13(12): 308.

Nikolakis, N., Maratos, V., Makris, S. 2019. A cyber physical system (CPS) approach for safe human-robot collaboration in a shared workplace. *Robotics and Computer-Integrated Manufacturing*, 56: 233–243.

Ortenzi, V., Adjigble, M., Kuo, J. A., Stolkin, R., Mistry, M. 2014. An experimental study of robot control during environmental contacts based on projected operational space dynamics. In *2014 IEEE-RAS International Conference on Humanoid Robots*, 407–412.

Pakistan Economic Survey, Ministry of Finance. 2022. Chapter 3: Manufacturing and mining. Available at: https://www.finance.gov.pk/survey/chapter_22/PES03-MANUFACTURING.pdf (accessed 30 August 2022).

Pedersen, M. R., Høilund, C., Krüger, V. 2012. Using human gestures and generic skills to instruct a mobile robot arm in a feeder filling scenario. In *2012 IEEE International Conference on Mechatronics and Automation*, 243–248.

Penazzi, S., Accorsi, R., Ferrari, E., Manzini, R., Dunstall, S., 2017. Design and control of food job-shop processing systems: A simulation analysis in the catering industry. *The International Journal of Logistics Management*, 28(3), 782–797.

Qi, X., Chen, G., Li, Y., Cheng, X., Li, C. 2019. Applying neural-network-based machine learning to additive manufacturing: Current applications, challenges, and future perspectives. *Engineering*, 5(4): 721–729.

Rex, J. W., Klemets, E. 2019. *Automated deliverance of goods by an automated guided vehicle*. Gothenburg, Sweden: Chalmers University of Technology.

Robotec. 2021. Open source automatic mobile robot (AMR) design. Available at: https://www.roboteq.com/roboamr-2021 (accessed 30 August 2022).

Rosenthal, S., Veloso, M. 2012. Mobile robot planning to seek help with spatially-situated tasks. In *Twenty-Sixth AAAI Conference on Artificial Intelligence*, 2067–2073.

Sato, F., Nishii, T., Takahashi, J., Yoshida, Y., Mitsuhashi, M., Nenchev, D. 2011. Experimental evaluation of a trajectory/force tracking controller for a humanoid robot cleaning a vertical surface. *In 2011 IEEE/RSJ International Conference on Intelligent Robots and Systems*, 3179–3184.

Sherwani, F., Asad, M. M., Ibrahim, B. S. K. K. 2020. Collaborative robots and industrial revolution 4.0 (IR 4.0). *In 2020 International Conference on Emerging Trends in Smart Technologies (ICETST)*, 1–5.

Sisbot, E. A., Marin-Urias, L. F., Alami, R., Simeon, T. 2007. A human aware mobile robot motion planner. *IEEE Transactions on Robotics*, 23(5): 874–883.

Wang, H. J., Xiong, W. 2009. Research on global path planning based on ant colony optimization for AUV. *Journal of Marine Science and Application*, 8(1): 58–64.

World Robotics Report. 2020. International Federation of Robotics. Available online: https://ifr.org/ifr-pressreleases/news/record-2.7-million-robots-work-in-factories-around-the-globe (accessed on 7 April 2021).

Xu, B., Cai, R., Zhang, Z., Yang, X., Hao, Z., Li, Z., Liang, Z. 2019. NADAQ: Natural language database querying based on deep learning. *IEEE Access*, 7: 35012–35017.

Yamazaki, K., Ueda, R., Nozawa, S., Mori, Y., Maki, T., Hatao, N., Okada, K., Inaba, M. 2010. System integration of a daily assistive robot and its application to tidying and cleaning rooms. In *2010 IEEE/RSJ International Conference on Intelligent Robots and Systems*, 1365–1371.

Zhang, H. Y., Lin, W. M., Chen, A. X. 2018. Path planning for the mobile robot: A review. *Symmetry*, 10(10): 450.

Zhao, M., Ansari, N., Hou, E. S. 1994. Mobile manipulator path planning by a genetic algorithm. *Journal of Robotic Systems*, 11(3): 143–153.

Zliobaite, I., Bifet, A., Gaber, M., Gabrys, B., Gama, J., Minku, L., Musial, K. 2012. Next challenges for adaptive learning systems. *ACM SIGKDD Explorations Newsletter*, 14(1): 48–55.

6 Collaborative Robots

Atal Anil Kumar
Department of Engineering, Faculty of Science, Technology, and Medicine, University of Luxembourg, Luxembourg

Uzair Khaleeq uz Zaman
Department of Mechatronics Engineering, College of Electrical and Mechanical Engineering, National University of Sciences and Technology, Islamabad, Pakistan

Peter Plapper
Department of Engineering, Faculty of Science, Technology, and Medicine, University of Luxembourg, Kirchberg, Luxembourg

6.1 INTRODUCTION

Market globalization and mass customization are pushing industries to move past the conventional mass production paradigm and focus on adopting innovative technologies with humans (as workers and customers) as the core of the production system (Kleindienst et al., 2016). Industry 4.0 is framed to help companies meet the new global demands and reform the industrial environment (Laudante, 2017). One of the main features of Industry 4.0 is to enable the factory of the future by including new types of intelligent information systems and automation as well as more flexible collaborative robots called "cobots." The purpose of this chapter is to elaborate on the role that "cobots" have in the well-known era of Industry 4.0.

6.1.1 What Are Collaborative Robots or "Cobots"?

According to ISO/TC 299, a collaborative robot or cobot is defined as a robot designed to interact directly with humans in a defined collaborative space (Faccio et al., 2022). The idea of collaborative robots was introduced by Colgate, Edward, Peshkin, and Wannasuphoprasit (Colgate et al., 1996), who focused on the simplest possible version with only a single joint, also known as a steerable wheel. Since then, several versions of cobots with different technologies embedded in them have been introduced in the market (Baumgartner et al., 2020).

According to the International Federation of Robotics (IFR), the market for cobots is still growing, and the end-users and engineers are still exploring the best configuration in terms of sensors, grippers, and intuitive programming interfaces for

DOI: 10.1201/9781003327523-8

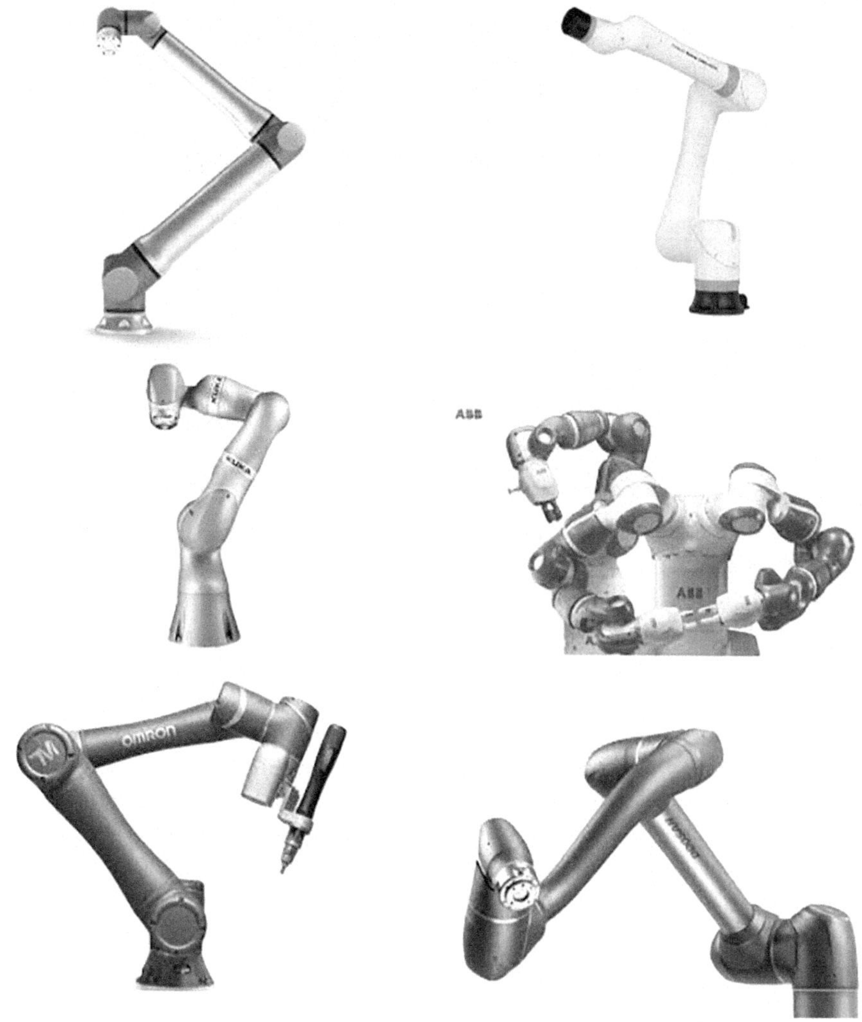

FIGURE 6.1 Cobots from different manufacturers (adapted from https://www.universal-robots.com/, https://crx.fanuc.eu/ch-fr/, https://www.kuka.com/en-de/products/robot-systems/industrial-robots/lbr-iiwa, https://new.abb.com/products/robotics/robots/collaborative-robots/yumi/irb-14000-yumi, https://industrial.omron.fr/fr/products/collaborative-robots, https://www.doosanrobotics.com/en/Index).

their efficient design and implementation in the manufacturing sector (IFR, 2020). Cobots from different manufacturers are shown in Figure 6.1.

The primary purpose of cobots is to be used in a shared workspace along with a human worker without fences to perform tasks involving varying levels of interaction. Different types of interactions are possible in a shared working environment, namely, coexistence, synchronization, cooperation, and collaboration (Baumgartner et al., 2020; Chiabert & Aliev, 2020; Fast-Berglund & Romero, 2019; Malik &

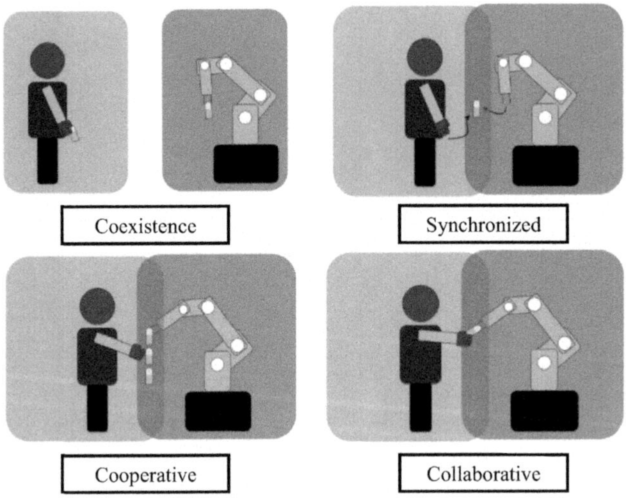

FIGURE 6.2 Schematic representation of different robot interactions (adapted from Dzedzickis et al., 2021).

Bilberg, 2019; Wang et al., 2019; Villani et al., 2018). Coexistence interaction exists when the worker and robot are close to each other without sharing the workplace. For synchronized interaction, the workspace is shared between the worker and robot but not at the same time. The task is completed in sequence, with the worker and the robot doing their steps one after the another. During cooperative interaction, the worker and the robot are in direct contact with each other; however, they work on different tasks. The collaboration interaction involves the worker and the robot having direct contact and working on the same tasks together. A schematic representation of different robot interactions is presented in Figure 6.2.

6.1.2 Increase in Demand for Cobots

The increase in the adoption of Industry 4.0 technologies in manufacturing and the rise in various customized products has significantly increased the demand for robots in general. Conventional industrial robots are unable to meet the current market demand efficiently due to their shortcomings as they are less intelligent and flexible, costly, and time-consuming. Cobots, on the other hand, can eliminate these disadvantages by working along with humans (Gervasi et al., 2020). A detailed comparison between conventional and collaborative robots is given in Table 6.1.

The primary market for cobots is small and medium-sized enterprises (SMEs) which account for about 90% of the world's enterprises and play a significant role in global economic growth and job creation (Muller et al., 2016). Since cobots are low-cost, safe, and have plug-and-play features, they can easily be installed in SMEs to produce low-volume, high-variant products with a faster return on investment (ROI). Almost all companies have reported increased productivity due to the implementation of cobots in their manufacturing setup.

TABLE 6.1
Difference Between Conventional and Collaborative Robots (Adapted from Villani et al., 2018; Bi et al., 2021; De Simone et al., 2022)

Characteristic Feature	Conventional Robots	Collaborative Robots
Proximity to humans	Prohibited	Allowed when the procedure of safety is followed
Safety barrier	Physical separation	Safety assurance mechanisms
Robot movement	Motion with separation of human workers	Simultaneous motions of worker and robot
Footprint of robot	Large for protection	Small for collaboration
Robot control	Pre-programmed	Can be modified in-line
Robot programming	Lead-through and off-line	On-line, off-line and multimodal interaction
Programming skill	Sophisticated	Intuitive
Complexity	Fixed programs for fixed tasks	Flexible programs to handle changes and uncertainties
Tasks	Repeatable, mostly fixed	Frequent changes
Structural feature	Heavy and rigid	Light-duty and easy to move
Payload	Medium to high	Low to medium
Workspace	Isolated	Shared
Use of workspace	High	Limited
Position	Fixed	Flexible
Productivity	High	Limited
Volume of products	Used for high volume	Used for lower volume and high variants
Adaptability	Hard automation by program	Soft automation by interactions

According to a study by market data forecast, the international cobot business is expected to grow to $12.48 billion by 2026 from its value of $0.65 billion in 2019, at a compound annual growth rate (CAGR) of 44.8% (Globe Newswire, 2020).

With the advancement in Industry 4.0 technologies, several factors are driving the need for using cobots in manufacturing, namely: market globalization, shortening product life cycles, high product customization, labor and social cost development, demographic changes, agility and changeability, digital transformation, and so on.

6.2 IMPORTANCE OF COBOTS IN MAKING MANUFACTURING SMART

The progress of Industry 4.0 has led to several innovative technologies, which, when combined, make manufacturing bright. In addition to cobots, a few such technologies are the Internet of Things (IoT) and artificial intelligence (AI). Using cobots supported by these technologies and humans has resulted in the development of augmented intelligence (a combination of cobots' AI and authentic human intelligence). This allows the manufacturing sector to be more efficient and accurate while enhancing creativity and introducing new diversity in the workplace. Implementing cobots has several benefits, such as:

- Improvement in productivity: Cobots are better in accuracy, perseverance, and reproducibility when compared to human workers (Villani et al., 2018). This makes them suitable for implementation in repetitive tasks where production volume also needs to be achieved.
- Increase in flexibility: Due to an increase in mass customization, manufacturing companies look for solutions without losing the classical advantages of a conventional production line (Y. Wang et al., 2017). Human-cobot teams have proven to be more efficient when compared to teams consisting of humans. The collaborative feature of cobots helps perform repetitive and physically demanding tasks while the human worker can focus on complex tasks requiring high cognitive skills.
- Increasing job attractiveness: Cobots are majorly used for performing repetitive and dull manual work. This allows the workers to engage themselves in new potential value-creating activities. This also helps them expand their skill set and take on new roles and responsibilities. Additionally, reducing repetitive and physically demanding tasks has helped improve the workers' working conditions and well-being.

Other key drivers for implementing cobots in the manufacturing sector (Bauer et al., 2016) are shown in Figure 6.3.

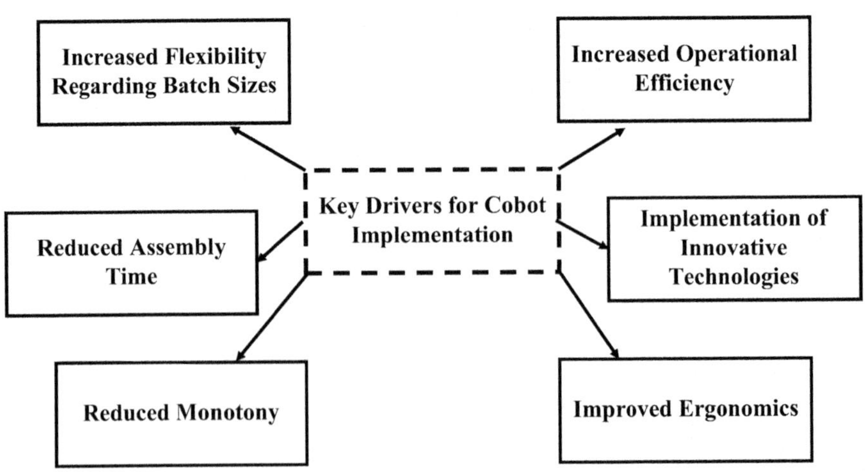

FIGURE 6.3 Key drivers for implementing cobots in manufacturing industries (adapted from Bauer et al., 2016).

6.3 NEED FOR COLLABORATION BETWEEN ROBOTS AND INDUSTRIAL WORKERS

There needs to be more clarity among the public about the introduction of robotic technology in the automation field. This is due to the common presumption that all the jobs and occupations of the workers will be replaced by robots, resulting in fear of job loss. However, a more accurate perception can be understood from the statistical study from the US Bureau of Labor and McKinsey Global Institute (McKinsey & Company, 2017), which states that 18% of the time consumed to accomplish a job is wasted in predicted activities that can be automated with a success rate of 78%. Also, 12% of the time consumed is spent on predictable activities, which have a low success rate (around 25%) of automation. In manufacturing, there are scenarios where the success rate of automating predictable activities is just 60%. Hence, manufacturing still has predictable and unpredictable tasks that cannot be automated fully and need human workers to accomplish them. Therefore, there is a definite need for an environment where workers and robots can collaborate to perform tasks efficiently.

6.3.1 Cobots as an Essential Tool to Achieve Industry 4.0

The shortage of skilled workers in recent years, especially in developed countries like Japan, the USA, and many countries in the EU, has affected manufacturing companies as they struggle to recruit qualified engineers and skilled production workers (McCarthy & Richter, 2020). According to a Deloitte report (2018), in 2018, it took companies 23 days longer to fill vacancies compared to 2015. Hence, employers compete for appropriate personnel in the labor market and make considerable efforts to increase their chances of recruiting suitable personnel. Large companies' reputation and publicity, and financial capabilities often allow them to offer higher salaries, making finding appropriate staff even more challenging for SMEs. Also, the younger generation has an opposing view on the manufacturing industry due to reports of offshoring manufacturing activities and jobs being dull, dangerous, dumb, and dirty (Skevi et al., 2014). All these factors have led to cobots being considered an attractive solution for relieving employees from physically and mentally stressful tasks, thereby increasing job attractiveness. Additionally, industries also believe that using cobots in their production line helps raise their reputation as innovative employers, which will attract a young and technology-driven workforce (Kopp et al., 2020).

6.3.2 Importance of Workers in Manufacturing

The ability of human workers to adapt dynamically to unpredictable tasks makes them irreplaceable in any manufacturing line involving a wide variety of products. One such example is building the interiors of an aircraft, where the tasks to be performed are more human-dependent than robots as they can adapt themselves much more efficiently. Human workers can use their natural senses intuitively to find instant solutions to complicated problems without wasting much time. For example, for the assembly of electronic devices, finding a standard way to connect

TABLE 6.2
Typical Strengths of Humans and Cobots (Adapted from Kopp et al., 2021)

Human Strengths	Cobot Strengths
Flexibility	Endurance
Perception	Power
Sensorimotor abilities	Reproducibility
Handling of different components (soft, moving, flexible etc)	Precision
Instant planning and action capability	Speed (except in collaboration mode)

some very tiny parts is extremely difficult to achieve with a robot compared to a human worker. Human workers' most significant advantage is their accumulated experience and logical judgment, which still needs to be achieved using robots.

A team of cobots and human workers provides a perfect synergy by combining the strengths of human beings and robots, which results in a superior working system (Selevsek & Köhler, 2018). Table 6.2 shows the typical strengths of humans and cobots (Kopp et al., 2021).

6.4 CHALLENGES IN IMPLEMENTING COBOTS IN MANUFACTURING

The introduction of cobots in any manufacturing environment faces challenges and difficulties. This chapter section aims to highlight these challenges and their necessary mitigation strategies.

One of the foremost vital challenges is the occupational safety issues when cobots are installed to work alongside the workers (Kopp et al., 2021). The reason for this is that the entire working system must be considered in the risk analysis in addition to the cobot itself (Kopp et al., 2020; Grahn et al., 2018). Such risk analysis is usually time-consuming and complex as it gives rise to many dynamic parameters that are difficult to envisage before its actual working in the environment. It is also important to note that the risks must be identified and eliminated (Mateus et al., 2019). In the work of Mateus et al., the authors have highlighted specific hazards grouped according to their hazard paths between the origin and the affected person (Mateus et al., 2019). These hazards might occur from (i) robots during human-robot collaboration, (ii) the process involved during the collaboration, or (iii) malfunctions in the control system of the robot during collaboration. The primary safety standards for applications of industrial robot systems are ISO standards 10218–1 (ISO, 2011a) and 10218–2 (ISO, 2011b).

Additionally, the guidelines for designing and implementing a collaborative workspace and its risk assessment are provided by ISO TS 15066 (ISO, 2016). However, more specified safety standards are needed as ISO TS 15066 is just an addition to the existing robot directive (Holm & Schnell, 2022). There is also the need to redefine safety in collaborative environments by considering the skill level of the workers and the tasks to be performed by the robots (Fast-Berglund et al., 2016).

The safety standards must also be redesigned due to the presence of conventional industrial robots and collaborative robots in the same manufacturing plant (Wang et al., 2019). This needs advanced risk assessments involving hazard identification, risk evaluation, and reductions during the early stages of the manufacturing layout planning (Gervasi et al., 2020; Poot et al., 2018).

The second challenge, a classical barrier in adopting cobots, is the fear among employees of being replaced by them (Richert et al., 2018). Industry specialists have already reported that full automation, a scenario where humans are removed from manufacturing, is not viable (Kolbeinsson et al., 2018) and that establishments should focus on creating shoulder-to-shoulder cooperation between the workers and different types of intelligent machines (Wurhofer et al., 2014). Hence, companies must present cobots as supportive colleagues to their workers rather than replacement machines. A study by Salvini et al. (2010) reports that this fear is one of the predominant factors hindering the acceptance of cobots among workers. One way to address this fear is by explaining the difference between conventional automation and cobots. They are not used to eliminate humans but to support them in collaborative working conditions.

The next challenge the companies face is getting workers' trust in cobots. This is very important during the installation and the operation phase to get the desired results as it indicates the relevance of worker-related aspects. Without trust, the worker may underutilize the robot, possibly reducing performance or even not using it (Gervasi et al., 2020). The workers' trust in cobots depends on social and emotional factors (Sadrfaridpour & Wang, 2017). A successful human-robot interaction conceptualizes trust as a prerequisite (Broadbent, 2017). Establishing trust begins before the actual physical implementation of the cobot. There is a difference between the initial trust prior to actual interaction and dynamic trust during interaction (Kopp et al., 2021).

According to one of the most comprehensive trust models in the domain of automation (French et al., 2018), trust is divided into dispositional, situation, and learned trust. Of these, dispositional and major situational factors can hardly be influenced. At the same time, the learned trust, which is a result of the worker's mental model of the robot, is dependent on the characteristics of the robot and is continuously modified by experiences (Ewart, 2019). Lewis and Walker reported that the actual reliability of the robot mainly determines the level of trust, which increases gradually when the worker becomes familiar with the robot through contact with it (Lewis et al., 2018). The initial trust can also be established by following efficient internal top-down communication strategies. These strategies work toward developing an appropriate mental model to allow positive experiences during an interaction. One thing to note here is that to reduce the impact of a single negative experience, many positive experiences are needed (Flook et al., 2019; Desai et al., 2013). The physical configuration of the robot (type of design) can also influence the trust built by the worker in the robot. For example, a giant robot may intimidate a worker into collaborating, while a more miniature robot may make the worker feel more comfortable (Gervasi et al., 2020). Three elements should be considered while assessing the worker's trust, namely (Charalambous et al., 2016), the robot, human, and external elements. The robot element considers performance (speed, movement, reliability, etc.) and physical aspects (dimensions, appearance, etc.). The human element includes safety, trust, previous experience with robots, and the human cognitive model. Finally, the external element is related to the complexity of the activity (De Simone et al., 2022).

Another challenge the decision makers face is choosing the appropriate robot configuration depending on the production process, tasks to be done by the worker and the selected cobot, and the positioning of the functional materials. This is because not all processes require a cobot to be used. It also means that the optimal workstation configuration should be planned (Turja & Oksanen, 2019). Due to the need for more well-defined standards, procedures, and steps, implementing cobot solutions require more work (Holm & Schnell, 2022). This requires the SMEs to have a precise robotics and automation strategy to implement a successful cobot application. The lack of expertise in this domain can result in choosing too complex automation tasks or involving too high interaction levels (Fast-Berglund & Romero, 2019).

One of the lesser challenges is the financial factors in introducing cobots. Generally, the costs are categorized as acquisition, maintenance, and operational costs. It has been reported that the impact of operational costs is more from a long-term perspective (Kopp et al., 2021). This is because the acknowledgment of value added to the industrial worker and the cobot needs to be clarified in teams consisting of workers and cobots, and the total costs exceed the one-time acquisition costs of the robot itself (Ranz et al., 2018).

6.5 COBOTS IN MANUFACTURING – CASE STUDIES

6.5.1 Case Study – Cobots in a High-variant Low-volume Manufacturing Line

One of the best applications of cobots has been in the high-variant low-volume manufacturing sector. One example is from Scott Fetzer Electrical Group (SFEG), an electronics manufacturer from Tennessee, USA. They have used robots to take over the workers‘ repetitive and potentially hazardous tasks (UR, 2016a). By doing so, they could use their experienced workers to perform more rewarding and value-added tasks, which in turn increased their employee satisfaction as they felt more valuable.

They used a mobile and flexible cobot fleet (cobots on rolling carts) which were easy to transport between different workstations. Since the cobots had safety features already embedded in them, the company did not need any additional safety fences or other surrounding safety sensors. This helped the high-variant low-volume company to use the robots where necessary and make the optimum use of the existing machinery, as shown in Figure 6.4.

6.5.2 Case Study – Cobots in Automotive Assembly Lines

One of the earliest implementations of cobot was done in the automotive assembly lines of an Indian automotive company (Bajaj Auto Ltd.) in 2010. Assembly lines in any manufacturing plant are highly labor intensive, require high precision, and face space challenges.

FIGURE 6.4 Use of cobot in low-volume, high-variant manufacturing line (adapted from UR, 2016a).

The specific features the company looked for in a cobot included flexibility in programming and installation, affordability, compactness, lightweight, and safe to work alongside humans (UR, 2016c). These features were chosen to improve the productivity, flexibility, and reliability of their assembly line along with the ergonomics for the company workers (finding the right kind of working position to eliminate work-related musculoskeletal disorders). They installed cobots from Universal Robots (UR) in their assembly line with a typical pitch dimension of one meter for assembling motorcycle engines, as shown in Figure 6.5.

FIGURE 6.5 Cobot setup used by Bajaj Auto for automotive assembly tasks (adapted from UR, 2016c).

Following the successful installation and integration of cobots in 2010, they currently have more than 100 cobots for various applications such as machine tending, material handling, and engine assembly. Their inclusion has increased the personnel productivity at the company from 507 to 804 vehicles per person per year. Cobots have also helped remove barriers for women in an assembly line and increased their participation to 50% of the total workforce. This contrasts with the widespread belief that the introduction of cobots leads to job losses.

6.5.3 Case Study – Integration of Cobots in Automotive Manufacturing

Continental automotive factories in Spain integrated cobots into their manufacturing line to increase market competitiveness and productivity (UR, 2016b). Having already been exposed to the advantages of automation and IoT due to the introduction of Industry 4.0, they were keen on using cobots for monotonous and repetitive tasks such as gluing, dispensing, and validating the printed circuit boards produced by the company.

Due to the salient safety features already available in cobots, the company was able to install them very quickly and was able to achieve a rapid return on investments (ROI) (less than 24 months). They also achieved a 50% reduction in their changeover time, which was reduced from 40 minutes before the introduction of cobots to 20 minutes afterwards. A cobot installed by the company is shown in Figure 6.6.

The easy programming methods available for cobots from UR helped the workers develop programs for the final products much faster than before, accelerating their implementation.

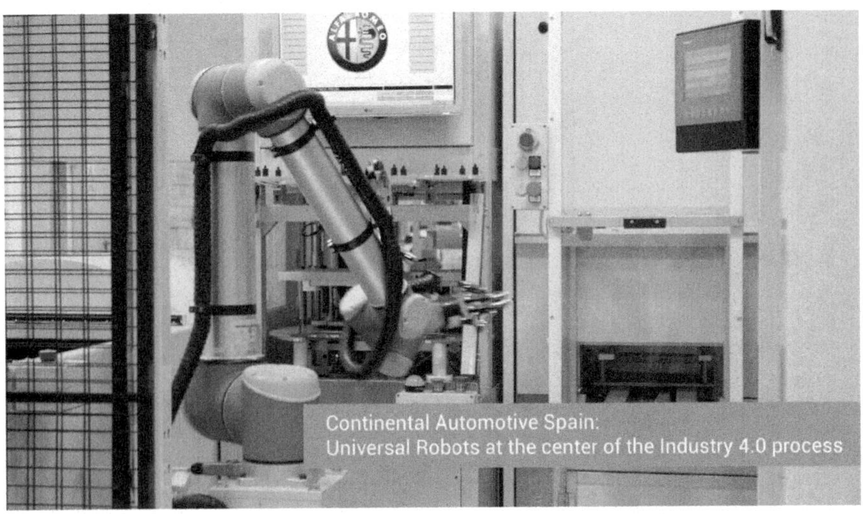

FIGURE 6.6 Cobot used by Continental Automotive for automating tasks (adapted from UR, 2016b).

6.5.4 Case Study – Assembling Fiat 500 Electric Car

Stellantis N.V., a Dutch company, installed 11 cobots from UR at its assembly plant in Turin, Italy, to automate complex assembly line operations and quality controls for the new Fiat 500 electric car (Crowe, 2022). They installed the cobots for a wide range of operations such as the application of waterproof liner to the vehicle doors, positioning of the soft-top using tracking and visual inspection, checking the soft-top frame dimensions, riveting of the tailgate, mounting the hood, tightening of the door hinges, and mudguard mounting. They used a mix of UR5 and UR10 cobots depending on the tasks. The setup used by the company is shown in Figure 6.7.

The company took steps in gradually introducing the cobots to help the operators fully understand their characteristics. They were first introduced in the company canteen, where their operators could interact with the cobot (which was tasked with the distribution of glasses of water during lunch break). This allowed the workers to observe the safety and collaborative features of the cobot by themselves and overcome inhibitions in accepting them in their working environment.

The company benefited from implementing cobots in all the applications they selected. Due to cobots' high repeatability and ability to follow complex paths easily within a confined space, they obtained high operating precision, increased productivity, and quality, along with better ergonomics and working conditions for their operators.

6.5.5 Case Studies from FANUC CRX Cobots

FANUC, being the largest maker of industrial robots in the world, has developed several models of cobots that have been implemented for different applications

FIGURE 6.7 Setup of the cobots used for assembling Fiat car (adapted from Crowe, 2022).

worldwide. Some of the recently implemented cobot solutions are presented here as case studies:

6.5.5.1 Lights-out Production Using FANUC CRX Cobots

Collaborative robots from FANUC, also known as FANUC CRX Cobot, have been used by Athena 3D Manufacturing company to achieve lights-out0 production (https://www.fanucamerica.com/case-studies/athena-3d-manufacturing). The printers that completed their jobs in the middle of the night were forced to sit idle until the next morning when the operator changed out the print bed to start the next job, which was highly time-consuming and impacted machine utilization and production deadlines. Using the collaborative robots from the FANUC CRX series, the company was able to quickly scale production from one to multiple printers, all running at the same time, all night long, because of which they were able to double their production with a 40% increase in machine utilization. This enabled the employees to focus more on product quality as well.

6.5.5.2 Case Study – Automated Welding Operation Using FANUC CRX Cobots

Last Arrow Manufacturing, an Ohio-based contract manufacturer, used FANUC CRX welding cobot to develop a flexible automated solution to perform simple and repetitive welding projects (https://www.fanucamerica.com/case-studies/all/last-arrow-manufacturing). This has helped them in increasing their productivity, freeing up their skilled welders to carry tasks requiring more expertise. They also achieved high employee satisfaction with improved flexibility and higher profits.

6.5.5.3 Case Study – Integrating Mobile Cobots with FANUC CRX Cobots

One application of integrating a collaborative mobile robot with a collaborative industrial robot can be seen at an Austrian company, GER4TECH (https://www.ger4tech.at/en-g4t4/). They have integrated G4T4 (a mobile cobot) with FANUC CRX-10iA/L. These mobile cobots can move autonomously between different workstations, transport materials, and switch from one job to another without any trouble. The company has used them to handle applications for machine tools, transportation of raw materials and finished parts, pick and place applications, warehouse picking, etc.

Further applications using FANUC CRX cobots can be found in Success Stories – YouTube.

6.6 CONCEPTUAL FRAMEWORK

Different frameworks and models are available for introducing cobots in a manufacturing line. Each framework follows different principles. Eekels and Roozenburg (1991) used an engineering design cycle, Ranz et al. (2018) proposed a framework using the quantitative and qualitative conceptual and technical aspects of human-robot interactions, and a generic two-dimensional design and implementation framework was proposed by Djuric et al. (2016). Further explanation of the models and framework can be found in Kopp et al. (2021).

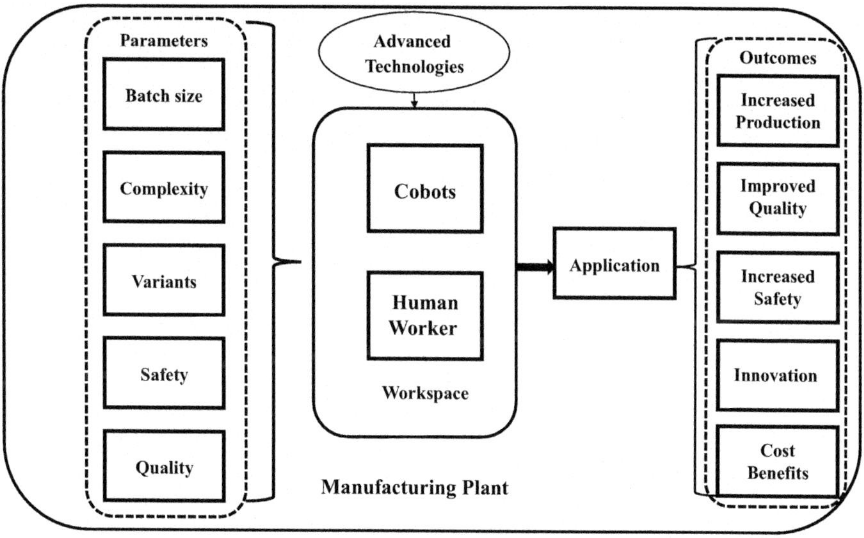

FIGURE 6.8 A simplified conceptual framework for implementing cobots in a manufacturing line.

A simplified conceptual framework is proposed for the readers as a starting point to check for the use of cobots in a manufacturing line. This framework is adapted from the work of (Kopp et al., 2021). The framework is shown in Figure 6.8. It takes into account the parameters that should be considered when the need for implementing a cobot for a manufacturing line is being planned.

6.7 CONCLUSION

As more SMEs and companies start implementing Industry 4.0 along with cobots for their production line, considerable modifications are expected in the production processes. Changes are foreseen on different levels, namely, work content, work organization, production management, and other organizational factors (Badri et al., 2018; Waschneck et al., 2016). These changes will result in the manufacturing systems being self-learning with the ability to make decisions by themselves. Such systems will incorporate the creativity and adaptability of human workers and help these workers in assuming more leadership and supervisory roles on the shop floor (Khalid et al., 2018).

Future research should explore more in detail which work can be usefully automated, what should be the optimum level and complexity of collaboration in such automated tasks and how it will affect the cognitive workloads of the human workers, the required skills and training needed to achieve it to reach the level needed for a booming Industry 4.0 implementation.

REFERENCES

Badri, A., Boudreau-Trudel, B., Souissi, A. S. 2018. Occupational health and safety in the industry 4.0 era: A cause for major concern? *Safety Science*, 109: 403–411.

Bauer, W., Bender, M., Braun, M., Rally, P., Scholtz, O. 2016. Lightweight robots in manual assembly–best to start simply. *Frauenhofer-Institut für Arbeitswirtschaft und Organisation IAO, Stuttgart.*

Baumgartner, M., Kopp, T., Kinkel, S. 2020. Industrial human-robot-collaboration in Smes–Smes underestimate the potential of human-robot-collaboration. *WT Werkstattstechnik*, 110(3): 146–150.

Bi, Z. M., Luo, C., Miao, Z., Zhang, B., Zhang, W. J., Wang, L. 2021. Safety assurance mechanisms of collaborative robotic systems in manufacturing. *Robotics and Computer-Integrated Manufacturing*, 67: 102022.

Broadbent, E. 2017. Interactions with robots: The truths we reveal about ourselves. *Annual Review of Psychology*, 68(1): 627–652.

Charalambous, G., Fletcher, S., Webb, P. 2016. The development of a scale to evaluate trust in industrial human-robot collaboration. *International Journal of Social Robotics*, 8: 193–209.

Chiabert, P., Aliev, K. 2020. Analyses and study of human operator monotonous tasks in small enterprises in the era of industry 4.0. In *IFIP International Conference on Product Lifecycle Management*, 83–97.

Colgate, J. E., Edward, J., Peshkin, M. A., Wannasuphoprasit, W. 1996. Cobots: Robots for collaboration with human operators. *Paper presented at Proceedings of the 1996 ASME International Mechanical Engineering Congress and Exposition*, Atlanta, GA, USA, 11/17/96 - 11/22/96, 433–439.

Crowe, S. 2022. 11 Cobots Assemble Fiat 500 Electric Car. https://www.Cobottrends.Com/11-Cobots-Assemble-Fiat-500-Electric-Car/. February 8, 2022.

Deloitte report. 2018. 2018 Deloitte skills gap and future of work in manufacturing study a deloitte series on the skills gap and future of work in manufacturing. Deloitte. https://www2.deloitte.com/content/dam/insights/us/articles/4736_2018-Deloitte-skills-gap-FoW-manufacturing/DI_2018-Deloitte-skills-gap-FoW-manufacturing-study.pdf (accessed on 09 June 2023)

Desai, M., Kaniarasu, P., Medvedev, M., Steinfeld, A., Yanco, H. 2013. Impact of robot failures and feedback on real-time trust. In *2013 8th ACM/IEEE International Conference on Human-Robot Interaction (HRI)*, 251–258.

De Simone, V., Di Pasquale, V., Giubileo, V., Miranda, S. 2022. Human-Robot collaboration: An analysis of worker's performance. *Procedia Computer Science*, 200: 1540–1549.

Djuric, A. M., Urbanic, R. J., Rickli, J. L. 2016. A framework for collaborative robot (CoBot) integration in advanced manufacturing systems. *SAE International Journal of Materials and Manufacturing*, 9(2): 457–464.

Dzedzickis, A., Subačiūtė-Žemaitienė, J., Šutinys, E., Samukaitė-Bubnienė, U., Bučinskas, V., 2021. Advanced applications of industrial robotics: New trends and possibilities. *Applied Sciences*, 12(1): 135.

Eekels, J., Roozenburg, N. F. 1991. A methodological comparison of the structures of scientific research and engineering design: Their similarities and differences. *Design Studies*, 12(4): 197–203.

Ewart, J., de Visser, E. J., Peeters, M. M. M., Jung, M. F., Kohn, S., Shaw, T. H., Pak, R., Neerincx, M. A. 2019. Towards a theory of longitudinal trust calibration in human-robot teams. *International Journal of Social Robotics*, 12(2): 459–478.

Faccio, M., Granata, I., Menini, A., Milanese, M., Rossato, C., Bottin, M., Minto, R., Pluchino, P., Gamberini, L., Boschetti, G., Rosati, G. 2022. Human factors in cobot era: A review of modern production systems features. *Journal of Intelligent Manufacturing*, 1–22.

Fast-Berglund, Å., Palmkvist, F., Nyqvist, P., Ekered, S., Åkerman, M. 2016. Evaluating cobots for final assembly. *Procedia CIRP*, 44: 175–180.

Fast-Berglund, Å., Romero, D. 2019. Strategies for implementing collaborative robot applications for the operator 4.0. In *IFIP International Conference on Advances in Production Management Systems*, 682–689.

Flook, R., Shrinah, A., Wijnen, L., Eder, K., Melhuish, C., Lemaignan, S. 2019. On the impact of different types of errors on trust in human-robot interaction: Are laboratory-based HRI experiments trustworthy?. *Interaction Studies*, 20(3): 455–486.

French, B., Duenser, A., Heathcote, A. 2018. Trust in automation–A literature review. In *CSIRO Report EP184082*. Australia: CSIRO.

Gervasi, R., Mastrogiacomo, L., Franceschini, F. 2020. A conceptual framework to evaluate human-robot collaboration. *The International Journal of Advanced Manufacturing Technology*, 108(3): 841–865.

Globe Newswire. 2020. Industrial Cobot Market to Grow $12.48 Billion by 2026: At 44.8% CAGR. Allied Market ReIndustrial Cobot Market to Grow $12.48 Billion by 2026: At 44.8% CAGR. *Globe Newswire*. https://www.Globenewswire.Com/En/Newsrelease/2020/06/10/2046305/0/En/Industrial-Cobot-Market-to-Grow-12-48-Billion-By2026-at-44-8-CAGR.Htm. June 10, 2020.

Grahn, S., Gopinath, V., Wang, X. V., Johansen, K. 2018. Exploring a model for production system design to utilize large robots in human-robot collaborative assembly cells. *Procedia Manufacturing*, 25: 612–619.

Holm, M., Schnell, M. 2022. Challenges for manufacturing SMEs in the introduction of collaborative robots. In *10th Swedish Production Symposium (SPS2022), Skövde, April* 26–29 2022, 173–183.

IFR. 2020. Demystifying collaborative industrial robots. In. *Positioning Paper*, IFR.

ISO, E. 2011a. Robots and Robotic Devices—Safety Requirements robotic devices—Safety requirements for Industrial Robots—Part industrial robots—Part 1: Robots. *ISO: Geneve, Switzerland*. 10218.

ISO, I. 2011b. Robots and Robotic Devices–Safety Requirements robotic devices–Safety requirements for Industrial Robots–Part industrial robots–Part 2: Robot Systems. *International Organization for Standardization: Geneva, Switzerland*, 10218- 2.

International Organization for Standardization. 2016. *Robots and Robotic Devices: Collaborative Robots*. ISO.

Khalid, A., Kirisci, P., Khan, Z. H., Ghrairi, Z., Thoben, K. D., Pannek, J. 2018. Security framework for industrial collaborative robotic cyber-physical systems. *Computers in Industry*, 97: 132–145.

Kleindienst, M., Wolf, M., Ramsauer, C., Pammer-Schindler, V. 2016. Industry 4.0: What workers need and what ICT can give-an analysis. In *i-Know 2016: 16th International Conference on Knowledge Technologies and Data-driven Business*.

Kolbeinsson, A., Lagerstedt, E., Lindblom, J. 2018. Classification of collaboration levels for human-robot cooperation in manufacturing. In *Advances in Manufacturing Technology XXXII*, 151–156.

Kopp, T., Schäfer, A., Kinkel, S. 2020. Kollaborierende oder kollaborationsfähige Roboter. *Welche Rolle spielt die Mensch-Roboter-Kollaboration in der Praxis*, 19–23.

Kopp, T., Baumgartner, M., Kinkel, S. 2021. Success factors for introducing industrial human-robot interaction in practice: an empirically driven framework. *The International Journal of Advanced Manufacturing Technology*, 112(3): 685–704.

Laudante, E. 2017. Industry 4.0, Innovation and Design. A new approach for ergonomic analysis in manufacturing system. *The Design Journal*, 20(sup1): S2724–S2734.

Lewis, M., Sycara, K., Walker, P. 2018. The role of trust in human-robot interaction. In *Foundations of trusted autonomy*, 135–159.

Malik, A. A., Bilberg, A. 2019. Developing a reference model for human–robot interaction. *International Journal on Interactive Design and Manufacturing (IJIDeM)*, 13(4): 1541–1547.

Mateus, J. C., Claeys, D., Limère, V., Cottyn, J., Aghezzaf, E. H. 2019. A structured methodology for the design of a human-robot collaborative assembly workplace. *The International Journal of Advanced Manufacturing Technology*, 102(5): 2663–2681.

McCarthy, N., Richter, F. 2020. Infographic: The countries polluting the oceans the most. *Statista Infographics*, 12.

McKinsey & Company. 2017. MGI-a-future-that-works-Executive-summary. Mckinsey Global Institute, https://www.mckinsey.com/~/media/mckinsey/featured%20insights/Digital%20Disruption/Harnessing%20automation%20for%20a%20future%20that%20works/MGI-A-future-that-works-Executive-summary.ashx (accessed on 9 June 2023).

Muller, P., Devnani, S., Julius, J., Gagliardi, D., Marzocchi, C. 2016. Annual report on European SMEs 2015/ 2016. *Technical Report*.

Poot, L., Johansen, K., Gopinath, V. 2018. Supporting risk assessment of human-robot collaborative production layouts: A proposed design automation framework. *Procedia Manufacturing*, 25: 543–548.

Ranz, F., Komenda, T., Reisinger, G., Hold, P., Hummel, V., Sihn, W. 2018. A morphology of human robot collaboration systems for industrial assembly. *Procedia CIRP*, 72: 99–104.

Richert, A., Müller, S., Schröder, S., Jeschke, S. 2018. Anthropomorphism in social robotics: Empirical results on human–robot interaction in hybrid production workplaces. *AI & Society*, 33(3): 413–424.

Sadrfaridpour, B., Wang, Y. 2017. Collaborative assembly in hybrid manufacturing cells: An integrated framework for human–robot interaction. *IEEE Transactions on Automation Science and Engineering*, 15(3):1178–1192.

Salvini, P., Laschi, C., Dario, P. 2010. Design for acceptability: Improving robots' coexistence in human society. *International Journal of Social Robotics*, 2(4): 451–460.

Selevsek, N., Köhler, C. 2018. Angepasste planungssystematik für MRK-systeme. *Zeitschrift für wirtschaftlichen Fabrikbetrieb*, 113(1-2): 55–58.

Skevi, A., Szigeti, H., Perini, S., Oliveira, M., Taisch, M., Kiritsis, D. 2014, September. Current skills gap in manufacturing: Towards a new skills framework for factories of the future. In *IFIP International Conference on Advances in Production Management Systems*, 175–183. Berlin, Heidelberg: Springer.

Turja, T., Oksanen, A. 2019. Robot acceptance at work: A multilevel analysis based on 27 EU countries. *International Journal of Social Robotics*, 11(4): 679–689.

UR. 2016a. Mobile robot colleagues on wheels increase productivity and worker safety at scott fetzer electrical group. https://www.Universal-Robots.Com/Case-Stories/Scott-Fetzer-Electrical-Group/. 2016.

UR. 2016b. UR10 at the center of the 4.0 industrialization process, reducing changeovers by 50%. https://www.Universal-Robots.Com/Case-Stories/Continental/. 2016.

UR. 2016c. Why the world's third largest motorcycle manufacturer opted for universal robots. https://www.Universal-Robots.Com/Case-Stories/Bajaj-Auto/. 2016.

Villani, V., Pini, F., Leali, F., Secchi, C. 2018. Survey on human–robot collaboration in industrial settings: Safety, intuitive interfaces and applications. *Mechatronics*, 55: 248–266.

Wang, L., Gao, R., Váncza, J., Krüger, J., Wang, X. V., Makris, S., Chryssolouris, G. 2019. Symbiotic human-robot collaborative assembly. *CIRP Annals*, 68(2): 701–726.

Wang, Y., Ma, H. S., Yang, J. H., Wang, K. S., 2017. Industry 4.0: A way from mass customization to mass personalization production. *Advances in Manufacturing*, 5(4): 311–320.

Waschneck, B., Altenmüller, T., Bauernhansl, T., Kyek, A. 2016. Production scheduling in complex job shops from an Industry 4.0 perspective: A review and challenges in the semiconductor industry. *SAMI@ iKNOW*, 1–12.

Wurhofer, D., Buchner, R., Tscheligi, M. 2014, June. Research in the semiconductor factory: Insights into experiences and contextual influences. In *2014 7th International Conference on Human System Interactions (HSI)*, 129–134. IEEE.

7 Control Strategies

A Pathway to Adaptive and Learning Techniques

Kanwal Naveed
Department of Mechatronics Engineering, College of Electrical and Mechanical Engineering, National University of Sciences and Technology, Islamabad, Pakistan

Atal Anil Kumar
Department of Engineering, Faculty of Science, Technology and Medicine, University of Luxembourg, Luxembourg

7.1 MACHINE LEARNING IN DESIGN AND MANUFACTURING: ADVANTAGES AND CHALLENGES

Machine learning (ML) is the future of manufacturing, and to comment that it will rapidly take over Industry 4.0 will not be an understatement. Manufacturing and its domains are experiencing a never-ending increase in the available analysis data related to production, environment, sensors, machine line parameters, etc. This data is utilized for parameter optimization, product customization, automation, and smart manufacturing (SM). That being stated, the more significant challenge effectively is utilizing this data. This calls for an intelligent and effective method to handle high dimensional dynamically complex data (Elangovan et al., 2015). This chapter, hence, aims to familiarize the reader with the basic concepts, advantages, challenges, and the deployment of machine learning concepts in some real manufacturing scenarios.

7.1.1 Variable Optimization

The manufacturing era from 2009 to 2014 has seen rapid developments in the integration of manufacturing optimization and learning techniques (Wuest et al., 2016). Variable optimization via machine learning is advantageous over normal control strategies (e.g., model predictive control) as it doesn't require any immediate feedback, thereby possessing the ability to operate as an open or closed-loop system. Machine learning will behave as a closed-loop technique if the quality

DOI: 10.1201/9781003327523-9

feedback data is immediately available. The available data for learning-based optimization can be classified into the following types:

1. Measured or simulated
2. Time-space or workspace
3. Qualitative or quantitative
4. Present or historical data
5. Observable or processable data

The workflow of the parameter optimization is shown in Figure 7.1.

7.1.2 Product Customization

Product customization can only reach the heights of engineering design with respect to variety, effectiveness, and responsiveness by finding the right balance between all three requirements (Tseng et al., 2018). The interdependence of these customization factors is displayed in Figure 7.2. The right balance among these customization factors is easier to attain in smart customization methods compared to mass customization techniques. This leads to the evolution of mass customization towards smart customization, where the system is more flexible and adaptable as per individual customer requirements. The following two machine learning-based techniques are the most popular when it comes to smart customization:

7.1.2.1 UGC Analysis Based on ML for Customer Research

UGC stands for user-generated content via internet resources because of any product consumption by the users. This is a promising alternative to manual surveys and other time-consuming and tedious data collection techniques. ML is utilized to analyze the UGC to identify the customer needs and bring on great product customization options which are not only fast but also cost-effective. Therefore, UCG can assist manufacturers in increasing product efficiency and variety through deep analysis of customer inputs (Suryadi et al., 2019).

7.1.2.2 ML-Based Classification and Formulation

ML-based classification of user requirements and feedback can assist the designers in better understanding the design priorities and allocation of product resources. Moreover, the designers can also utilize ML to predict the future performance of a newly designed product based on the analysis of performance and feedback data of a previously designed product (Zhang et al., 2021).

7.1.3 Automated Experimentation

The manufacturing of consumer products (CP) is designed based on the product lifecycle. The overall flow of the testing site from a product from design to validation to product testing is shown in Figure 7.3. The product design is completed by a product design team (Chavhan & Ugale, 2016). This design is later validated via a

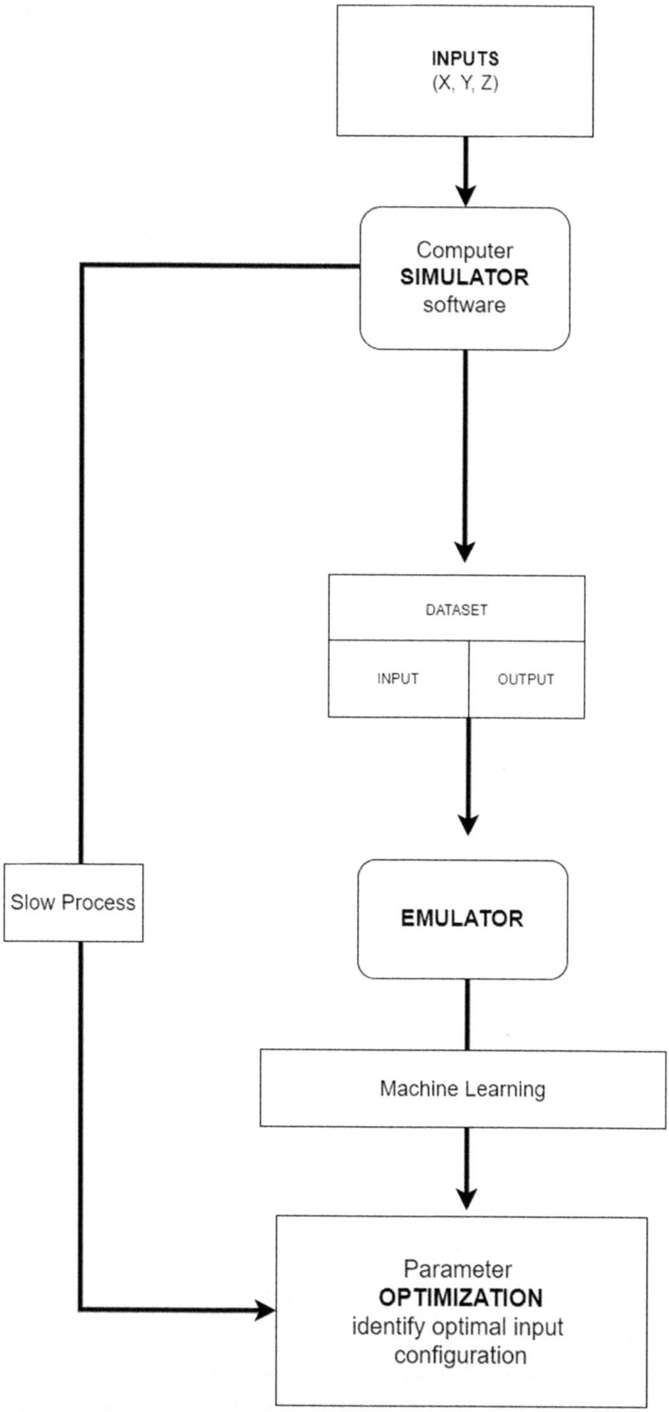

FIGURE 7.1 Workflow of parameter optimization.

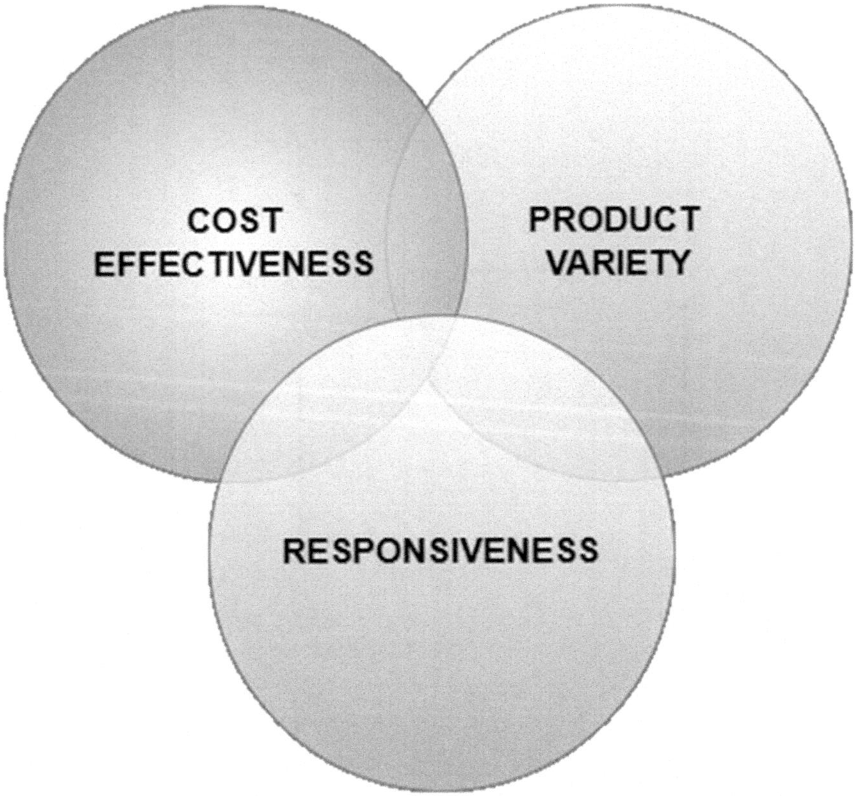

FIGURE 7.2 Interdependence of customization requirements.

validation process that is product specific. The validation process might be weeks, months, or even year-long, depending upon the type of product under consideration. After successful validation, the product is manufactured and passed to the testing phase. The testing phase is carried out to test the manufactured product to test out any manufacturing faults and the sufficiency of the product as per the consumer requirements. To carry out successful product testing, consumer product companies analyze the collected data via consumer feedback and previous analysis data. The collected data for product testing can be classified into two categories.

1. Performance key indicators or product
2. Analysis of data specific to a product

As discussed in section 7.1.2, a machine learning method can be used as an analysis framework. The work presented by Chavhan and Ugale (2016) employed LabVIEW as a testing site for testing and analysis of an induction motor. Similarly, a fully automated testing site was presented by Ramaprasad et al. (2018) for testing an

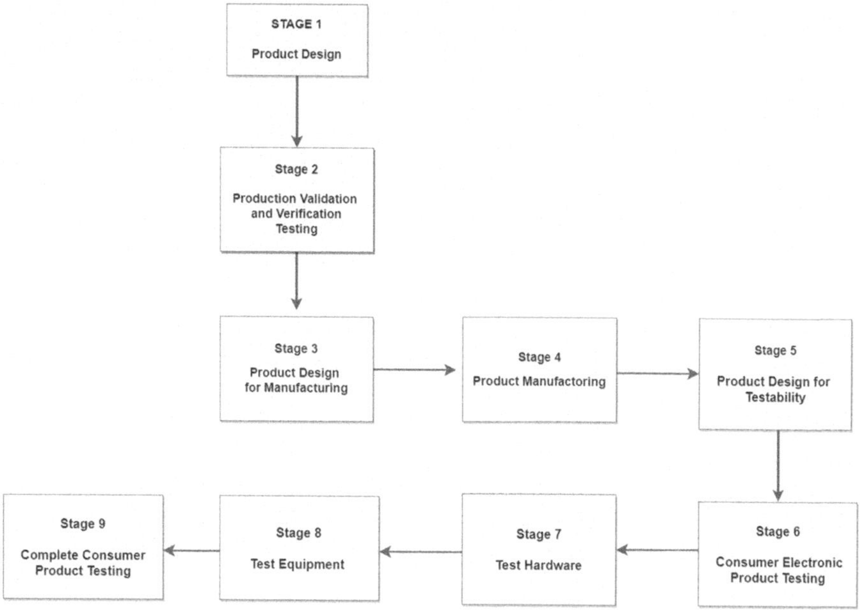

FIGURE 7.3 Complete product design to validation process.

aerospace product. One of the major drawbacks the consumer product manufacturing industry faces is the design and development of standalone testing sites that can only be applied to a specific product. Moreover, only a single centrally dedicated data collection and analysis system is available, which might lead to hours of manual fault findings and rectification processes. This gives rise to the importance of a universal data collection and analysis system that can classify useful data and efficiently manage product testing sites. The various types of variable data analysis and classification can be seen in Table 7.1.

TABLE 7.1
Variable Optimization – Data Classification

Data Classification			
Qualitative Vs Quantitative	**Time Vs Workspace**	**Controlled Vs Uncontrolled**	**Measured Vs Simulated**
• Corners are rounded up to avoid edges • Intelligent collision sensors	• Requires minimal human assistance • Uses a vision monitoring system to keep an eye on safe proximity area	• Similar characteristics to Speed and Separation • Can sense and stop when a human is in proximity	• Easier to program and allows a human to teach • Can achieve the minimum downtime

7.1.4 SM/Maintenance

SM is defined as an interconnection of a fully automated machine network that can communicate wirelessly. Such a network is monitored via various sensors and can be controlled with the help of different control algorithms, such as machine learning, for a reliable manufacturing process. This not only improves the quality of the product but also offers great help in increasing the product life cycle and sustainability and reducing production costs. Advanced data collection techniques like cyber-physical systems, the Internet of Things (IoT), and cloud computing support advanced manufacturing methods. The collected data is then processed by advanced machine learning algorithms. These algorithms can relate large volumes of highly nonlinear information and present it as useful information for SM processes, as shown in Figure 7.4. Applications of various machine learning algorithms in different areas of industrial manufacturing aspects are also discussed in Table 7.2. An important aspect here is to understand the relationship and differences between the so-called intelligent manufacturing and SM and artificial intelligence (AI). Before the early 2000s, the use of AI in industrial manufacturing applications was mostly referred to as intelligent manufacturing; however, this term was mostly replaced by

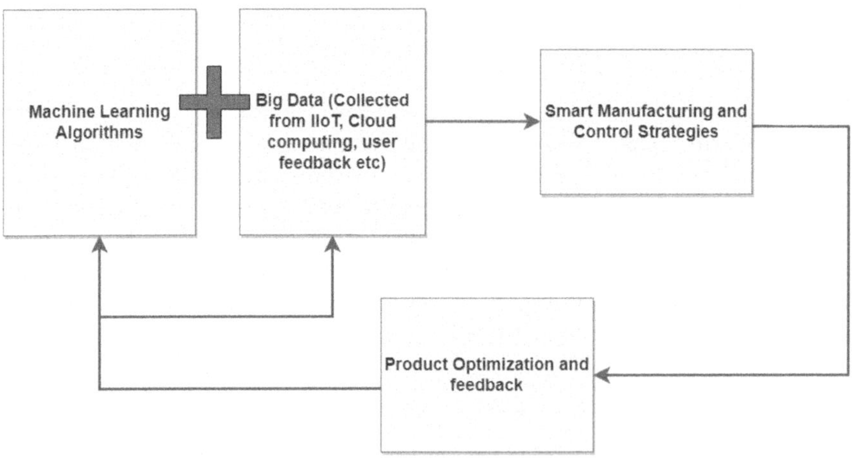

FIGURE 7.4 Workflow of smart manufacturing.

TABLE 7.2
Application of Machine Learning Algorithm in Manufacturing Industries (Adapted from Wang et al., 2018)

Machine Learning Module	Application Areas in Manufacturing Industry
• Convolutional Neural Networks	• Surface Integrations, Fault diagnosis of machine parts
• Deep Belief Networks	• Predictive Analysis and Fault diagnosis of machine parts
• Auto Encoders	• Fault diagnosis
• Recurrent Neural Networks	• Predictive Analysis and Fault diagnosis of machine part

SM. SM includes many sophisticated techniques, for example, cloud computing. The term "smart" in SM refers to utilizing available analysis data throughout the product cycle. This assists in a more efficient and flexible manufacturing method that can alter design and manufacturing processes on demand.

7.2 PREDICTIVE MAINTENANCE DURING MANUFACTURING PROCESSES

A promising maintenance technique is essential to reduce equipment failures and improve overall equipment health condition. Such maintenance techniques can be divided into four different categories as follows:

a. **RUN to failure:** This maintenance is not planned and is also called corrective action maintenance. It is applied only if the equipment fails.
b. **Preventive maintenance:** This type goes one step ahead of the former and performs maintenance on equipment after some intervals. The disadvantage of this method is that it can add up unnecessary maintenance without the need, thereby increasing the overall maintenance cost.
c. **Conditional maintenance:** This technique follows a specific criterion of degradation thresholds before performing maintenance on the equipment. Thus, this way, conditional maintenance only performs maintenance when necessary.
d. **Preventive maintenance:** This technique monitors the machine and schedules maintenance activity when necessary. It can predict a machine failure occurrence based on available data analysis collected by continuous monitoring of the machine.

The predictive maintenance approach performs the best of all four types discussed above and offers the least operation cost (Jezzini et al., 2013). The main task of predictive maintenance is to process and analyze the collected data and classify it as equipment health data, failure detection-related data, and so on. On the other hand, machine learning and adaptive algorithms are defined as techniques that can learn from little to no support and constantly update their parameters based on new available information/data. Some of the famous learning techniques being implemented in manufacturing industries are discussed next.

7.2.1 Artificial Neural Network

Artificial neural networks (ANNs) are preferred for scenarios involving a large amount of complex data and unclear machine structure. ANNs are mostly used in industrial applications such as soft sensing and predictive control methods (Shin et al., 2018). In 2019, Hesser and Markert proposed an ANN-based technique to monitor the wearing of a milling machine. Another approach was presented by Scalabrini et al. (2019), where a predictive failure technique for vibrating machines was developed. For that purpose, the dataset was collected based on simulations of a vibrating motor and was classified as frequency and amplitude data to predict future failures.

7.2.2 State Vector Machine

State vector machine (SVM) is based on adaptive computational learning methods and is applied for classification and regression analysis because of its highly accurate results. This technique was first introduced in 1995 by Vapnik, and since then, it has been ever-growing and has found many applications in SM. Falamarzi et al. (2019) applied an ML-based SVM for degradation measurements of the curved and straight rail tracks. This study also compared SVM-based and ANN predictions for the same tracks using mean squared error as the performance parameter, proving, that the SVM-based adaptive method performed far better than the ANN approach. A data diagnostic-based predictive analysis and maintenance technique was presented by Xiang et al. (2018) to reduce the overall maintenance cost.

7.2.3 Linear Regression

Linear regression methods are popular in industrial predictive maintenance applications (Zenisek et al., 2019). It is mainly because linear regression is simpler and deterministic, the parameter calculation is always less, and no extra model training adjustments are required. According to Zenisek et al. (2019), the linear regression method was deployed in the industry to model the healthy machinery and perform predictive analysis over the concept drift of the machine using the continuously changing data. The results of this method proved to be highly accurate from detection and predictive perspectives.

7.2.4 Extreme Gradient Trees

This state-of-the-art technique is mainly utilized by data scientists. Successful implementation of extreme gradient trees (XGTs) for predictive application is the deployment of such an algorithm for predicting the downtime of a printing machine based on previous failures (Binding et al., 2019). Quiroz (2018) presented a method to diagnose a synchronous motor's rotor bar failures.

Similarly, other famous adaptive and machine learning techniques involve logistic regressions, random forest (RF) and gradient-boosting methods. The work presented by Quiroz (2018) shows a comparison of linear regression, RF and gradient-based techniques for predicting motor disorders, and it was seen that the RF method outperforms most of these methods with an accuracy of 98.4%.

7.2.5 ML and Manufacturing Industry – Case Study I

It is imperative for manufacturing industries to not be able to manufacture the products efficiently but be able to proactively predict and fix any potential defect in the machinery. Therefore, predictive maintenance algorithms should be designed and tested in industries that can collect and store sensor data, analyze and process the data and predict any real-time malfunction or wear-tear effects that may result in a machine's downtime. Zenisek et al. (2019) presented one such case study, where a

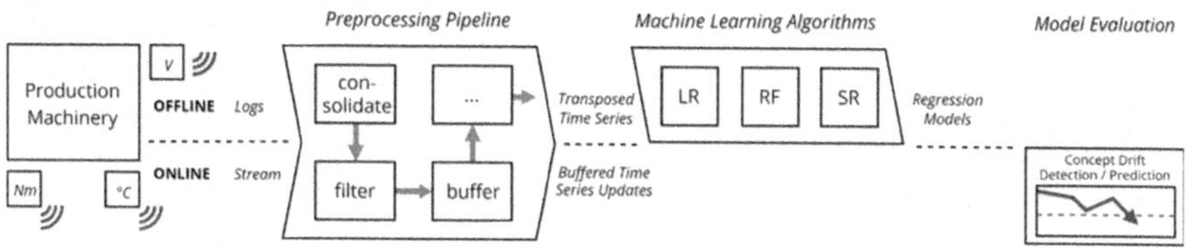

FIGURE 7.5 Machine learning algorithm approach to detect process faults (adapted from Zenisek et al., 2019).

machine learning algorithm was developed to predict defective system behavior by detecting the concept drift in the data streams. The primary aim of this case study is to model the state of a healthy machine via a machine learning algorithm where the machine state is labelled as "normal" until it is signalled as "defective." For this purpose, two methods have been considered:

a. Maintain a large, supervised pool of datasets and compare the machine data with it to label it as normal or defective.
b. Use regression models to predict the upcoming data stream values and match them with sensed values. If the error is higher, it might indicate a defect.

This case study uses machine learning models to predict the data stream values using regression models. The developed algorithm is tested in real-time with industrial radial fans. In this regard, this study collaborated with Scheuch GmbH, an industrial enterprise based in Austria. The industry provided a new radial fan for data collection and later used it for run-to-failure predictive analysis. This study used various machine learning approaches, including regression models and gradient trees, to predict the next values in the incoming data stream and compared them with the on-ground sensed value. If the difference in the values of these quantities is high, this might lead to potential concept drift, thus indicating that a fault might occur, as shown in Figure 7.5. The comparative analysis of all the applied techniques proved that RF outperformed every other method in accuracy.

7.2.6 ML and Manufacturing Industry – Case Study II

Condition monitoring and failure prediction of industrial equipment is an essential part of the predictive maintenance process. ML has a vast number of applications when it comes to predictive maintenance. It can be used in failure diagnostics (Liu et al., 2017), failure predictions (Rietman et al., 1998), remaining useful life (Zhao et al., 2016), etc. Currently, there is a strong inclination towards using machine learning algorithms for failure predictions and predictive maintenance.

In this case study presented by ABB corporations (Amihai et al., 2018), an attempt was made to predict future failures by collecting vibration data for two and a half years. The equipment selected for data collection is low-voltage motors and rotating mechanisms like voltage pumps. The vibration data is collected over the years via a commercial monitoring equipment called *ABB WiMon100*. The position of sensors is selected while keeping in mind the effects of their position on the data accuracy. The WiMon system itself has an acceleration and a temperature sensor. The collected data is sent to a computer software called WiMon data manager. This software is the primary medium for the data provided to the appropriate machine learning algorithm.

This study considers failures like looseness, cavitation, blade problems, imbalance, and misalignment. For each of the considered failures, the monitoring system calculates key condition indices (KCIs), a mathematical representation of the severity of the failure mode of the equipment under consideration. The KCI and vibration data are fed as inputs to the machine learning algorithm for the appropriate analysis. The gathered data is then passed through a data-cleaning process for noise and spike removals, as represented in Figure 7.7. The successful completion of this stage leads to the data preparation stage, where the data is formatted and labelled as per analysis requirements. Further to this step, it is processed by the learning algorithm. The choice of the algorithm in this study is a RF learning algorithm. Once the model is trained over a specific look-ahead time, it is tested over a selected testing dataset. The expected and the predicted KCIs are then compared to verify the accuracy of the trained model, which resulted in the fact that this case study could predict the KCIs up to 7 days ahead of time.

7.3 EXPANDING DESIGN POSSIBILITIES: GENERATIVE DESIGNS FOR SM

The term "generative design" is defined as a design proposed by an AI technique based on all the previously collected data, old product samples, customer inputs, cost and weight estimation etc. An AI technique can predict many designs based on all the information being transferred. Such design techniques are employed by various industries, including AirBus, Stanley Black, and Under Armour (Akella, 2018). One of the major impacts of generative designs is energy saving and waste production. The integration of generative designs with additive manufacturing will help in the production of lesser waste and thus leading towards fewer adversarial effects on the environment (Korner et al., 2020).

7.3.1 Parametric and Generative Techniques: A Comparison

Parametric and generative design techniques are high-level design methods that utilize ML to develop reliable and high-quality solutions. Both techniques rely on high-end computer software programs, offering greater design speed and reliability.

> Parametric design is an online design technique where users can make real-time changes and reuse the data in other relevant projects.

> Generative design is an iterative method where a solution is provided in several iterations. In each iteration, useful data is separated from the overall collected data, and then a scoring feature is used to make the next iteration.

The parametric design method is faster than the traditional design methods. It provides the user with the ability to reuse the old data. Fast speed is another significant advantage of this technique. Furthermore, parametric design methods use optimization techniques to achieve high resource efficiency. However, this method might suffer some problems, including facing clashes when dealing with design-imposed constraints. This might lead to a product that may not only be ineffective or, in some cases, entirely unusable. Moreover, while dealing with many design constraints, the overall process might slow down and lead to an expensive product.

The generative designs consider the designer-generated parameters to develop the solutions. Based on the provided data, generative designs can explore many smart solutions quickly, thereby saving design time. The biggest disadvantage these designs suffer from is creating too many solutions, thus rendering human input/labour redundant. Moreover, it also requires a large amount of computational power to manage solution development over a short period.

The best application of the parametric and generative approach is in decision-making processes where it is important to maintain a relationship between various kinds of information/data (Shea et al., 2005). It is interesting to note that parametric and generative designs complement each other to allow a more efficient method of information selection.

7.3.2 Generative Design for Mass Customization: A Case Study

The term "mass customization" refers to the development of a product as per user requirements while maintaining a balance between cost and variety. Under such circumstances, generative designs exploit advanced manufacturing opportunities to develop a better and improved product according to user satisfaction. In a data-driven approach, the user-to-product survey data feeds the design requirements instead of designer inputs. A detailed framework for a data-driven generative design is given in Figure 7.6 (Jiang et al., 2022). The framework declares three main levels of design. First is the high level, where the user input directly determines the element under design. Second is a weak level where the element is designed based on manual design specifications by the designer. Third is the medium level, which consists of design strategy as a mixture of the above two types. This case study has specifically used the data-driven approach to design the bike saddle. To design the shape of such a customized product, input is required from the already present designs in the market and user insights. The user-related data can be easily collected via sensors, e.g., a force sensor can be used to gather the data to lay down the relationship between the pressure distributions and bike saddles. The framework designed in Figure 7.6 is being tested on a bike saddle where the target is to design a bike saddle that can continuously interact with the user for data collection and is designed as per the user's needs and demands. The ML algorithm for the design of such a saddle used pre-existing data for training, the user needs were determined via various questionnaires.

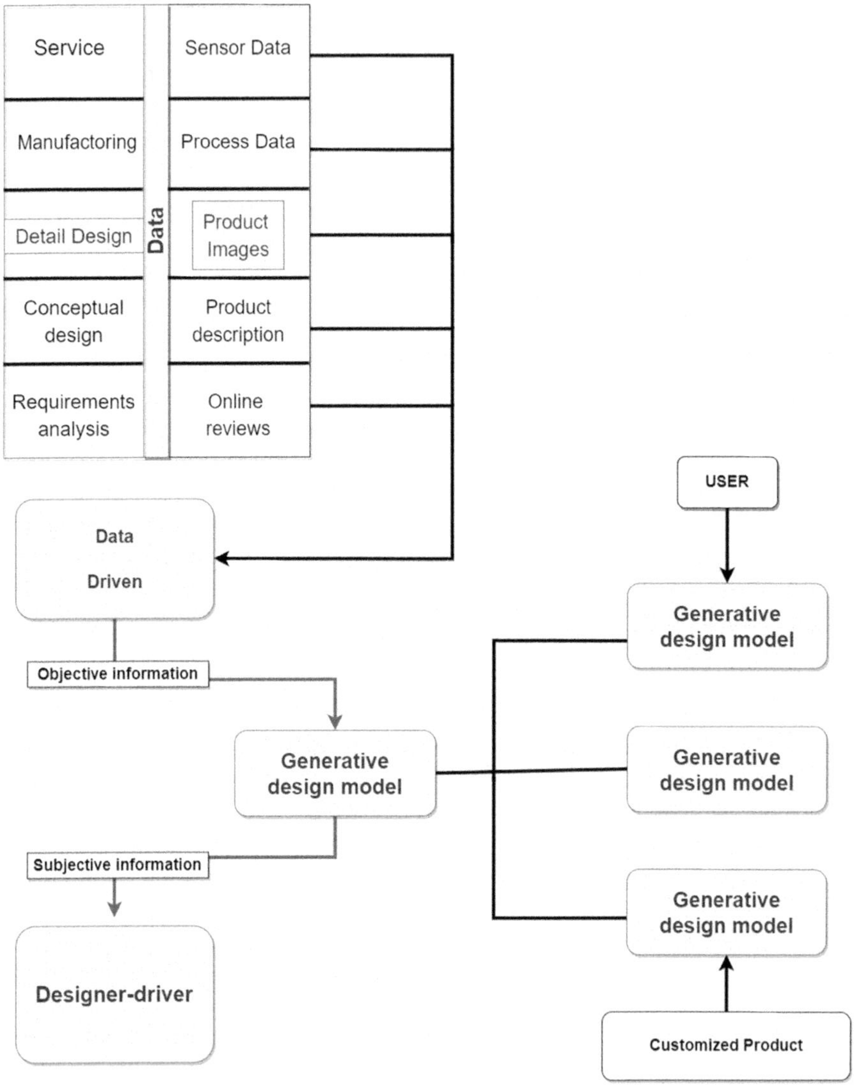

FIGURE 7.6 Framework for data-based generative design (adapted from Jiang et al., 2022).

The saddle images were mostly gathered from amazon.com. This allowed the computer software to produce a generative design for the bike saddle, which is personalized in design as per the customer's requirements. This design is then manufactured using additive manufacturing (AM). After successful testing, it was found that a more in-depth user-to-product analysis is required. Moreover, it is also concluded that the internal structure of the saddle overlooked the relationship between the geometrical factors and the saddle materials.

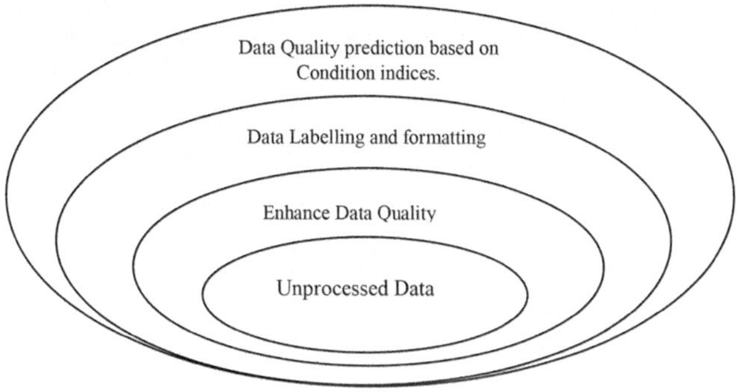

FIGURE 7.7 Data preparation for product/process failure prediction.

7.4 COMPREHENSIVE PLAN FOR UTILIZING THE BEST LEARNING MODULE

As per the discussions carried out in the previous sections, there is no doubt in the fact that learning-based techniques and designs are the future of the so-called advanced Industry 4.0. Any new proposed study for the development of a new structural layout for a manufacturing process is based upon smart decisions and learning one way or another. The ability to fuse the data-driven techniques and ML processes allows for a more sophisticated and learned design for a product which is not only efficient but also fulfils the user's demands.

7.4.1 Future Trends and Conceptual Framework

Currently, many manufacturing industries are still utilizing typical machine systems to accomplish manufacturing tasks. This machinery mostly relies on manual instructions or paper-based working. This might lead to a highly non-efficient system. This is because the interaction, selection and operation stages are very time-consuming at the designer's end, where many human interactions are involved. Secondly, the process of data collection is usually manual, whether it is from the workers in the industry or the customer feedback; people are always reluctant to write down data over pieces of paper, and this data is usually very important for a new enhanced design of the product. Unavailability or incorrect data might lead to a faulty product design. Therefore, special attention is required towards the data collection stage. This process can be automated by incorporating IoT concepts and smart sensors to collect real-time data. This will lead the overall manufacturing process toward the concept of smart designs and smart monitoring. Typical manufacturing design methods are being upgraded via various advanced technologies, for example, virtual realities, generative designs, and augmented reality. The successful manufacturing of smart designs is carried out with the help of smart monitoring, where a web of widespread sensors can help sense and visualization of data to indicate any future abnormalities or failures.

Another challenge faced by manufacturing systems is manufacturing control which requires decision-making at certain stages of manufacturing. The challenge mainly arises due to the presence of large interconnected data. Therefore, a typical

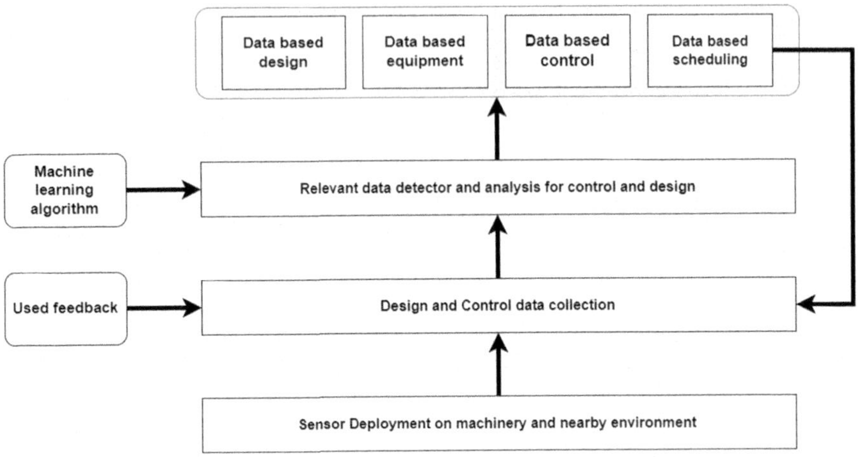

FIGURE 7.8 Future framework for control strategies and product design towards smart manufacturing.

control model may take a long time to come up with any manufacturing-related decision. This may also affect the decision-making to schedule the product optimization. This leads to a need for an SM and control approach which can differentiate relevant information for the task at hand and can produce a learned and well-trained design model for the product. The ML algorithms for this purpose can be deployed over cloud computing to keep track of all previously incorporated upgrades and make the designs and models available readily for daily decision-making. With respect to the dire need for these future trends in manufacturing, a detailed conceptual framework to achieve this technology process is shown in Figure 7.8.

Smart design, as the first step refers to the utilization of relevant data to produce a smart product design that can interact with computer software like computer-aided design (CAD) in real-time (Zheng et al., 2018). This helps realize a smart product design paradigm for AM. As a second step, the designed product is passed to the manufacturing stage, which is governed by smart machinery, product manufacture monitoring, and control techniques. The deployment of smart robots, for example, cobots that can communicate with one another and humans around them, helps take the manufacturing industry to the next level by creating a much easier and faster control system for the manufacturing of any product. The smart machinery is equipped with smart sensors which can interact with the environment and collect useful data which is further utilized in generative product designs and implementation of an effective control strategy for the machinery responsible for product manufacturing. The data collected over the smart monitoring stage is used by the smart scheduling stage to study the product behavior according to the available data and schedule procedures related to product maintenance and designs. With all these advanced features, the presented framework is very much capable of being applied in the manufacturing industry, thereby revolutionizing the Industry 4.0 concept and bringing out the best combination of smart product design and advanced learning techniques for smart control of manufacturing systems.

REFERENCES

Akella, R. 2018. What generative design really is, what it isn't, and why it's the future of manufacturing. *New Equipment Digest*.

Amihai, I., Gitzel, R., Kotriwala, A. M., Pareschi, D., Subbiah, S., Sosale, G. 2018. An industrial case study using vibration data and machine learning to predict asset health. *In 2018 IEEE 20th Conference on Business Informatics (CBI)*, 1: 178–185.

Binding, A., Dykeman, N., Pang, S. 2019. Machine learning predictive maintenance on data in the wild. *In 2019 IEEE 5th World Forum on Internet of Things (WF-IoT)*, 507–512.

Chavhan, K. B., Ugale, R. T. 2016. Automated test bench for an induction motor using LabVIEW. *In 2016 IEEE 1st International Conference on Power Electronics, Intelligent Control and Energy Systems (ICPEICES)*, 1–6.

Elangovan, M., Sakthivel, N. R., Saravanamurugan, S., Nair, B. B., Sugumaran, V. 2015. Machine learning approach to the prediction of surface roughness using statistical features of vibration signal acquired in turning. *Procedia Computer Science*, 50: 282–288.

Falamarzi, A., Moridpour, S., Nazem, M., Cheraghi, S. 2019. Prediction of tram track gauge deviation using artificial neural network and support vector regression. *Australian Journal of Civil Engineering*, 17(1): 63–71.

Hesser, D. F., Markert, B. 2019. Tool wear monitoring of a retrofitted CNC milling machine using artificial neural networks. *Manufacturing letters*, 19: 1–4.

Jezzini, A., Ayache, M., Elkhansa, L., Makki, B., Zein, M. 2013. Effects of predictive maintenance (PdM), Proactive maintenace (PoM) & Preventive maintenance (PM) on minimizing the faults in medical instruments. *In 2013 2nd International conference on advances in biomedical engineering*, 53–56.

Jiang, Z., Wen, H., Han, F., Tang, Y., Xiong, Y. 2022. Data-driven generative design for mass customization: A case study. *Advanced Engineering Informatics*, 54: 101786. https://doi.org/10.1016/j.aei.2022.101786

Korner, H. M. E., Lambán, M. P., Albajez, J. A., Santolaria, J., Ng Corrales, L. D. C., Royo, J., 2020. Systematic literature review: Integration of additive manufacturing and industry 4.0. *Metals*, 10(8): 1061.

Liu, Z., Jia, Z., Vong, C. M., Bu, S., Han, J., Tang, X., 2017. Capturing high-discriminative fault features for electronics-rich analog system via deep learning. *IEEE Transactions on Industrial Informatics*, 13(3): 1213–1226.

Quiroz, J. C., Mariun, N., Mehrjou, M. R., Izadi, M., Misron, N., Radzi, M. A. M. 2018. Fault detection of broken rotor bar in LS-PMSM using random forests. *Measurement*, 116: 273–280.

Ramaprasad, S. S., Rajesh, G. N., Kumar, K. S., Prasad, P. R. 2018. Fully automated PCB testing and analysis of SIM module for aircrafts. In *2018 3rd IEEE International Conference on Recent Trends in Electronics, Information & Communication Technology (RTEICT)*, 2016–2020.

Rietman, E. A., Beachy, M. 1998. A study on failure prediction in a plasma reactor. *IEEE Transactions on Semiconductor Manufacturing*, 11(4): 670–680.

Scalabrini, S. G., Filho, A. R. D. A., Santos da Silva, L., Augusto da Silva, L. 2019. Prediction of motor failure time using an artificial neural network. *Sensors*, 19(19): 4342.

Shea, K., Aish, R., Gourtovaia, M. 2005. Towards integrated performance-driven generative design tools. *Automation in Construction*, 14(2): 253–264.

Shin, J. H., Jun, H. B., Kim, J. G. 2018. Dynamic control of intelligent parking guidance using neural network predictive control. *Computers & Industrial Engineering*, 120:15–30.

Suryadi, D., Kim, H. M. 2019. A data-driven approach to product usage context identification from online customer reviews. *Journal of Mechanical Design*, 141(12).

Tseng, M. M., Wang, Y., Jiao, R. J., In, M. D., Chatti, S., Laperrière, L., Reinhart, G., Tolio, T. (Eds.) 2018. Modular Design. In*CIRP Encyclopedia of Production Engineering* (2018 edition). Springer, Berlin, Heidelberg

Wang, J., Ma, Y., Zhang, L., Gao, R. X., Wu, D. 2018. Deep learning for smart manufacturing: Methods and applications. *Journal of Manufacturing Systems*, 48: 144–156.

Wuest, T., Weimer, D., Irgens, C., Thoben, K. D. 2016. Machine learning in manufacturing: Advantages, challenges, and applications. *Production & Manufacturing Research*, 4(1): 23–45.

Xiang, S., Huang, D., Li, X. 2018. A generalized predictive framework for data driven prognostics and diagnostics using machine logs. In *TENCON 2018-2018 IEEE Region 10 Conference*, 0695–0700.

Zenisek, J., Holzinger, F., Affenzeller, M. 2019. Machine learning based concept drift detection for predictive maintenance. *Computers & Industrial Engineering*, 137: 106031.

Zhang, J., Chu, X., Simeone, A., Gu, P., 2021. Machine learning-based design features decision support tool via customers purchasing data analysis. *Concurrent Engineering*, 29(2): 124–141.

Zhao, R., Wang, J., Yan, R., Mao, K. 2016. Machine health monitoring with LSTM networks. In *2016 10th International Conference on Sensing Technology (ICST)*, 1–6.

Zheng, P., Sang, Z., Zhong, R. Y., Liu, Y., Liu, C., Mubarok, K., Yu, S., Xu, X. 2018. Smart manufacturing systems for Industry 4.0: Conceptual framework, scenarios, and future perspectives. *Frontiers of Mechanical Engineering*, 13(2): 137–150.

8 Mixed Reality for Industry 4.0 in Manufacturing Systems

Sri Kolla, Atal Anil Kumar, and Peter Plapper
Department of Engineering, Faculty of Science, Technology and Medicine, University of Luxembourg, Kirchberg, Luxembourg

8.1 INTRODUCTION

Human interactions still play a significant role despite the advancements in the manufacturing landscape with Industry 4.0 technologies. Moreover, the aim of Industry 4.0 is not to replace machines with humans but to enhance human capability with the support of advanced technologies. The market trends suggest an increased demand for customized products, increasing the product variants and hence more demand on the shopfloor (Kolla et al., 2020). Therefore, operators need systems that can assist them in completing their tasks quicker, more precisely, and with a lesser workload. Extended reality (XR) technologies can be employed in the industry to give operators an enhanced perception of the instructions in a visual and interactive manner (Doolani et al., 2020; Fast-Berglund et al., 2018). The XR spectrum includes technologies such as augmented reality (AR), mixed reality (MR), and virtual reality (VR). Figure 8.1 illustrates the XR spectrum, which encompasses AR, MR, and VR technologies.

The virtual extreme of the XR continuum is VR. In VR, the user is completely immersed in the virtual world and isolated from the real environment (Mujber et al., 2004). In a VR environment, the user is surrounded by virtual objects and is unaware of reality. On the other hand, the real world is the other extreme, consisting exclusively of physical objects and free of any virtual objects to improve the operator's perception. AR and MR enhance the user experience by improving reality with virtual elements. In an AR environment, the virtual information is overlaid in the natural environment. However, in an MR environment, it is possible to develop interactions between real and virtual objects. Therefore, MR can be considered an umbrella term for AR and VR. In a nutshell, AR and MR can be differentiated based on the extent of interactions between real and virtual elements.

Numerous advantages of XR technology include traceability, increased intuitive learning, decreased error rates, and reliability. AR and MR can help shop floor employees in their daily jobs by enabling digital poka-yoke systems and reducing

DOI: 10.1201/9781003327523-10

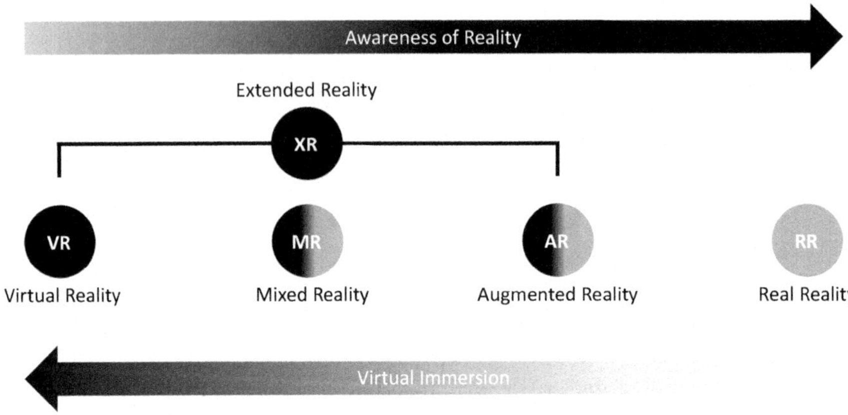

FIGURE 8.1 Extended reality (XR) spectrum illustrating both virtual and real extremes of awareness.

manual errors and rework of complex and unfamiliar assembly tasks. VR can be effectively employed in training activities in any manufacturing enterprise. A new way of interacting with machines and the shop floor environment is also introduced by XR, opening the door for future solutions for human-machine interfaces (HMI).

This chapter is structured as follows: Section 8.2 presents basic components and different types of AR systems, followed by various interaction modalities in AR systems in Section 8.3. AR systems architectures and case studies are detailed in Section 8.4. This is followed by a brief introduction to VR technology in Section 8.5. The important applications of VR technology in manufacturing are presented in Section 8.6 and the case studies are highlighted in Section 8.7. A conceptual framework of MR is shown in Section 8.8, followed by the conclusion and discussion in Section 8.9.

8.2 BASIC COMPONENTS AND TYPES OF AR SYSTEMS

With the support of several existing technologies, AR overlays virtual information in the real world. As a result, AR improves the operator's perception of Reality that has been supplemented with digital data. The way operators interact with their environment and carry out their daily activities will shift because of this paradigm shift.

There are five essential components in an AR system: capturing technology (CT), visualization technology (VT), processing unit (PU), tracking system, and user interface (UI). CT includes a camera or a sensor to capture the real world. VTs are the essential components of an AR system as they superimpose virtual elements in the real world. Example VTs include a mobile phone or a HoloLens®. A PU in an AR system analyses the data input and aids in visual rendering. A tracking system includes a marker, such as a QR code, used to orient the virtual information in relation to the real environment. Some AR systems can use advanced approaches such as spatial mapping and bypass a tracking system. AR systems with markers are often utilized in the industry in contrast with marker-less technologies,

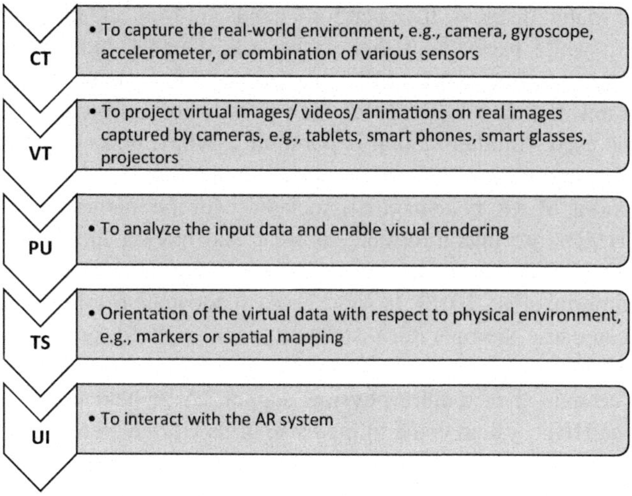

FIGURE 8.2 Basic components of an AR system.

as markers could significantly reduce the computing power in rendering virtual information. Finally, the user interacts with the AR system supported by a UI (Fraga-Lamas et al., 2018; Lee et al., 2017; Wang et al., 2016). Figure 8.2. illustrates the basic components of an AR system.

8.2.1 Types of AR Devices

Three AR devices are used: handheld devices (HHDs), head-mounted devices (HMDs), and spatial AR devices. HHDs are devices that a user can manipulate using their hands. The most common example of HHDs is mobile phones and tablets. These devices employ all the necessary components, such as a camera, display unit, processing unit, and other additional sensors. HMD is a device to wear on the head of the user, and these days they are available as part of a helmet. While using the device, the user is not entirely immersed in the virtual world as they can see both virtual and real environments intact. HHDs employ image sensors, depth cameras, holographic optical elements, holographic processing units, and microphones to support the development of AR applications. Spatial AR devices display virtual information in a real environment using video projectors, beamers, and lights. Currently, logistics and warehouse optimization projects use spatial AR technology extensively.

8.2.2 Benefits and Limitations of AR Systems

In general, HMDs are favored for assembly applications because they give the user a hands-free experience and enhance the user's perception of the real world because the view through an HMD is nearly identical to the real world. On the other hand, HHDs are used in everyday life (e.g., tablets or a smartphone). Using an HHD lessens the need for training and enables an operator to use the system directly with little to no training. HHDs are superior to HMDs from an ergonomic and financial standpoint. In spatial displays, the operator's hands are free, and they do not need to carry anything while performing their tasks. AR systems with markers are less computationally intensive compared to marker-less AR systems. Because AR applications do not require a lot of processing power, both mid-range and high-end devices can be used. Another benefit is portability, which makes it simple to move AR applications and hardware from one place to another.

The limitations of AR systems are associated with the maturity of the available technology. HMDs are uncomfortable to wear and have a limited field of view (FoV). Limited FoV compromises the operators' safety and can lead to workplace accidents (Palmarini et al., 2018). In some cases, inadequate resolution of the virtual information can cause the operator's sickness when exposed for extended periods. HMDs are heavy to wear and work for a prolonged period. Furthermore, HHDs are not portable because they require physical support to be held in place. A further drawback of an HHD system is the limited size of its UI, which restricts the amount of information that can be displayed on it. AR systems that use markers need special care concerning the maintenance of the markers. Marker-less AR systems require more computational power for object tracking in real-time.

8.3 INTERACTION MODALITIES IN AR SYSTEMS

Interaction modality is a way of sensing such as touch (haptic), voice (acoustical or auditive), vision (optical), etc. In AR, interaction modalities help users to interact with AR interfaces (Esteves et al., 2017). Currently, there are three ways to interact with AR systems: touch, touchless, and wearables. These interaction modalities are divided into input modalities and output modalities. Input modalities are for the user to give instructions to the AR system, and output modalities are for the user to receive feedback from AR systems. The AR modalities are shown in Figure 8.3.

Touch modality is not new in the manufacturing industry. Most of the existing operator panels on the mechanical systems have touch modality enabled, either with buttons or with touch screen displays.

With the development of AR technology, especially with HMDs, a new type of deictic gesture to interact with AR devices is possible (Williams et al., 2019). These gestures are mostly given mid-air with the operator's hands (Mistry & Maes, 2009). For example, raising an index finger in the FoV of smart glass can indicate that the operator is ready for a selection, and pinching the index finger with the thumb can follow the selection of a particular target.

Voice, a natural interaction modality, has been the most widely used in our daily communication. Yet, this modality is intercepted by machine noises in an industrial environment, especially on shop floors. The voice interaction modality enabled by speech recognition techniques allows the user to interact with the AR systems using voice commands (Nizam et al., 2018). Eye tracking (Liu et al., 2019) is a recent development in AR that is derived from complex algorithms that estimate the movement of the cornea and pupil of the eye in space. In an HMD powered by eye tracking interaction modality, cameras that operate at a high frame rate capture the images of the eyes. As the name suggests, screen dwelling (Qian et al., 2020) makes interaction with a system possible by gazing at the target for a specified amount

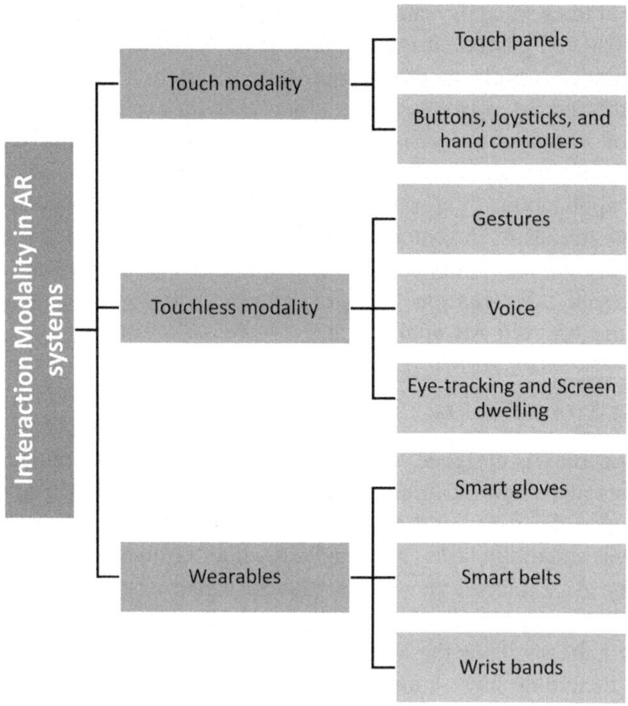

FIGURE 8.3 Interaction modalities in AR systems.

of time. The gyroscope and motion sensors of the system facilitate the dwelling interaction. Of course, dwelling is only possible by combining it with another modality, such as head, hand, or eye tracking.

Even though wearables found their way into the fitness industry and health tech, they aren't matured to be used along with AR devices in an industrial environment. However, research on wearables such as smart gloves (Hsieh et al., 2016), belts (Dobbelstein et al., 2015), and wristbands (Hu et al., 2020) is rapidly progressing for better-enhanced interactions with smart devices. This chapter presents two case studies where touch, gesture, and voice interaction modalities are employed in the AR application.

8.4 AR CASE STUDIES

All-in-one AR solutions are readily available in the market to meet a wide range of industrial needs. These solutions are expensive and, to an extent, overkill for smaller applications in manufacturing enterprises. In this section, the authors demonstrated the development of AR applications from scratch using inexpensive hardware, open-source software, and plugins. The workflow process presented in this section is attractive to the needs of SMEs as the capital expenditure required is minimal compared to off-the-shelf solutions on the market.

Both case studies solve the same problem of digitalizing the work instructions to assemble a planetary gearbox using two alternative hardware setups and features on each. In the first case study, an affordable Android system is used to develop an AR solution; in the second, a HoloLens® is proposed with expanded capabilities.

In terms of software, Unity® is the primary development engine used to create interactive 3D experiences. Besides Unity®, C# programming is used for interactions in AR applications, Vuforia® for marker tracking and orientation of virtual elements, and Autodesk® inventor for 3D modeling. Various open-source plugins are used from the Unity® store for specific operations in developing AR applications. For example, the lean touch plugin from the Unity® store enabled the touch modality in the Android AR application.

8.4.1 Case Study Problem

A planetary gearbox is designed by Autodesk® inventor, which contains a diversity of components and requires a number of steps to be assembled. The case studies need to simulate the real-world shopfloor scenario. Therefore, care was taken to have different assembly tasks such as screwing, fastening, circlip fixing, and bearing fitting. An exploded view of the planetary gear is shown in Figure 8.4. All the parts are 3D printed, and components such as bearings, circlips, nuts, and bolts are bought off the shelf for the assembly operation. The usage of several tools is required by including such components, which increases the complexity of the assembly process and simulates the shopfloor experience. In Sections 8.4.2 and 8.4.3, two AR applications are developed to digitalize the work instruction to assemble a planetary gearbox.

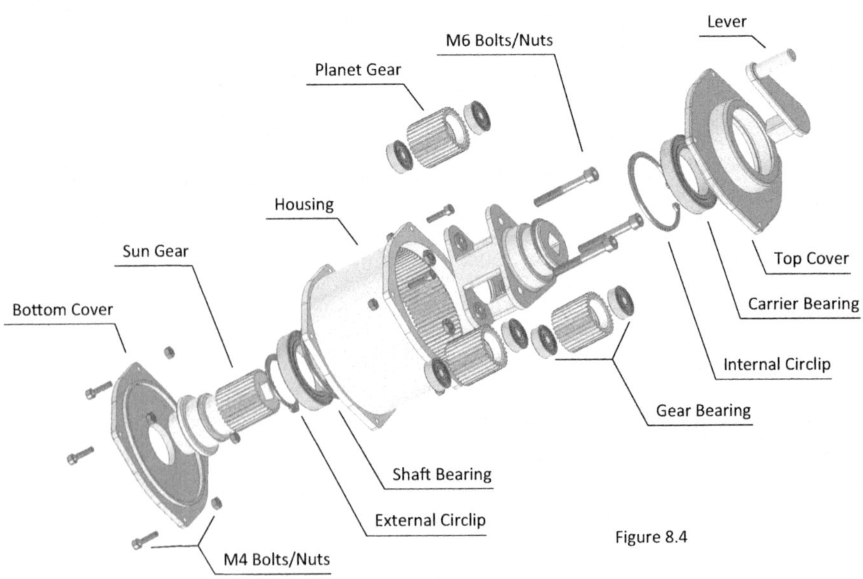

FIGURE 8.4 Exploded view of planetary gear.

8.4.2 Case Study 1 – Digitalizing Assembly Instructions Using a Handheld AR Device

This section presents the development process of digitalizing the assembly instructions to assemble a planetary gear using an HHD (e.g., a mobile phone). From the hardware perspective, a Samsung® A7 device running on the Android operating system is chosen. The device comes with all the necessary AR elements, such as a camera, processing unit, touch interface, and other useful sensors. Moreover, mobile phones are readily available, and most users feel comfortable using these devices.

From a software perspective, a framework is developed to model the abstraction first approach. As the manufacturing domain is adapting software to assist their daily operations, it is important to understand and adapt the workflow of software developers. In software development, the C4 framework (Brown, 2019) is very popular for visualizing the software elements in a simple using a lean graphical representation technique. Figure 8.5 illustrates the software framework used to design the Android

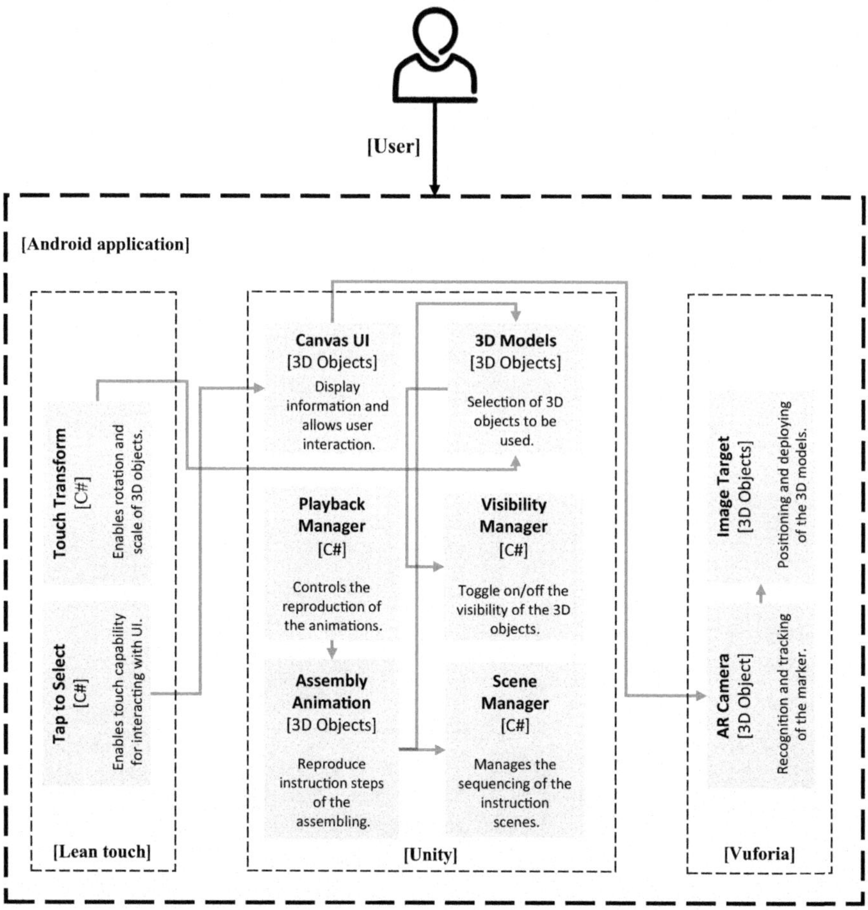

FIGURE 8.5 Software framework for mobile device (HHD).

AR application on a mobile device. The C4 framework has three layers: software system, containers, and components. The idea of a user interacting with the AR application forms the software system. Containers are executable applications, and in developing the Android application, the following containers are used: Vuforia®, Unity®, and Lean Touch. Components provide functionality, and are not separately deployable without a well-designed interface. In this application, C# programming and 3D files are examples of components.

The virtual instructions to assemble the planetary gear are shown using animations created by 3D models. The user can use the interactive buttons on the UI to reproduce, pause, replay, and navigate to the next/ previous instructions. The UI is designed such that the user can get auxiliary information about the assembly parts by tapping each part. The auxiliary information is available on demand, as having it on the UI by default is not ideal for the usability of the application. The user also can scale and rotate the animations by pinching the display unit with two fingers. Touch input modality was applied in the development of this AR application. The program is developed from the Unity® scenes containing all the necessary elements to run on it. AR Software Development Kit (SDK) from PTC (PTC, 2020) is used to convert the Unity® scenes to an executable Android application using an inbuilt plugin. In the development of the application, a fiducial marker (QR code) optimized by Vuforia® is used to track image target and orientation. Figure 8.6 illustrates the UI of the Android application.

In summary, the AR application is developed for Android devices using see-through video technology using a marker-based tracking system. Touch modality is the only input modality used for the development of this AR application. The application is

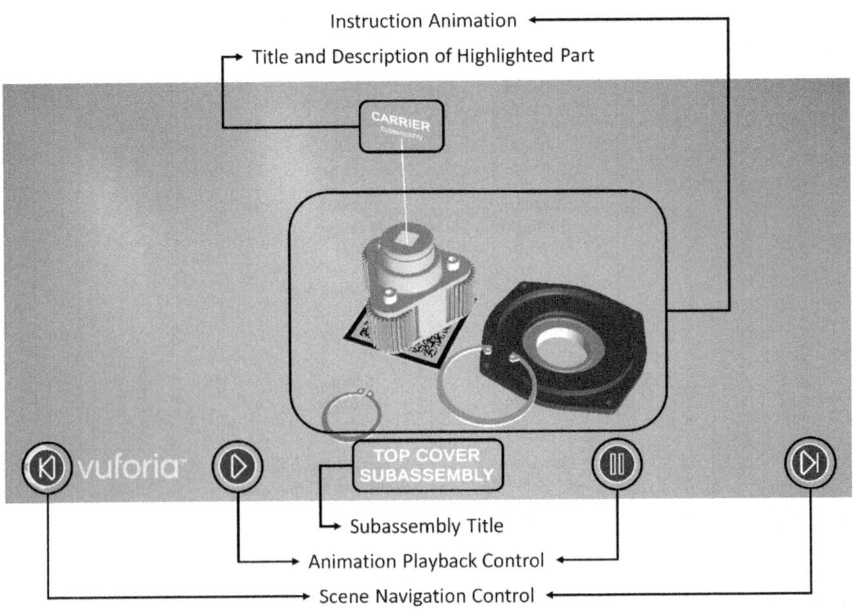

FIGURE 8.6 User Interface (UI) on mobile device (HHD).

portable to different locations on the shop floor. However, the FoV of the UI is limited by the size of the display unit of mobile devices. Visual cues such as part and tool pointers are not included in the application as it requires much more computing power and creates lags while running the application.

8.4.3 Case Study 2 – Digitalizing Assembly Instructions Using a Head-Mounted AR Device

Like the HHD application in Section 8.4.2, the application on HMD (e.g., ®HoloLens) delivers the digital instructions to assemble a planetary gear through several steps. The hardware used in this development process is an HMD from Microsoft® running on the Windows® operating system. HoloLens® 1 is packed with several sensors and a processing unit and supports input modalities such as gaze tracking, gesture recognition, and voice recognition. The FoV in HMD devices can be varied by head movement. This new freedom allows the implementation of several new features compared to the HHD application to help the user navigate the application. However, the user needs to be trained before using the HoloLens® as it is relatively new and unfamiliar to most users.

From a software perspective, a framework is developed just like in case study 1. Figure 8.7 illustrates the software framework used to design the Android AR application on a HoloLens®. The three layers of the C4 framework represented here are the software system (HoloLens® application), containers (Unity and Mixed Reality Tool Kit), and components. The spatial mapping feature of HoloLens® is used to detect the workplace and display instructions instead of a QR code. The voice commands from the user are translated to inputs for the HoloLens®, and this is enabled by the inbuilt Voice Manager feature of the Mixed Reality Tool Kit.

When the user interacts with the HoloLens® for assembly instructions, by default, a scene is loaded without any supplementing information. This will keep the visual and cognitive load of the user to a minimum level. However, the user can command the device using voice inputs such as "Parts," "Tools," "Next," and "Previous" to get the information needed or to navigate through different scenes. The hand gesture is also enabled in the application and is limited to positioning the digital workstation on top of the real one. This is done to situate the origin of the digital system on the correct location of the spatial map; after that, it is anchored and will stay locked in position even when the application is restarted. This initial setup is done when the application is used for the first time or when the workstation is physically moved somewhere other than the initial place. This also enables the device's portability for different workstations for diverse operations. Figure 8.8 illustrates the UI of the application on HoloLens®.

In summary, the AR application is developed for HoloLens® using see-through optical technology using spatial mapping as a tracking system. Speech and gesture modalities are incorporated into the application, validating the multimodal approach in interacting with AR devices. The application is portable to different locations on the shop floor. Moreover, the FoV of the UI is variable enabled by the head tracking feature of the device. Visual cues such as part and tool pointers are included in the application as an increased field of view allows multiple feature implementations.

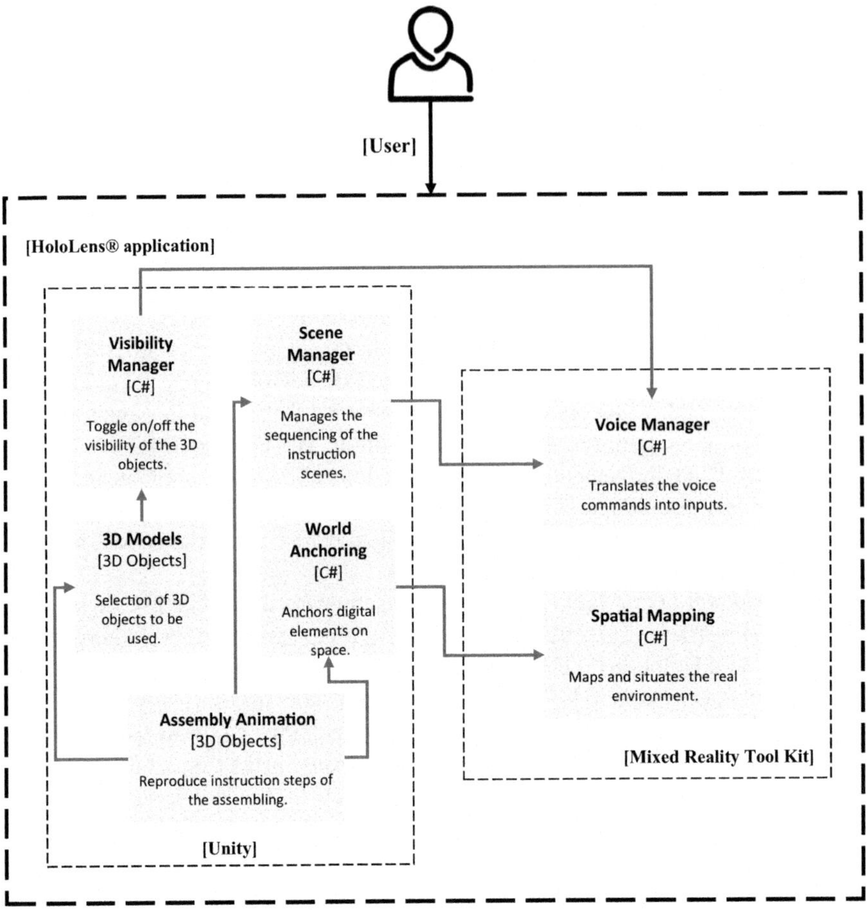

FIGURE 8.7 Software framework for HoloLens® (HMD).

8.5 VR TECHNOLOGY IN MANUFACTURING

The availability of powerful, immersive hardware and software systems has made the visualization of complex systems in a realistic virtual environment very efficient. VR is defined as an immersive computing technology that provides a unique way to interact with the ever-growing digital landscape (Berg et al., 2017). This interaction helps in depicting the behavior of complex and abstract systems in a holistic way to both experts and non-experts. The ability of VR technology to revolutionize traditional processes and working methods has been identified by large companies in the automotive and aircraft sector, especially in the topics of knowledge transfer, professional training, and new product development.

The objective of the immersive technology tools developed to experience the virtual environment is mainly to deliver information to our senses, particularly sight, hearing, and touch. Some of the most used VR tools are VR-Powerwall, cave virtual environment, and a VR-head-mounted display.

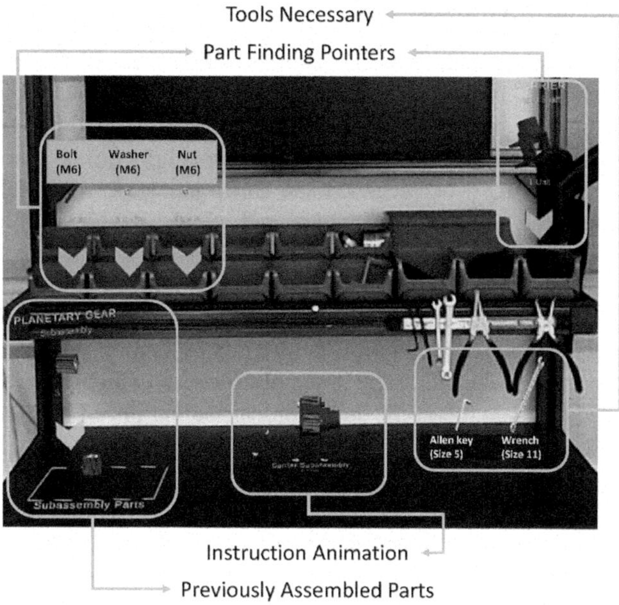

FIGURE 8.8 User Interface (UI) on HoloLens (HMD).

The rest of the sections will focus on the important applications of VR in manufacturing and the various case studies which have shown the potential of VR in manufacturing.

8.6 VR APPLICATIONS IN MANUFACTURING

VR is being used increasingly in many manufacturing applications. The various applications are explained in this section to provide the reader with the possible opportunities for using VR (Choi et al., 2015).

8.6.1 VR in Design

VR technology plays a significant role in designing and prototyping new products. The best use of VR is in the conceptual design stage, where it provides the designers with a virtual environment for designing a new product in 3D. This feature helps them in analyzing the performance of different mechanical features such as hinges, assembly, etc., which in turn will aid in evaluating the conceptual design.

Another important application of VR is seen in virtual prototyping, where the engineers can use the virtual prototype before fabricating the physical prototype to suggest improvements in the design, evaluate specific characteristics of the prototype, help in manufacturing planning, and get feedback from the potential customers in case of a new product. Generally, such applications incorporate all the different functionalities needed to realistically simulate the dynamic behavior of the product, including human interaction with the virtual environment, and consist of both offline and real-time virtual simulation.

8.6.2 VR in Manufacturing Processes

VR technology helps in understanding three different areas of the manufacturing process, namely, machining, assembly, and inspection. Machining using VR technology is used in processes such as turning, milling, drilling, grinding, and so on. They help the engineers in simulating the manufacturing processes where the user can mount the workpiece on the machine, perform direct machining operations using the corresponding tool and verify if the desired output is obtained.

Virtual assembly using VR technologies helps in making assembly-related engineering decisions as it helps the designers to analyze, visualize and implement different models without realization of the product or support process. On a larger scale, it can help in investigating the assembly process in an assembly line where the location, sequence, and type of operation can be optimized beforehand.

8.6.3 VR in Planning, Simulation, and Training

The virtual environment developed using different VR tools can lead to optimal planning of a manufacturing system with the best possible arrangement of shop floor and different machines. The comparison of different layouts/solutions considering human experiences and facts has resulted in the rapid start of several manufacturing processes.

VR-based training is among the most advanced method for training workers with the necessary manufacturing skills and processes. The virtual model of the product, plant layout, and assembly process helps the workers understand the different features associated with each field. Since the visualization and simulation are done realistically, the workers feel it is easier and faster to learn than with the conventional methods.

8.7 VR CASE STUDIES

8.7.1 Case Study 1 – VR for the Development of a Virtual Robotic Cell

Darmoul et al. (2015) developed a virtual model of a robotic cell and analyzed its functioning in a semi-immersive virtual environment to provide a feasible layout planning for setting up the real robotic cell. Through this work, the authors demonstrated that saving time and effort while designing such robotic cells using VR technology was possible. They were able to assess various layouts and configurations without the risk of damaging the system. They used the virtual robotic cell for layout planning and set up a real robotic cell in an advanced manufacturing institute, King Saud University, Saudi Arabia.

To achieve their goals, they used a combination of hardware equipment and software architecture consisting of CATIA for the CAD model generation, InterSense IS-900 wired tracking devices for head and hand tracking, pro-engineer, division mockup, and so on. The developed robotic cell was used for training industrial workers and academic researchers to demonstrate the advantages of VR technology and its potential for the manufacturing industry.

8.7.2 Case Study 2 – VR-Based Configurator of Interior Accessories for Fire Truck

Researchers from the University of Applied Sciences Karlsruhe, Germany, in collaboration with company Rosenbauer Karlsruhe GmbH & Co. KG, used VR technology to develop a configurator of interior accessories for a fire truck (Bellalouna, 2020). This helped the company decouple the determination of the firefighting equipment position from the production of the fire truck. The author identified the necessary functions to be implemented in the VR application by organizing several workshops with the sales engineering departments of the company and different fire departments. Some of the necessary functions were VR libraries containing all the firefighting accessories, different fire truck models, the configuration of the equipment shed, a function to determine the equipment weight, center of gravity, and axle load, and finally, an option to create the bill of materials. The VR application developed was evaluated to check its usability for further business processes, and the following points were inferred:

a. The users found the functions to be intuitive and also allowed a cognitive configuration process.
b. The application helped in creating a realistic VR environment with a high emersion ability without any restriction.
c. The application helped in identifying conflicts between design and customer requirements during the early product lifecycle phase.
d. One weak point of the system was that the successful VR-based configurator only allowed a single-user application. In contrast, successful deployment of the VR application requires multi-user involvement.

8.7.3 Case Study – Implications of VR Technology on Environmental Sustainability in Manufacturing Industries

An interesting case study was carried out by Chen et al. (2021) with an international automotive company with a research and development center in Sweden and manufacturing plants in China to explore the implications of VR technology on environmental sustainability in manufacturing industries. They developed a VR demo of the welding process in a manufacturing environment in China to be used by the users from the development center in Sweden. In addition to the VR demo, they used various tools such as customer experience feedback, interviews, and questionnaires.

They showed that the users were able to communicate more efficiently about the product design as the VR application offered them the product view from different angles and with more details. This improved communication efficiency enabled constant and frequent dialogue between developers and implementation sites and potentially reduced the number of travels between China and Sweden.

Due to the possibility of performing product and process design verification from Sweden, this demo application could contribute to environmental sustainability with less CO_2 emissions generated from international trips.

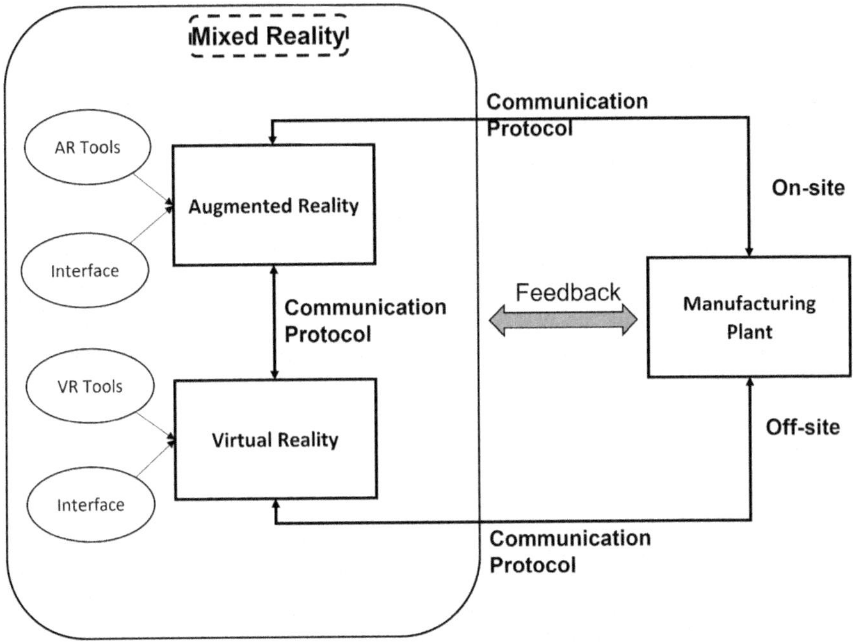

FIGURE 8.9 Conceptual framework for integrating MR technology in a manufacturing plant.

8.8 CONCEPTUAL FRAMEWORK FOR MR TECHNOLOGIES

MR, like several other Industry 4.0 solutions, draws in data from the tools available to create virtual realities the workers can use to augment their work. Implementing MR technologies by using VR and AR for both on-site and off-site operations in a manufacturing sector can help improve the working conditions for the workers, have better control over the plant operations, and integrate Industry 4.0 technologies for their production activities. A simplified conceptual framework for MR technology in manufacturing is shown in Figure 8.9.

8.9 DISCUSSION AND CONCLUSION

MR technologies play a very significant role in complex product development and production processes. Due to the advancement in different immersive technologies, MR has found its way from many high-tech industry sectors to SMEs as well. However, there are several challenges that need to be addressed to reach its actual potential in the field of manufacturing.

One of the topics that will strongly influence the future of MR applications development is spatial registration technology, which can combine the virtual world and the real world through a proper relationship of relative positions. This will help the user in efficient localization in the MR environment and improve the performance of the MR application. The second challenge is the development of a user-friendly interface for MR applications that are affected by aesthetics and ergonomics. One big problem in using MR technology at a large scale is the heavy display equipment, which is unsuitable for a production line. The alternative solution of using mobiles or smartphones also has a limitation: the workers must perform their work simultaneously using the MR application. During such times, fixing the mobile phone in the workplace or using a lightweight advanced HMD could be a good option. The next challenge is in developing MR applications for multiuser collaboration. Generally, a fully functional manufacturing line consists of a line manager, engineers from different disciplines, operators, and technicians. In such cases, it is of utmost importance to provide an MR platform enabling helpful collaboration, information sharing, and real-time communication among the different users.

Despite their limitations, the potential of MR technology for manufacturing is enormous and is becoming increasingly mature through the innovation of related technologies. Combining the benefits of mixed Reality with other Industry 4.0 tools will enable the manufacturing sector to bring different teams together to increase their performance, efficiency, and market share.

REFERENCES

Bellalouna, F. 2020. Industrial case studies for digital transformation of engineering processes using the virtual reality technology. *Procedia CIRP*, 90: 636–641.

Berg, L. P., Vance, J. M. 2017. Industry use of virtual reality in product design and manufacturing: a survey. *Virtual Reality*, 21(1): 1–17.

Brown, S. 2019. *The C4 model for visualising software architecture.* https://c4model.com/

Chen, X., Gong, L., Berce, A., Johansson, B., Despeisse, M. 2021. Implications of virtual Reality on environmental sustainability in manufacturing industry: A case study. *Procedia CIRP*, 104: 464–469.

Choi, S., Jung, K., Noh, S. D. 2015. Virtual reality applications in manufacturing industries: Past research, present findings, and future directions. *Concurrent Engineering*, 23(1): 40–63.

Darmoul, S., Abidi, M. H., Ahmad, A., Al-Ahmari, A. M., Darwish, S. M., Hussein, H. M. 2015. Virtual reality for manufacturing: A robotic cell case study. In *2015 International Conference on Industrial Engineering and Operations Management (IEOM)*, 1–7. IEEE.

Dobbelstein, D., Hock, P., Rukzio, E. 2015. Belt: An unobtrusive touch input device for head-worn displays. *Proceedings of the 33rd Annual ACM Conference on Human Factors in Computing Systems*, 2135–2138.

Doolani, S., Wessels, C., Kanal, V., Sevastopoulos, C., Jaiswal, A., Nambiappan, H., Makedon, F. 2020. A review of extended reality (XR) technologies for manufacturing training. *Technologies*, 8(4): 77.

Esteves, A., Verweij, D., Suraiya, L., Islam, R., Lee, Y., Oakley, I. 2017. SmoothMoves: Smooth pursuits head movements for augmented reality. *Proceedings of the 30th Annual ACM Symposium on User Interface Software and Technology*, 167–178.

Fast-Berglund, Å., Gong, L., Li, D. 2018. Testing and validating Extended Reality (XR) technologies in manufacturing. *Procedia Manufacturing*, 25: 31–38.

Fraga-Lamas, P., Fernández-Caramés, T. M., Blanco-Novoa, Ó., Vilar-Montesinos, M. A. 2018. A review on industrial augmented reality systems for the Industry 4.0 Shipyard. *IEEE Access*, 6: 13358–13375.

Hsieh, Y.-T., Jylhä, A., Orso, V., Gamberini, L., Jacucci, G. 2016. Designing a willing-to-use-in-public hand gestural interaction technique for smart glasses. *Proceedings of the 2016 CHI Conference on Human Factors in Computing Systems*, 4203–4215.

Hu, F., He, P., Xu, S., Li, Y., Zhang, C. 2020. FingerTrak: Continuous 3D hand pose tracking by deep learning hand silhouettes captured by miniature thermal cameras on wrist. *Proceedings of the ACM on Interactive, Mobile, Wearable and Ubiquitous Technologies*, 4(2): 1–24.

Kolla, S. S. V. K., Sanchez, A., Minoufekr, M., Plapper, P. 2020. Augmented reality in manual assembly processes. *Augmented Reality in Manual Assembly Processes*, 121–128.

Lee, J., Jun, S., Chang, T.-W., Park, J. 2017. A smartness assessment framework for smart factories using analytic network process. *Sustainability*, 9(5): 794.

Liu, C., Berkovich, A., Chen, S., Reyserhove, H., Sarwar, S. S., Tsai, T.-H. 2019. Intelligent vision systems–Bringing human-machine interface to AR/VR. *2019 IEEE International Electron Devices Meeting (IEDM)*, 10–15.

Mistry, P., Maes, P. 2009. SixthSense: A wearable gestural interface. In *ACM SIGGRAPH ASIA 2009 Art Gallery & Emerging Technologies: Adaptation*, 85.

Mujber, T. S., Szecsi, T., Hashmi, M. S. J. 2004. Virtual reality applications in manufacturing process simulation. *Journal of Materials Processing Technology*, 155: 1834–1838.

Nizam, S. S. M., Abidin, R. Z., Hashim, N. C., Lam, M. C., Arshad, H., Majid, N. A. A. 2018. A review of multimodal interaction technique in augmented reality environment. *International Journal of Advance Science and Engineering Information Technology*, 8(4–2): 4–8.

Palmarini, R., Erkoyuncu, J. A., Roy, R., Torabmostaedi, H. 2018. A systematic review of augmented reality applications in maintenance. *Robotics and Computer-Integrated Manufacturing*, 49: 215–228.

PTC. (2020). *Develop AR Experiences with Vuforia Engine*. https://www.ptc.com/en/products/vuforia/vuforia-engine

Qian, J., Shamma, D. A., Avrahami, D., Biehl, J. 2020. Modality and depth in touchless smartphone augmented reality intERACTIOns. *ACM International Conference on Interactive Media Experiences*, 74–81.

Wang, X., Ong, S. K., Nee, A. Y. C. 2016. A comprehensive survey of augmented reality assembly research. *Advances in Manufacturing*, 4(1): 1–22.

Williams, T., Bussing, M., Cabrol, S., Lau, I., Boyle, E., Tran, N. 2019. Investigating the potential effectiveness of allocentric mixed reality deictic gesture. *International Conference on Human-Computer Interaction*, 178–198.

9 Inferential Modeling and Soft Sensors

Muhammad Asad Ullah Khalid
School of Mechanical Engineering, Chung-Ang University, Seoul, South Korea

Shahid Aziz
Jeju National University, South Korea

Uzair Khaleeq uz Zaman
Department of Mechatronics Engineering, College of Electrical and Mechanical Engineering, National University of Sciences and Technology, Islamabad, Pakistan

9.1 INTRODUCTION TO SOFT SENSORS

Soft sensors are essential analytical tools for industrial process monitoring and control. The word "soft" is short for software, which indicates that these sensors are software-based. Since these soft sensors are developed using software, they are less expensive to create than expensive hardware-based sensors. They can complement the existing hardware-based sensors, allowing the implementation of smart monitoring networks to provide better process control. The added advantage of real-time data analysis without time delay issues, preliminarily found in hardware-based sensors, can significantly improve control strategies. Historical data of different process variables, both dependent and independent, is very beneficial for soft sensor design and modeling. This chapter is quite different from the previously reviewed topic on soft sensor models, as we will briefly present the data-related context for soft sensor modeling and focus more on a case study-based approach for the applications of soft sensors in bioprocess monitoring and control. In the end, it will be concluded with the key recommendations of soft sensors integration for smart process monitoring and control.

There are many different approaches for modeling a soft sensor which can be mainly categorized into mechanistic, multivariate statistical, and artificial intelligence-based modeling (Ghosh et al., 2020). The data-driven models need accurate data of the plant or process variables to make accurate measurements and decisions for control, which means that the choice of data is crucial in modeling the soft sensors. Modeling a soft sensor might involve typically following steps (Luigi et al., 2007):

1. Data classification and pretreatment
2. Assigning variables

DOI: 10.1201/9781003327523-11

3. Soft sensor model structuring and identification
4. Validation of the soft sensor model

Data should be correctly categorized into different classes after collection based on the information required so that feature identification is easy. Data contamination is a serious concern, affecting the model's performance, and the soft sensor model's reliability is compromised. This suggests that the data pretreatment is important before passing it to a model structure. The data pretreatment might involve normalization, filtration, etc. The characteristics of process data have been discussed by Ghosh et al. (2020) include problems like missing data, data outliers, and data collinearity and how to deal with them. They are also briefly discussed here to provide a complete understanding of the concepts. The typical process flow for soft sensor modeling is given in Figure 9.1.

Before reviewing the data characteristics, let's look at the flow diagram of soft sensor-integrated process monitoring and control shown in Figure 9.2. The diagram shows that the soft sensor depends on a hardware sensor's data to determine the estimated variables. The process generates process variables, which pass through this soft sensor module comprising a hardware sensor and an estimator. This estimator is our soft sensor design model, which estimates new parameters by analyzing the historical process data of the hardware sensor to determine new estimated variables. These new estimated variables can then be used as process monitoring parameters as well as control parameters. To keep the consistency of the concepts being discussed, it is important to mention that the estimator or soft sensor model is

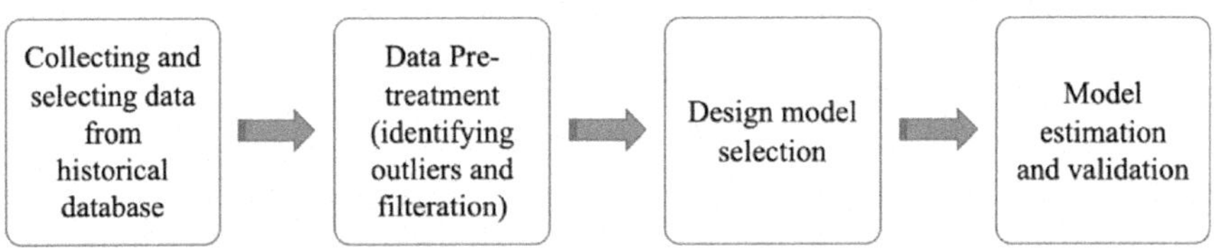

FIGURE 9.1 A block diagram of soft sensor modeling process.

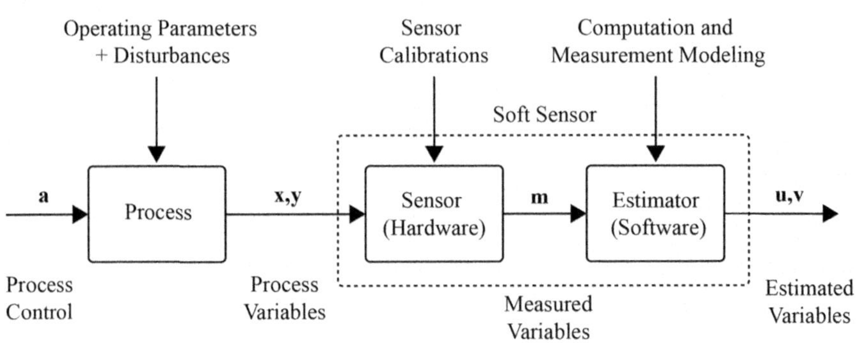

FIGURE 9.2 A flow diagram of a soft sensor integrated process monitoring and control.

entirely dependent on the single/multiple hardware sensors' data. We previously described that this data is available in the raw form and is pretreated due to certain problems. One major problem is data collinearity, which makes it very time-consuming to select the appropriate data bank from huge data with poor information management. If not tackled properly, it could seriously affect the performance of data-driven, for example, artificial neural network (ANN)-based sensors. The plant personnel working might help acquire the right data due to their working experience with that particular process control. Then the problem of missing data can also emerge, which can happen because of power outages due to maintenance, repairs, and certain other reasons. So, these missing data need to be accommodated. Sometimes this missing data is ignored within a chronological stream, but at other times, it is replaced with a filler data set to complete a chronological data stream. Data outliers are also a prevalent problem. They are mostly the process variable data points, which are outside a defined or normal range of interest, and are caused by instrumental errors, plant disturbances, etc. They can negatively affect the soft sensor model. They can be categorized into different types and handled accordingly using various data reconciliation and proximity-based methods (Ghosh et al., 2020).

Virtual instruments (VIs) play an important role in modeling soft sensors. These instruments are software-based measurement and computational blocks that utilize the existing hardware instruments and computer systems with programmed user interfaces. These VIs can help design soft sensors as they provide customization capabilities. They can be utilized according to application-based requirements to measure and process data. All these VIs are available in a software facility store as packages for custom soft sensor modeling. LabVIEW by National Instruments (NI) is a powerful computer program used in many industrial process monitoring and control applications. It combines with various data acquisition hardware or other instruments for real-time variables' data collection. This raw data is then computationally evaluated, combined with other data, or processed according to certain logical and mathematical constraints within a LabVIEW program. This LabVIEW program provides a convenient front panel interface for user data display and control, while a block diagram represents a backend software that runs the algorithm to perform tasks in the front panel according to user design applications. This provides a great degree of comfort to the designers and users as well. The users are mostly concerned with these front panel tasks, which are relatable to the actual physical process as it is designed to do so, whereas all the logical and mathematical operations are carried out by the implemented algorithm using VIs in the block diagram. Examples of a typical front panel and block diagram are provided in Figure 9.3.

9.2 SOFT SENSOR INFERENTIAL MODELS

Modern-day manufacturing plants are very different from the old ones. They consist of old equipment, which was built in the 1920s as well as new equipment up to the 2020s. Smart manufacturing (SM) is utilizing data science to improve efficiency at all levels of the manufacturing cycle. Since it is not possible to get real-time feedback from all important processes, there is a need to use soft sensors and inferential models to realize the feedback. Improvements in sensors embedded in

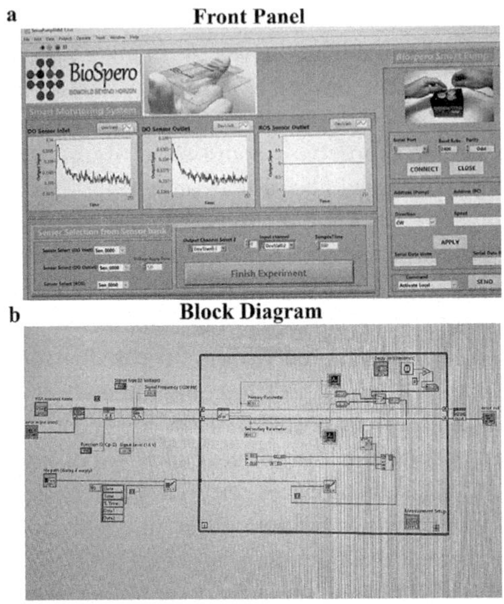

FIGURE 9.3 Images of a front panel and block diagram for LabVIEW-based VIs for sensor-based measurements in a bioprocess (Khalid et al., 2022).

the manufacturing processes enable inferential models that simulate manufacturability characteristics that affect quality, cost, and ease of assembly. Inferential modeling techniques automate the process of generating digital twin models through direct observation, thereby maximizing the benefits of Industry 4.0 technologies. It enables the automatic modeling of the processes using computational models and automatically updates models due to the change in environment. Inferential modeling relies on the input data from various sensors to estimate the output, which lies at the center of modeling techniques to provide robust, repeatable, and sustainable results. Data can be efficiently utilized for modeling when available or refined in a meaningful manner; hence, data collection and training are critical to infer effective soft sensor models in the manufacturing plant. There are various techniques for modeling soft sensors. In this section, we will discuss principal component analysis (PCA), partial least squares (PLS), and ANN techniques to model input data from soft sensors.

9.2.1 Principal Component Analysis

SM requires smart measurement methods to monitor the advanced process control parameters. Since it is not always possible to measure them, soft sensors are used to model these processes. PCA is one of the methods to model the huge amount of data inferred from the soft sensors, which relies on reducing the dimension of the data set while increasing the probability of retaining the variance of the parent data. PCA method is preferable when the manufacturing process is linear; otherwise, it is not recommended. PCA has been widely used in various soft sensor applications (Shi & Liu, 2006; Pe et al., 2018; Farsang et al., 2015).

In order to model sensor data, consider a data matrix M = {(X, Y)$_p$; p = 1, 2, …, n}, where n is the number of sample data points, and X = (x_1, x_2, x_3, … , x_l), Y = (y_1, y_2, y_3, … , y_m), where l and m represent the number of input and output variables of soft sensors.

All input variable data X is selected by PCA to filter the irrelevant input variables from going into the multiscale decomposition stage. Based on the experimental data and trial error, the selected new variables are scaled down to several components that best explain the data variation. At this stage, new pairs are formed in the training process, and then the PCA model reconstructs the trained data in the multiscale reconstruction stage before it is obtained as a model output Y.

9.2.2 Partial Least Squares

Partial least squares (PLS) is another simple method for inferential modeling of soft sensor data (Hasnen et al., 2019; Wang et al., 2015). PLS is used to evaluate the performance of the variable selection methods. Just like PCA, PLS is also limited only to linear processes. However, some researchers have modified PLS algorithms to account for the non-linearities subjected to time-varying characteristics experienced by most manufacturing processes in industries (Fu et al., 2017). Since PLS is based on the variables' scores instead of the original values of variables, it is considered more stable than those models that use the original variables.

The mathematical model of the PLS has been explained in detail by some researchers (Fu et al., 2017). The basic idea is to start modeling by centralizing and normalizing the input and output data matrices. Then the PLS algorithm is used to decompose the input and output matrices into external models having a component matrix that consists of score matrices. The detailed breakdown of the PLS matrices can be found in Fu et al. (2017), Chen et al. (2009), and Poerio and Brown (2020).

9.2.3 Artificial Neural Networks

ANNs are algorithms that help deep learning to solve advanced data-driven problems. Deep learning makes use of ANN that behave similarly to the neural networks in a human brain. A neural network functions when some input data is fed to it. This data is then processed via layers of perceptrons to produce the desired output. During the training process, the ANN-based model can learn the complex dynamics of the manufacturing process through variable examples readily available through previous experimental data. Unlike PCA and PLS models, the ANN model is well known to deal with complex nonlinearities, which play an essential role in SM processes. ANN is used extensively in SM for Industry 4.0 to predict important process parameters critical for mechanical components. In a recent work (Kazi et al., 2021), ANN has been used to predict the required cotton fiber filler content to increase the mechanical properties in a cotton fiber/thermoplastic composite, i.e., polypropylene. The polyethylene terephthalate production process has been monitored via soft sensors developed through a feed-forward ANN model (Gonzaga et al., 2009). ANN-based soft sensors have been utilized to estimate the apparent viscosity of water-based drilling fluids from the available data about the concentration of additives and temperature of that fluid (Bispo et al., 2017). The modeling of ANN-based soft sensors can be found in many research works ("Soft Sensor Model Based on Improved Elman Neural.Pdf"; Ghosh et al., 2020).

9.2.4 Support Vector Machines

ANN-based modeling has been successfully implemented in soft sensor modeling, but they lack the ability for generalization. Therefore, support vector machines (SVM)-based soft sensor modeling is needed. The basis of SVM is the statistical learning theory. SVMs are more robust to raw data and have a better ability for generalization because of the natural enforcement of regularization. To predict key production indicators in the complex grinding process, a modification of SVM, i.e., least squares SVM, has been used as a soft sensor model with strong generalization ability (Xie et al., 2020). SVM has also been used to develop soft sensor models for continually assessing quality indicators in the refinery isomerization process of 2,3-dimethylbutane and 2-methyl-pentane mole percentage (Herceg et al., 2021). The advantages of SVM over ANN are the convergence of SVM-based soft sensors to the global optimal more efficiently and its ability to be interpreted based on the training data, which cannot be done in the case of ANN-based soft sensors.

9.3 SOFT SENSORS IN MULTISTAGE SM

SM processes have become more complex and multistage due to several quality control parameters and customer requirements. In most multistage manufacturing processes, the process monitoring techniques provide prediction models at the end of the manufacturing process, which is less practical and costly. Soft sensor models discussed in the previous section can be useful in challenging modeling problems like multistage SM processes. This requires continuous online monitoring of processes, which involves big data handling to predict anomalies in manufacturing processes early in the multistage SM system. Several soft sensor modeling approaches, such as ANN surrogate (Bambach et al., 2021) and hybrid sequence-to-sequence recurrent neural network-deep neural network (Seq2Seq RNN–DNN) (Hong et al., 2022) can be adapted to model multistage manufacturing processes in which each stage can be monitored through a different soft sensing approach along with embedded sensors throughout the processes. ANN surrogate model-based soft sensing approach has been demonstrated by Bambach et al. (2021) to model the multistage hot forming process. A hybrid model (Hong et al., 2022) combining several soft sensing models can be integrated to enhance quality monitoring in multistage SM systems.

9.4 HARD AND SOFT SENSORS FOR SMART BIOREACTORS

This is an important and new topic that distinguishes the contents of this book chapter from others in similar contexts. Bioreactors and their different types will be briefly discussed here, most importantly focusing on the hardware and software sensors used for bioprocess monitoring (Biechele et al., 2015; Busse et al., 2017).

Then this topic will briefly discuss some soft sensors implemented-bioreactors with a case study.

Bioreactors are engineered spaces to allow safer and controlled bioreactions. To briefly describe the main purpose of bioreactor designing is to realize perfect conditions for these bioreactions, which may involve cell culture and microorganism growth for various applications like the production of biofuels, metabolic products, biowaste management etc. There are many types of bioreactors depending on the application, to name a few, they are stirred-tank reactor, bubble reactor, airlift reactor, loop reactor, reactor with immobilized cells, fluidized reactor with recycling of cells, solid-phase tray reactor, rotary drum bioreactor as discussed by Carl (2016). They have discussed the advantages and shortcomings of these bioreactors' designs. In this text, only specific sensor-integrated bioreactors will be discussed. When we talk about key parameters of bioprocess monitoring, they mostly include some physical and/or chemical environmental parameters, e.g., temperature, CO_2, and culture media pH. However, biosensors also play a very important role in determining the state and output of the bioreactor systems. These biosensors can detect metabolic products, proteins, DNA/RNA structures, or the microorganisms themselves with very high specificity. For example, a review was conducted on the usefulness of soft sensing systems in online monitoring of upstream industrial bioprocess, focusing on the sensing performance and process' outputs in terms of production economy (Randek & Mandenius, 2018). A basic overview of the soft sensing requirements and applications in various industrial bioprocesses, like recombinant protein production and production of baker's yeast has been presented. Similarly, different online analytical techniques like near and, mid-infrared (N/MIR) spectroscopy with high potential for soft sensor development have been discussed, which can measure several parameters like metabolic products' concentrations or cell numbers. Other discussed techniques include *in-situ* microscopy (ISM), fluorescence microscopy, high-performance liquid chromatography (HPLC), and biosensors. Similarly, they reviewed various upstream bioprocess monitoring by soft sensors' examples, followed by suggesting a step-wise soft sensor development general methodology for this purpose.

9.4.1 Case Study – Soft Sensor for Bioprocess Monitoring

Metabolic heat-based soft sensor has been used in a fed-batch *Escherichia coli* cultivation-based recombinant protein production (green fluorescent protein – GFP) bioprocess for estimating the specific growth rate and biomass concentration (Paulsson et al., 2014) as shown in Figure 9.4a. The sequential filtering further improved the methods like moving average filtering methods, such as low-pass filters, Savitzky-Golay filters, and extended Kalman filters. They successfully controlled the feed rate based on the soft sensor's output estimates. After sequential filtering, the final model equation obtained was

$$q_{filter\ 2,n} = \{(q_{filter\ 1}[n-180+1:n]) + min(q_{filter\ 1}[n-180+1:n])\}/2 \ \ldots \quad (9.1)$$

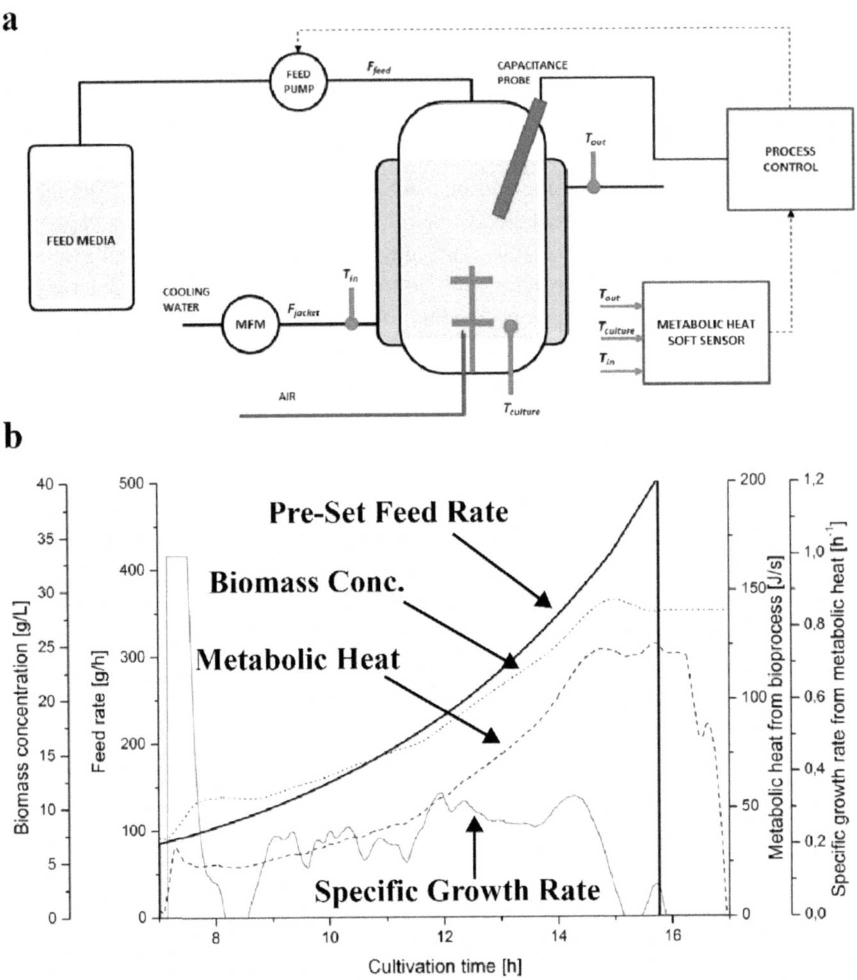

FIGURE 9.4 **(a)** Metabolic heat soft sensor for bioprocess control, and **(b)** Estimated metabolic heat, specific growth rate ($\mu_{metabol}$) and biomass concentration ($X_{metabol,n}$) with passing cultivation time in pre-set feed rate (Paulsson et al., 2014).

Similarly, further mathematical modeling was done to derive specific growth rate $\mu_{metabol}$ and biomass concentration $X_{metabol,n}$ and the feed rate was controlled using a conventional proportional-integral (PI) controller. Further details can be found in the original work (Paulsson et al., 2014). Figure 9.4b shows the estimated outputs of the soft sensing system for uncontrolled fed-batch cultivation. In the end, the soft sensor's performance with hardware-based sensor probes and capacitance on-line sensor, measuring the same parameters, were in high agreement.

Apart from this, many other soft sensors have been developed for bioprocess monitoring and control applications. A few of them have been presented in Table 9.1 for a brief review.

TABLE 9.1
Soft Sensors Developed for Various Bioprocesses' Monitoring and Control

Soft Sensors	Purpose & Parameters Estimated	Reference
a. Biomass concentration from an online NIR probe, b. Biomass concentration from titrant addition, c. Specific growth rate from titrant addition, d. Specific growth rate from the NIR probe, and e. Specific substrate uptake rate and by-product rate from online HPLC and NIR probe signals	Recombination protein (GFP) production in E. Coli cultivation, mainly for a. Biomass concentration estimation b. Estimation of specific growth rates	Warth et al., 2010
Feed-forward ANN, support vector, and relevance vector regression soft sensor mode based on glucose/lactose feed rate and oxygen uptake	Biomass and product concentrations	Simutis et al., 2013
Biomass soft sensor in Eppendorf DASbox® Mini Bioreactor System using the DASware® control software using mass balancing	Biomass production and biomass specific nutrient uptake estimation (e.g., substrate uptake rate)	Kager et al., 2017
Multiphase Artificial Neural Network (MANN) based dynamic soft sensor having Nonlinear Auto Regressive with eXogenous input (NARX) models to capture the complete dynamics of lag, log, and stationary phases of the microbe	For biomass concentration of *Trichoderma* estimated from online sensors data like pH, substrate concentration and agitation speed.	Murugan and Natarajan, 2019
The end of the growth phase in a thraustochytrid Industry 4.0 (I4.0) fermentation process	Dynamic control of compressor based on dissolved oxygen (DO) levels	Alarcon and Shene, 2021

9.4.2 Case Study – Advanced Sensing Strategies with Soft Sensing Potential for Biochip Systems

The authors have worked on state-of-the-art microfluidic organs-on-chips, which are engineered microchips – sometimes also referred to as biochips, containing living human organ microtissues in a dynamic culture environment to recapitulate the in-vivo-like organ functionalities in-vitro (Wu et al., 2020; Leung et al., 2022). They are recently becoming popular in different fields of study, e.g., toxicology, drug development (Caplin et al., 2015; Ching et al., 2021), disease modeling (Wang et al., 2017; Farooqi et al., 2021; Asif et al., 2021), and personalized medicine (Ingber, 2022). While they have the potential to replace animal testing models in the traditional drug development pipeline, they still need to improve on continuous processing monitoring and control strategies due to a lack of technological advancements. The authors have developed an embedded-sensors gut-bilayer microchip for real-time assessment of cellular dissolved oxygen (DO) and reactive oxygen species concentrations to analyze developmental and induced hypoxia (Khalid et al., 2022). The expanded and cross-sectional views of this biochip are shown in Figures 9.5a and b. Similarly, the bioprocess schematic demonstration and real-time data of the sensors are given in Figure 9.5c and d, respectively. While this biochip system generates very useful data of sensors for process monitoring, for a comprehensive analysis, the soft sensors to estimate metabolic products like glucose uptake and lactate release can also be implemented using the existing hardware sensors data as inputs to the model in future studies.

Another study on the microfluidic biochip was conducted by the authors to analyze the effect of chemotherapeutic drugs (docetaxel and doxorubicin) on a lung cancer microtissue using the integrated hardware sensors (optical pH sensor,

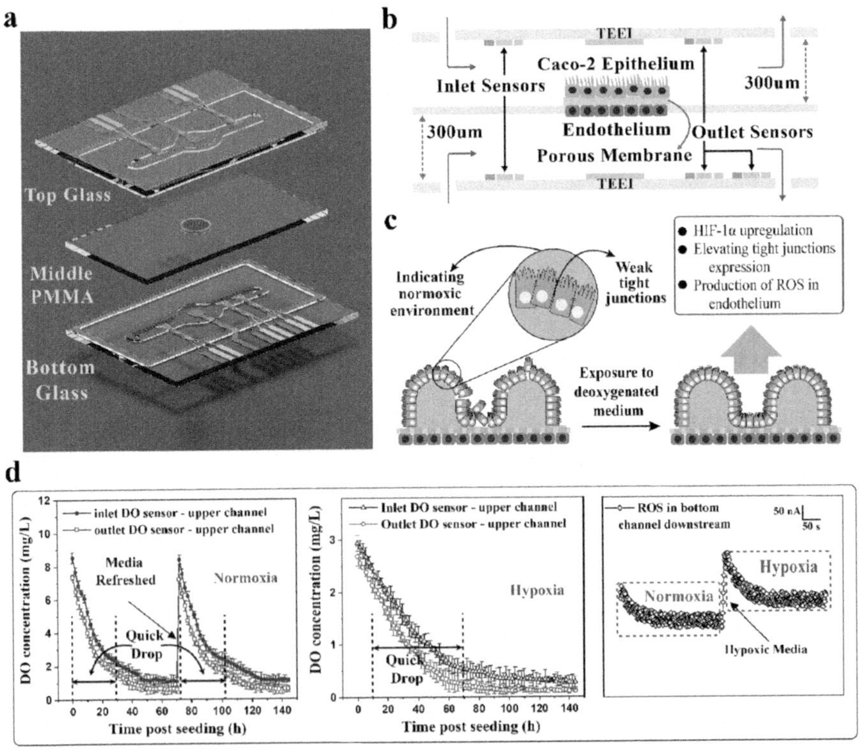

FIGURE 9.5 A gut bilayer organ on chip for hypoxia bioprocess monitoring: **(a)** Sensors (DO, reactive oxygen species, and TEEI) integrated microfluidic biochip expanded view, **(b)** Schematic cross-section indicating different components like printed electrochemical (DO and reactive oxygen species) sensors, TEEI sensor, cell culture area and cell types etc., **(c)** Hypoxia bioprocess schematic, **(d)** Response of the sensors in real-time for both normoxia and hypoxia bioprocesses (Khalid et al., 2022).

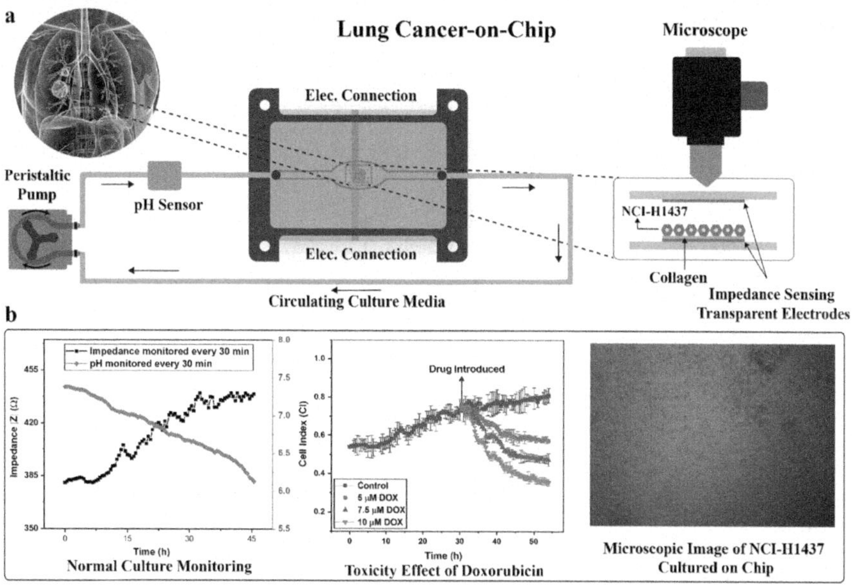

FIGURE 9.6 A lung cancer on chip with integrated sensors for process monitoring and chemotherapeutic assessment: **(a)** Schematic representation of the sensor (optical pH sensor, TEER sensor, and a portable microscope) integrated process, **(b)** Results of the sensors in normal process monitoring and response analysis to chemotherapeutic drug Doxorubicin (Khalid et al., 2020).

Transepithelial electrical resistance [TEER], and portable microscope) approach (Khalid et al., 2020). The sensor data were collected for normal growth monitoring and after treatment with different drug concentrations in real-time. A process schematic is shown in Figure 9.6a, whereas the sensor output response is shown in Figure 9.6b.

This system can produce enormous data banks of the process parameters using integrated online monitoring sensors, which could again be used for estimating relevant secondary parameters. These secondary parameters could be metabolic products, as previously described, as well as some pathological protein biomarkers or even DNA or RNA. The visual monitoring capability can allow for geometrical growth predictions of computer vision-based soft sensor models. The use of soft sensor models can be extremely beneficial for the complex multi-organ on-chip systems, where hardware usage is plentiful. The complex multi-organ on-chip systems provide more data, and data-driven soft sensors can benefit from this for state estimation and process control.

9.4.3 Conceptual Framework for Future Smart Bioprocess Monitoring Systems

A Hubka-Eder map (Hubka & Eder, 1988) can help build smart, complex bioreactor systems, as shown in Figure 9.7, by identifying the contributing individual

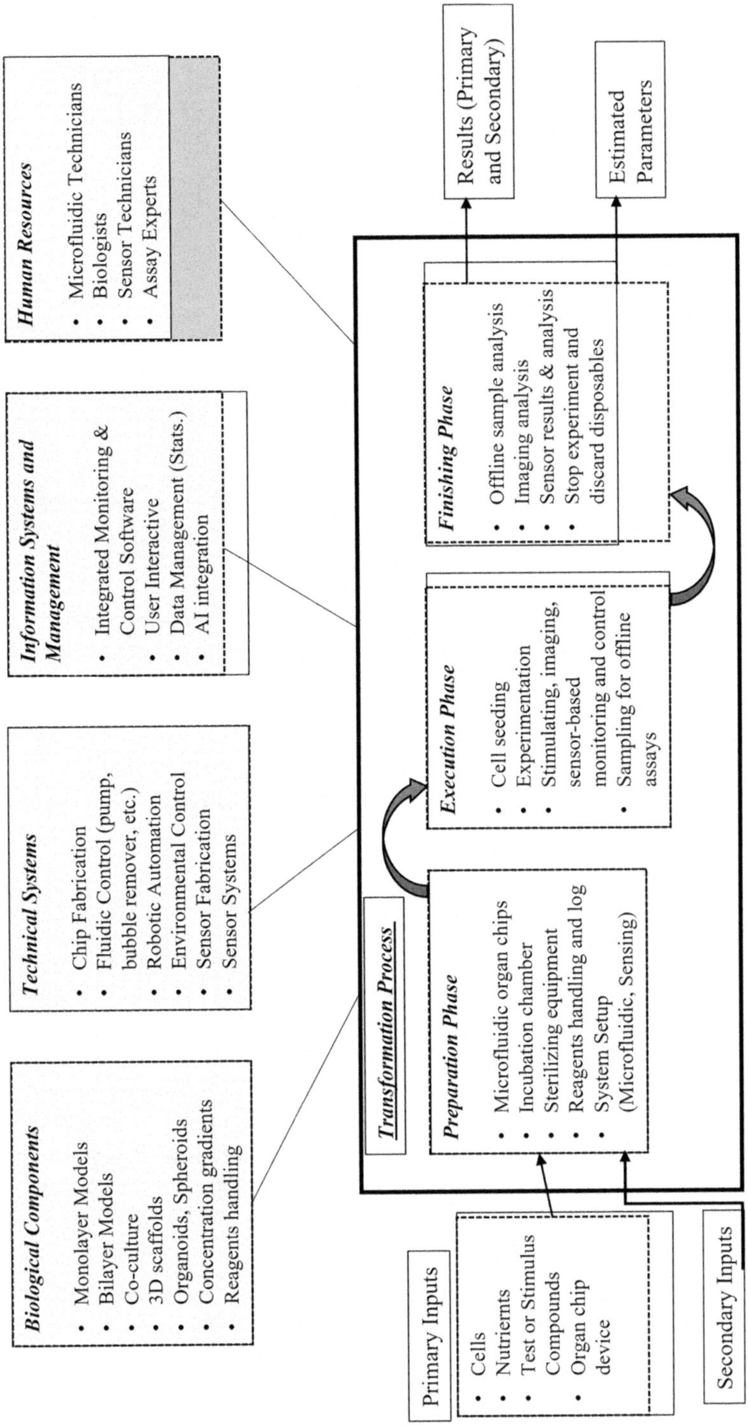

FIGURE 9.7 The Hubka-Eder map for smart bioreactor on chip systems.

functional units like biological components, technical systems, Information management systems, and human resources. These subunits contribute towards the transformation of primary inputs into results and estimated parameters. The transformation process can be divided into various execution phases, which convert the inputs to outputs using hardware and soft sensors.

9.5 CONCLUSION

Data-driven soft sensor models expand the capabilities of existing hardware sensors and systems in traditional integrated process monitoring and control applications. They can estimate the parameters which are not compatible with hardware measurement strategies. However, data selection is the most crucial step in modeling soft sensors. This data should be high-quality and free of noise, disturbances, and misinformation, for which different data treatment strategies are applied. A few inferential models have been briefly reviewed in this chapter. And some recent literature on soft sensors in smart multi-stage manufacturing systems and smart bioreactor systems has also been discussed, which suggest the more widespread use of soft sensor models in complex bioprocess monitoring and control with a case study. Similarly, emerging state-of-the-art bioreactor on-chip systems, which use online monitoring strategies, were explored with the potential for soft sensor integration to enhance different biotechnological aspects of these systems. Some conceptual soft sensor integration frameworks have also been presented in this regard.

9.5.1 Future Trends (Big Data and Deep Learning Revolution)

Since the data volumes are increasing with the complexity and abundance of the available hardware, there is a great need to utilize deep learning algorithms. However, the general rules of data selection apply here as well. But due to enormous instrumentation and sensing devices in modern-day plants, the data is generated in big volumes and variety. This big data needs large storage and a large number of data processing units to extract useful data features or parametric information. This could also lead to longer data processing times. Thankfully, the current data processing systems have high computational capabilities and can process big data in a reasonable time frame. The second important aspect is to use intelligent deep learning algorithms to model the soft sensors. A few example models have been discussed in the references (Ma et al., 2020; Wang et al., 2020) for the reader's choice.

REFERENCES

Alarcon, C., Shene, C. 2021. Fermentation 4.0, a Case Study on Computer Vision, Soft Sensor, Connectivity, and Control Applied to the Fermentation of a Thraustochytrid. *Computers in Industry*, 128: 103431.

Asif, A., Park, S. H., Soomro, A. M., Khalid, M. A. U., Salih, A. B. C., Kang, B., Ahmed, F., Kim, K. H., Choi, K. H. 2021. Microphysiological System with Continuous Analysis of Albumin for Hepatotoxicity Modeling and Drug Screening. *Journal of Industrial and Engineering Chemistry*, 98: 318–326. doi:10.1016/j.jiec.2021.03.035

Bambach, M., Imran, M., Sizova, I., Buhl, J., Gerster, S., Herty, M. 2021. A Soft Sensor for Property Control in Multi-Stage Hot Forming Based on a Level Set Formulation of Grain Size Evolution and Machine Learning. *Advances in Industrial and Manufacturing Engineering*, 2: 100041.

Biechele, P., Busse, C., Solle, D., Scheper, T., Reardon, K. 2015. Sensor Systems for Bioprocess Monitoring. *Engineering in Life Sciences*, 15(5): 469–488.

Bispo, V. D. da Silva, Scheid, C. M., Calçada, L. A., Meleiro, L. A. C. 2017. Development of an ANN-Based Soft-Sensor to Estimate the Apparent Viscosity of Water-Based Drilling Fluids. *Journal of Petroleum Science and Engineering*, 150: 69–73.

Busse, C., Biechele, P., Vries, I., Reardon, K. F., Solle, D., Scheper, T. 2017. Sensors for Disposable Bioreactors. *Engineering in Life Sciences*, 17(8): 940–952.

Caplin, J. D., Granados, N. G., James, M. R., Montazami, R., Hashemi, N. 2015. Microfluidic Organ-on-a-Chip Technology for Advancement of Drug Development and Toxicology. *Advanced Healthcare Materials*, 4(10): 1426–1450.

Carl, F. M. 2016. *Bioreactors*. Edited by Carl-Fredrik Mandenius. Weinheim, Germany: Wiley-VCH Verlag GmbH & Co. KGaA.

Chen, K., He, M., Zhang, D. 2009. Sliding-Window Recursive PLS Based Soft Sensing Model and Its Application to the Quality Control of Rubber Mixing Process. *Communications in Computer and Information Science*, 51: 16–24.

Ching, T., Toh, Y. C., Hashimoto, M., Zhang, Y. S. 2021. Bridging the Academia-to-Industry Gap: Organ-on-a-Chip Platforms for Safety and Toxicology Assessment. *Trends in Pharmacological Sciences*, 42(9): 715–728.

Farooqi, H. M. U., Kang, B., Khalid, M. A. U., Salih, A. R. C., Hyun, K., Park, S. H., Huh, D., Choi, K. H. 2021. Real-Time Monitoring of Liver Fibrosis through Embedded Sensors in a Microphysiological System. *Nano Convergence*, 8(1). Springer Singapore.

Farsang, B., Balogh, I., Németh, S., Székvölgyi, Z., Abonyi, J. 2015. PCA Based Data Reconciliation in Soft Sensor Development – Application for Melt Flow Index Estimation. In *Chemical Engineering Transactions*, 43: 1555–1560.

Fu, Y., Yang, W., Xu, O., Zhou, L., Wang, J. 2017. Soft Sensor Modelling by Time Difference, Recursive Partial Least Squares and Adaptive Model Updating. *Measurement Science and Technology*, 28(4).

Ghosh, S., Yang, S., Bequette, B. W. 2020. Inferential Modeling and Soft Sensors. In *Smart Manufacturing: Concepts and Methods*, 323–351. Elsevier Inc.

Gonzaga, J. C. B., Meleiro, L. A. C., Kiang, C., Filho, R. M. 2009. ANN-Based Soft-Sensor for Real-Time Process Monitoring and Control of an Industrial Polymerization Process. *Computers and Chemical Engineering*, 33(1): 43–49.

Hasnen, S. H., Zabiri, H., Prakash, K. K., Mat, T. T. 2019. Adaptive PLS Inferential Soft Sensor for Continuous Online Estimation of NOx Emission in Industrial Water-Tube Boiler. *IOP Conference Series: Materials Science and Engineering*, 702(1).

Herceg, S., Andrijić, U., Bolf, N. 2021. Support Vector Machine-Based Soft Sensors in the Isomerisation Process. *Chemical and Biochemical Engineering Quarterly*, 34(4): 243–255.

Hong, S., An, N., Cho, H., Lim, J., Han, I. S., Moon, I., Kim, J. 2022. A Dynamic Doft Sensor Based on Hybrid Neural Networks to Improve Early Off-Spec Detection. *Engineering with Computers*, 0123456789. Springer London.

Hubka, V., Eder, W. E. 1988. *Theory of Technical Systems*. Berlin, Heidelberg: Springer Berlin Heidelberg.

Ingber, D. E. 2022. Human Organs-on-Chips for Disease Modelling, Drug Development and Personalized Medicine. *Nature Reviews Genetics*, 0123456789. Springer US.

Kager, J., Fricke, J., Herwig, C. 2017. A Generic Biomass Soft Sensor and Its Application in Bioprocess Development. *Eppendorf*, 357: 1–8, https://www.eppendorf.com/product-media/doc/en/228755/Fermentors-Bioreactors_Application-Note_357_DASware_A-Generic-Biomass-Soft-Sensor-Its-Application-Bioprocess-Development.pdf (accessed on 7 June 2023).

Kazi, M. K., Eljack, F., Mahdi, E. 2021. Data-Driven Modeling to Predict the Load vs. Displacement Curves of Targeted Composite Materials for Industry 4.0 and Smart Manufacturing. *Composite Structures*, 258: 113207.

Khalid, M. A. U., Kim, K. H., Salih, A. B. C., Hyun, K., Park, S. H., Kang, B., Soomro, A. M. 2022. High Performance Inkjet Printed Embedded Electrochemical Sensors for Monitoring Hypoxia in a Gut Bilayer Microfluidic Chip. *Lab on a Chip*, 22(9): 1764–1778.

Khalid, M. A. U., Kim, Y. S., Ali, M., Lee, B. G., Cho, Y.-J., Choi, K. H. 2020. A Lung Cancer-on-Chip Platform with Integrated Biosensors for Physiological Monitoring and Toxicity Assessment. *Biochemical Engineering Journal*, 155: 107469.

Leung, C. M., Haan, P., Bouchard, K. R., Kim, G. A., Ko, J., Rho, H. S., Chen, Z., et al. 2022. A Guide to the Organ-on-a-Chip. *Nature Reviews Methods Primers*, 2(1). Springer US.

Luigi, F., Graziani, S., Rizzo, A., Xibilia, M. G. 2007. *Soft Sensors for Monitoring and Control of Industrial Processes*. Advances in Industrial Control. London: Springer London.

Ma, Y., Liu, S., Xue, G., Gong, D. 2020. Soft Sensor with Deep Learning for Functional Region Detection in Urban Environments. *Sensors*, 20(12): 1–17.

Murugan, C., Natarajan, P. 2019. Estimation of Fungal Biomass Using Multiphase Artificial Neural Network Based Dynamic Soft Sensor. *Journal of Microbiological Methods*, 159: 5–11.

Paulsson, D., Gustavsson, R., Mandenius, C. F. 2014. A Soft Sensor for Bioprocess Control Based on Sequential Filtering of Metabolic Heat Signals. *Sensors*, 14(10): 17864–17882.

Pe, E. A. G., Oliveira, V. A., Cruvinel, P. E. 2018. Soft-Sensor Approach Based on Principal Components Analysis to Improve the Quality of the Application of Pesticides in Agricultural Pest Control. 95–100.

Poerio, D. V., Brown, S. D. 2020. Localized and Adaptive Soft Sensor Based on an Extreme Learning Machine with Automated Self-Correction Strategies. *Journal of Chemometrics*, 34(7): 0–2.

Randek, J., Mandenius, C. F. 2018. On-Line Soft Sensing in Upstream Bioprocessing. *Critical Reviews in Biotechnology*, 38(1): 106–121.

Shi, J., Liu, X.-G. 2006. Product Quality Prediction by a Neural Soft-Sensor Based on MSA and PCA. *International Journal of Automation and Computing*, 3(1): 17–22.

Simutis, R., Galvanauskas, V., Levisauskas, D., Repsyte, J., Vaitkus, V. 2013. Comparative Study of Intelligent Soft-Sensors for Bioprocess State Estimation. *Journal of Life Sciences and Technologies*, 1(3): 163–167.

Wang, B., Wang, Z., Chen, T., Zhao, X. 2020. Development of Novel Bioreactor Control Systems Based on Smart Sensors and Actuators. *Frontiers in Bioengineering and Biotechnology*, 8: 7.

Wang, L., Tao, T., Su, W., Yu, H., Yu, Y., Qin, J. 2017. A Disease Model of Diabetic Nephropathy in a Glomerulus-on-a-Chip Microdevice. *Lab on a Chip*, 17(10): 1749–1760.

Wang, Z. X., He, Q. P., Wang, J. 2015. Comparison of Variable Selection Methods for PLS-Based Soft Sensor Modeling. *Journal of Process Control*, 26: 56–72.

Warth, B., Rajkai, G., Mandenius, C.-F. 2010. Evaluation of Software Sensors for On-line Estimation of Culture Conditions in an Escherichia Coli Cultivation Expressing a Recombinant Protein. *Journal of Biotechnology*, 147(1): 37–45.

Wu, Q., Liu, J., Wang, X., Feng, L., Wu, J., Zhu, X., Wen, W., Gong, X. 2020. Organ-on-a-Chip: Recent Breakthroughs and Future Prospects. *BioMedical Engineering Online*, 19(1): 1–19.

Xie, W., Wang, J. S., Xing, C., Guo, S. S., Guo, M. W., Zhu, L. F. 2020. Adaptive Hybrid Soft-Sensor Model of Grinding Process Based on Regularized Extreme Learning Machine and Least Squares Support Vector Machine Optimized by Golden Sine Harris Hawk Optimization Algorithm. *Complexity*, 2020.

10 Energy Harvesting and Industry 4.0

Rashid Naseer
Department of Mechanical Engineering, College of Electrical and Mechanical Engineering, National University of Sciences and Technology, Islamabad, Pakistan

10.1 INTRODUCTION

In the last few years, Industry 4.0 has gained more attention from researchers around the world as progress in modern technologies is enabling the combination of internet of things (IoT), use of big data, and the concept of cyber-physical systems (CPS) to achieve higher levels of productivity, operational efficiency, and automation. Here IoT means merging the physical world with digital space through a huge quantity of web-connected smart devices and sensors routinely sharing data (Kumar & Yadav, 2022). It is estimated that factories use about 35 percent of the electricity supply worldwide while producing around 20 percent of the global carbon emissions (da Silva et al., 2020). Developments in Energy Harvesting technologies are making it feasible for smart factories to be self-sustainable and renewable in terms of energy which is necessary for achieving long-term environmental goals. Continuous data acquisition is essential for establishing smart factories in the Industry 4.0 era, and wireless sensor networks (WSNs) are important in data acquisition. Reliable power supply to these WSNs is a significant challenge in this digital transformation, whereas energy harvesting has surfaced as a powerful solution (Díez et al., 2018). Hence, the concept of Industry 4.0 has an inbuilt inclination toward integrating energy harvesting technologies.

10.1.1 Energy Harvesting and Its Types

"Energy harvesting" is a broad term encompassing various systems capable of extracting electricity from ambient energy sources like light, heat, sound, radio frequency (RF), wind, work function, or mechanical vibrations. A graphical representation of energy harvesting sources is shown in Figure 10.1. Solar energy devices have been extensively discussed in the literature, and photovoltaic (PV) solar cells are widely used for power generation at domestic and commercial levels. However, their use in powering wireless sensors in an indoor industry environment is restricted due to limitations like low light intensity, the need for the correct exposure angle, and continuous maintenance like cleaning. On the other hand, solar thermal power plants also produce a handsome amount of electric power. Still, the requirement of large areas, light intensity and angles, and environmental concerns for

DOI: 10.1201/9781003327523-12

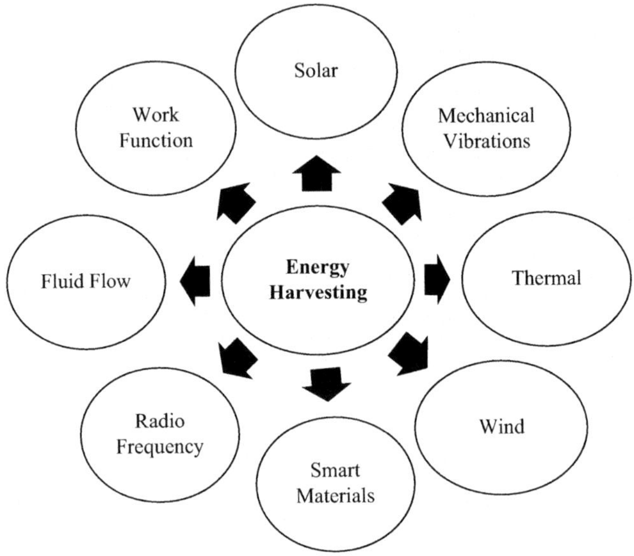

FIGURE 10.1 Types of ambient sources for energy harvesting (adapted from Williams et al., 2021).

birds and insects restrict its widespread use. Another type, the pyro-electric energy harvester, is primarily used in temperature sensors because pyro-electric materials generate an electric potential difference with a temperature change. Similarly, using flow-induced vibrational harvesters, the wind blowing due to environmental events can be used to harvest energy (Soyata et al., 2016). Vibrational energy harvesters produce electric power converting mechanical vibrational energy into electric current. These energy harvesters are more relevant to industrial sensor applications as ambient vibrations are common in industrial environments. This chapter will focus on various kinds of energy harvesters and their applications in industry.

10.1.2 Transduction Mechanisms and Activation/Excitation Phenomena

Extracting energy from ambient sources is a process where energy is converted from one form to the other, generally requiring a transduction mechanism, as shown in Figure 10.2. Furthermore, a summary of the various energy harvesting methods

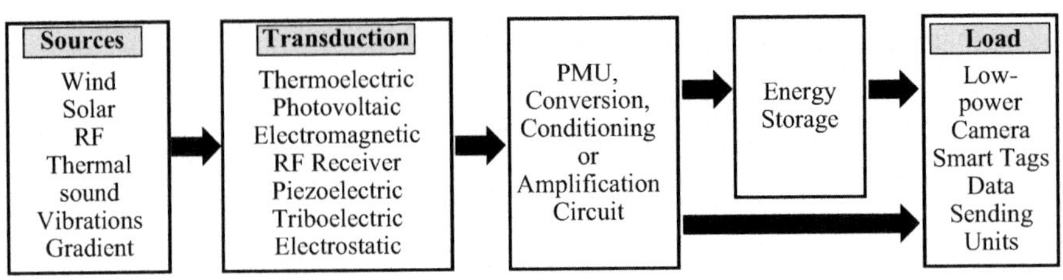

FIGURE 10.2 Energy harvesting process (modified from Divakaran & Krishna, 2019).

along with their applications, conversion factors, and advantages and disadvantages are tabulated in Table 10.1 (Williams et al., 2021).

10.1.2.1 Piezoelectric Transduction

Piezoelectric materials are smart materials used for structural actuation applications because they are simple, lightweight, low-cost, easy-to-install, and easy-to-control materials with high adaptability. Piezoelectric materials can be transformed into shapes, such as patches, thin films, discs, cylinders, and fibers (Aabid et al., 2021). Piezoelectric materials exhibit electric polarization once they undergo mechanical strain and vice versa. Such electric polarizations happen in monocrystalline materials, polycrystalline materials, and ferroelectric ceramics. With the developments in thin-film technologies, more materials and their variants are being discovered which could be used in a piezoelectric energy harvester. Piezoelectric performance, design flexibility, application frequency, available volume, and cost depend on selecting a particular piezoelectric material for a specific application. More than 200 piezoelectric materials are presently being used for energy harvesting applications. Piezoelectric materials can be broadly divided into three categories: ceramics such as lead zirconate, single crystals like lead magnesium niobate-lead titanate (PMN-PT), and semi-crystalline polymers like polyvinylidene difluoride called PVDF (Elvin & Erturk, 2013). Furthermore, Barium titanate ($BaTiO_3$) is another significant inorganic ceramic compound which exhibits ferroelectric, pyroelectric, and piezoelectric effects (Baek et al., 2017). Further refinement of lead-free variants/combinations like $BiFeO_3$ –$BaTiO_3$ (BF-BT) is being explored for better piezoelectric properties (Lee et al., 2022). Some commonly used piezoelectric materials in automotive and aerospace engineering are lead titanate (LT), lead zirconate titanate (PZT), lead magnesium niobate (PMN), leadmetaniobate (LMN), and sodium potassium niobate (SPN). PZT is the most used and studied ferroelectric material, famous for its range of applications like vibration control, repair of cracks, shape control, structural health monitoring, etc. (Aabid et al., 2021). Some materials with a higher piezoelectric response, like relaxor ferroelectric $(Pb, Mg, Nb)O_3$-$PbTiO_3$ (PMN-PT), can function as sensors/actuators and energy harvesters (Chen et al., 2017).

Piezoelectric transduction has emerged as a preferred choice for vibrational energy harvesters due to its large power density, ease of application, flexibility, and ability to work with high-frequency resonators (Zhou et al., 2011). A low-stiffness substrate beam is used to harvest energy from vibrations with low frequency. A typical piezoelectric vibrational energy harvester consists of a cantilever beam with an attached tip mass and a layer of piezoelectric material pasted on one or both sides of the substrate beam near the fixed end to achieve maximum strain. Such a typical bimorph piezoelectric-based energy harvester using flow-induced vibrations (FIV) is presented in Figure 10.3.

Excitation phenomena like base excitation, galloping, vortex-induced vibrations, wake oscillations, or their combinations can be used to vibrate the beam. Once the beam vibrates, compressive and tensile stains are produced on opposite sides of the beam alternately. Hence, by pasting the piezoelectric material on one side (unimorph) or both sides (bimorph), electric energy can be harvested. The strains are more pronounced near the attached end of the beam; hence generally, beams are

TABLE 10.1
Summary of Energy Harvesting Methods

Method	Application Environment	Energy Conversion Factors	Advantages/Disadvantages
Electromagnetic	Industrial machinery, Transportation, Human activity, Roads and infrastructure	Vibration frequency, vibration acceleration	**Advantages:** Operate at low frequencies **Disadvantages:** Requires resonant frequency matching; Moving parts; AC only; Low voltage; Interference with microelectronics
Electrostatic			**Disadvantages:** Requires resonant frequency matching; Moving parts; AC only; Requires input voltage to activate
Piezoelectric			**Advantages:** High voltage output; High power density **Disadvantages:** Requires resonant frequency matching; Moving parts; Delicate materials; AC only; Not receptive at low frequencies
Thermoelectric	Industrial waste heat; Household water; Combustion engines;	Spatial temperature gradient	**Advantages:** Long life due to stationary parts; High reliability **Disadvantages:** Requires constant thermal gradient; Low conversion efficiency; Performs poorly on small gradients
Pyroelectric	Domestic heaters; Body heat; Water vapor	Temporal temperature gradient; Cycle frequency	**Advantages:** Room temperature operation; Wide spectral response; Piezoelectric properties; Flexible and thin form factor **Disadvantages:** Requires cyclical temperature variation; AC only
Photovoltaic	Natural light; Brightly lit indoor spaces	Light intensity; Temperature gradient; Material properties	**Advantages:** Mature technology; Predictable; Scalable; Simple form factor; No moving parts **Disadvantages:** Long periods of natural absence; Natural prediction limited
Aerodynamic	Outdoor environments; Ventilation ducts	Flow speed; Turbine design	**Advantages:** Mature technology **Disadvantages:** Difficult to apply form factor; Hostile application environments
Hydrodynamic	Rivers / oceans; Water pipes		
Magnetic	Power delivery infrastructure	Magnetic field strength	**Advantages:** Predictable energy source **Disadvantages:** Requires resonant frequency matching; Moving parts; Limited range; AC only
Radiofrequency	(semi-)urban environments; Dedicated transmitter setup	Source transmission power; Distance from source; Antenna gain; Antenna design	**Advantages:** Ambient or dedicated techniques; High conversion efficiency; Flexible form factor; No moving parts; AC or DC; **Disadvantages:** Requires tuning to frequency bands; Energy availability limited by safety; Inconsistency if non-dedicated

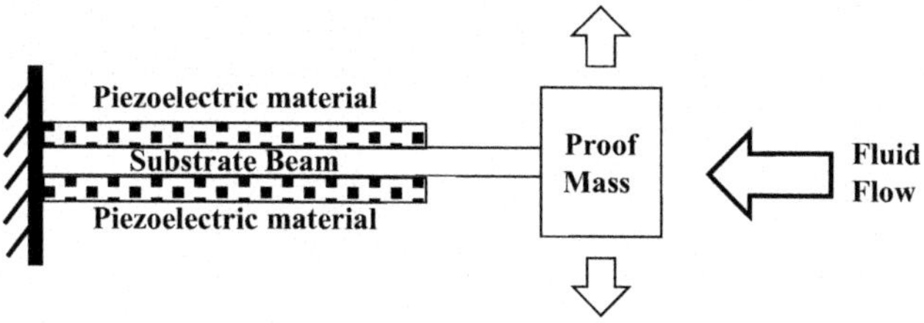

FIGURE 10.3 A typical piezoelectric energy harvester working under flow-induced vibrations (modified from Naseer & Abdelkefi, 2022).

only partially covered with piezoelectric material. The produced current depends on the area of the pasted piezoelectric material and the strain generated, thereby creating higher amplitudes of vibration and causing more electricity. In base excitation, the base where the fixed end of the cantilever beam is attached is forced to vibrate, making the whole beam vibrate. In base excitation, the beam's amplitude is dependent on the amplitude of the vibrating base. If the proof mass is a circular cylinder, vortices are shed after regular intervals at the trailing edge of the cylinder. When the shedding frequency becomes closer to the system's natural frequency, the cylinder starts vibrating due to the resonance making the whole beam vibrate. Such excitation is called vortex-induced vibrations (VIV). Alternately, suppose the cylinder working as a proof mass is a prismatic cylinder. In that case, the cylinder will start vibrating after the onset speed of galloping, and the amplitude of vibrations will be proportional to the fluid speed, a phenomenon called galloping. Moreover, suppose the cantilever beam has a significant width, making it closer to a wing profile or flat surface with a significant torsional mode of vibrations. In that case, the phenomenon will be referred to as flutter. If a combination of the excitation exists in the system, the excitation is termed "hybrid." When base excitation is coupled with any other excitation, quenching can be observed. Suppose vortex-induced vibrations and galloping are considered together in a cylinder. In that case, the phenomenon is termed unsteady galloping with the obvious advantage of harvesting energy over a wide range of fluid velocities. If the beam buckles under the axial compressive force, the central static position becomes unstable, causing the system to vibrate around one or both non-central static potions. Hence, the system is called a bistable system. To harvest energy from mechanical vibrations, various orientations/configurations are possible. Five commonly used configurations are shown in Figure 10.4.

10.1.2.2 Electromagnetic Transduction

The most popular transduction mechanism used to convert mechanical energy to electric energy is based on the principle of electromagnetism. Electromagnetic generators rely on electromagnetic induction from relative motion between the conductor and magnetic flux (Beeby et al., 2006). In 1831, Faraday discovered the principle of electromagnetic induction, which not only generates electricity in a conductor (in the form of a coil) placed within a magnetic field but also generates electricity because of the change in the magnetic flux caused by the relative motion between the coil and the magnets. Examples include generators and alternators. The amount of electric energy

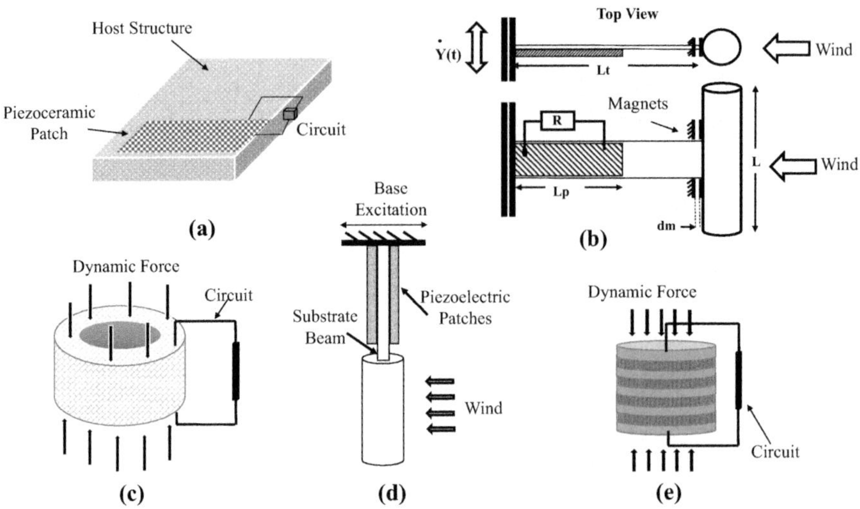

FIGURE 10.4 Five commonly used configurations for harvesting energy from mechanical vibrations (a) a piezoelectric patch attached at that location of the host structure or substrate where maximum strain occurs, (b) a unimorph bistable piezoelectric energy harvester working under vortex-induced vibrations and base excitation along with a nonlinear axial compressive magnetic force, (c) a hollow cylinder connected with an external circuit and compressive force is applied at its end faces, (d) a bimorph cantilever beam working under fluid-induced vibrations and / or base excitation, and (e) a combination of substrate and piezoelectric materials arranged together to enhance output power (modified from Zhang et al., 2017 and Naseer et al., 2019).

generated depends on the turns in a conductor coil, the magnetic field's strength, and the conductor coil's relative velocity with respect to the magnet. The mechanism is generally simple and rugged and does not require any voltage source or use of any smart or expensive materials; however, electromagnetic (EM) energy harvesters are less flexible and lightweight than piezoelectric harvesters. A typical EM energy harvester consists of a coil, pair of permanent magnets, and a mechanical system designed to excite the relative movement of the magnets against the coil aimed to achieve electromagnetic induction. Such an EM generator adapted by Muscat et al., 2022 is shown in Figure 10.5.

The significant advantage of EM energy harvesters over their piezoelectric competitors is their ability to generate high power, which can be replicated economically on a large scale, making them suitable for many electro-mechanical industries. The principle used in the electromagnetic generator can be easily used in reverse, where a shaft is rotated by providing current to the coils. Electric motors and alternators/generators based on electromagnetic induction are widely used in almost every mechanical industry and countless daily appliances. For harvesting energy using electromagnetic transduction, a magnet needs to be carefully moved through a coil, thereby causing a change in magnetic flux, which, in turn, generates a flow of current through the coil. Different configurations can achieve such a

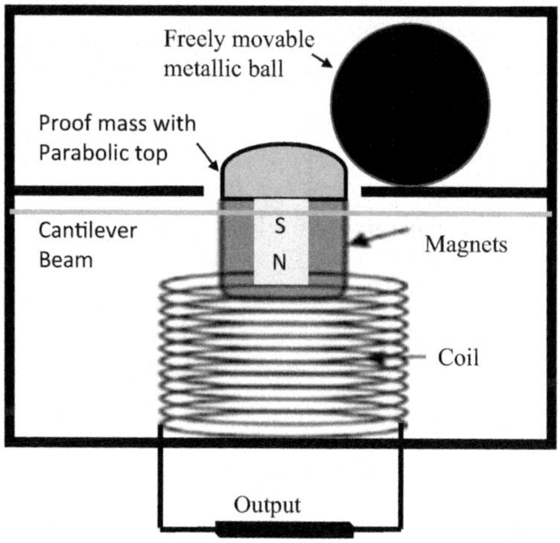

FIGURE 10.5 Schematic of an EM energy harvester (adapted from Muscat et al., 2022).

change of flux through the coil, including a vibrating cantilever beam, clamped beam, rotating shaft, linearly moving resonator, and many more. Liu et al. (2018) used an electromagnetic motor to harvest energy from the walking of humans on a paver. The schematic of a generic electromagnetic energy harvester using base excitation is depicted in Figure 10.6 (Maamer et al., 2019).

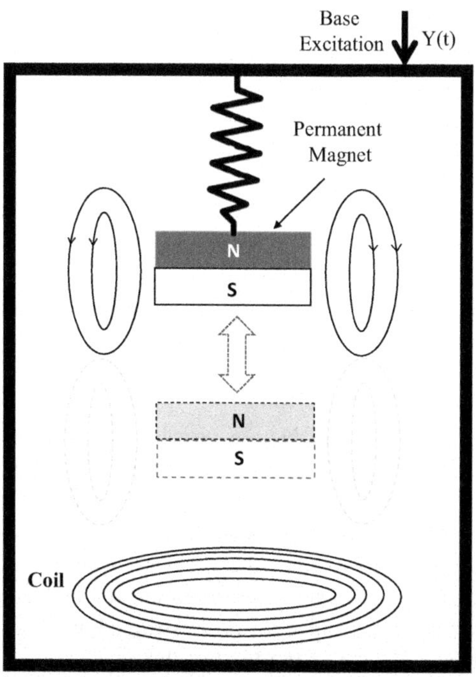

FIGURE 10.6 Electromagnetic energy harvester using base excitation (adapted from Maamer et al., 2019).

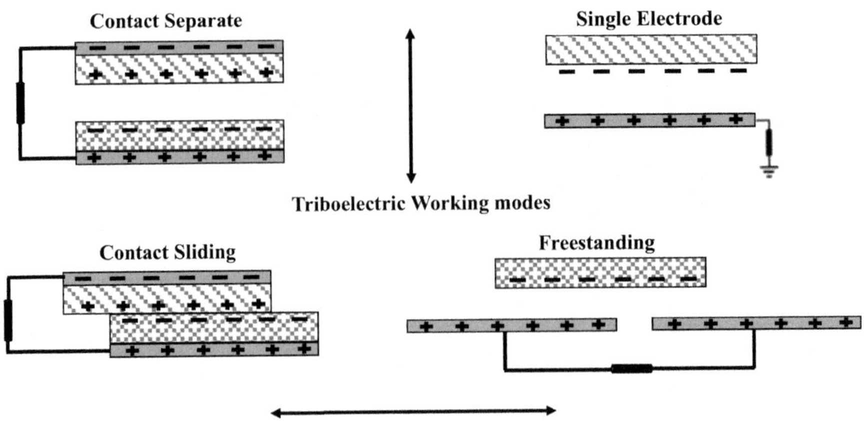

FIGURE 10.7 Main modes of triboelectric nanogenerators (adapted from Ahmad & Khan, 2021).

10.1.2.3 Triboelectric Transduction

In tribology, triboelectric is one of the earliest effects studied. Over 2000 years old worth of history, widespread use of this phenomenon is still a challenge due to the low charge density, complexity of the mathematical model, and lack of reliable theoretical interpretations of the phenomenon (Pan & Zhang, 2019). Despite these issues and difficulties, research articles after 2012 [when the first triboelectric nanogenerator (TENG) was reported] have observed an unprecedented surge in triboelectric-based energy harvesting devices (Zhang & Olin, 2020). This surge can be attributed to significant developments in low-power devices, standalone sensors, wireless smart tags (Pan & Zhang, 2019), and other enabling technologies for the realization of industrial IoT (IIoT)-based smart industries (Aabid et al., 2021). In a pictorial schematic representation, Figure 10.7 shows the main modes of TENGs depicting separate contact mode, contact sliding mode, single electrode mode, and free-standing mode. In triboelectric energy harvesting, electron donors and acceptors play a key role. In this context, Zhang and Olin (2020) selected 100 articles to compile the data to rank the acceptors and donors per fraction percentage of their usage as donors or acceptors category. The collected data is reproduced as Figure 10.8.

10.1.2.4 Electrostatic Transduction

Electrostatic transduction is a transduction process in which mechanical energy is converted into electrical energy. Variable capacitors are the basis of electrostatic induction and electrostatic energy conversion. The structure for variable capacitance is fabricated using micro-electromechanical system (MEMS) technology. Mechanical vibrations are used to make the structure oscillate between the values of a maximum capacitance (C_{max}) and a minimum capacitance (C_{min}).

When the voltage across the capacitor is constrained, the charge will move from the capacitor to a storage device or the load as the capacitance is decreased. When the charge on the capacitor is constrained, the voltage will increase with the

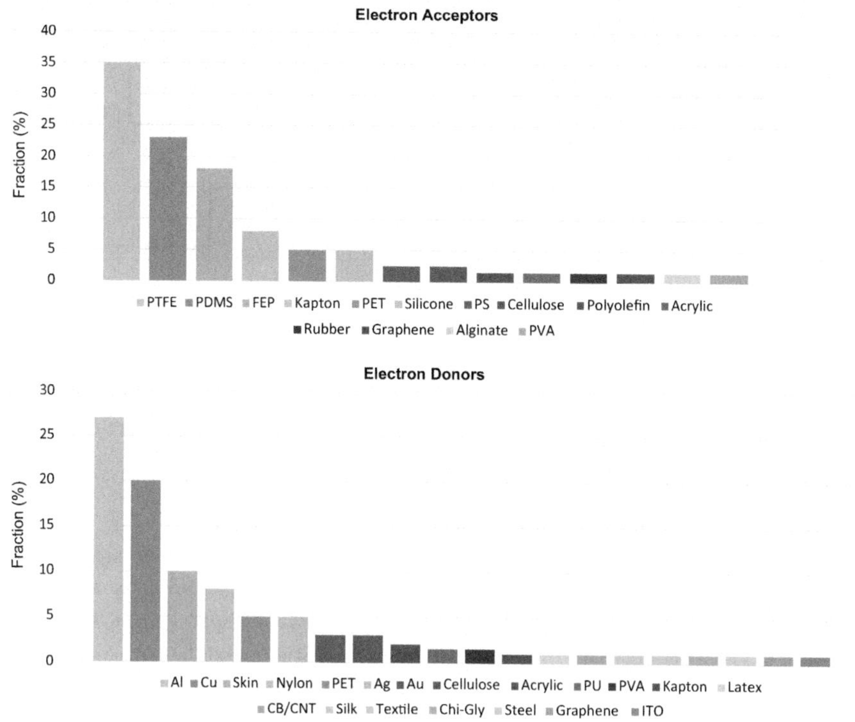

FIGURE 10.8 (a) Electron donors and (b) electron accepters (adapted from Zhang & Olin, 2020).

decrease in the capacitance. The electrostatic transduction process requires an initial charge on the plates, which is a disadvantage from the energy harvesting perspective; however, an electret can be used to supply this initial charge. For this purpose, one or both plates are coated by the electret substance at the time of manufacturing which will provide polarization voltage. Air or a dielectric can be used as the medium between the two plates (Ahmad et al., 2021).

The advantages of electrostatic energy harvesters as compared to vibrating energy harvesting include wide tuning range, low noise, high-quality factor Q, and contained size. However, electrostatic harvesters generate less energy than other kinetic harvesters, and their range of application is restricted due to operational issues (Pozo et al., 2019). The schematic of an electrostatic transduction principle is illustrated in Figure 10.9.

10.1.2.5 Work Function-Based Energy Harvesting

The work function is the minimum thermodynamic work (i.e., energy) required to remove an electron from a solid to a point in the vacuum immediately outside the solid surface. A work function is a property of the material's surface, not the bulk material. It depends on the material crystal face and the presence of contaminants. The charging effect between two bodies having different work functions is well-characterized, showing that when two different materials having different work functions are

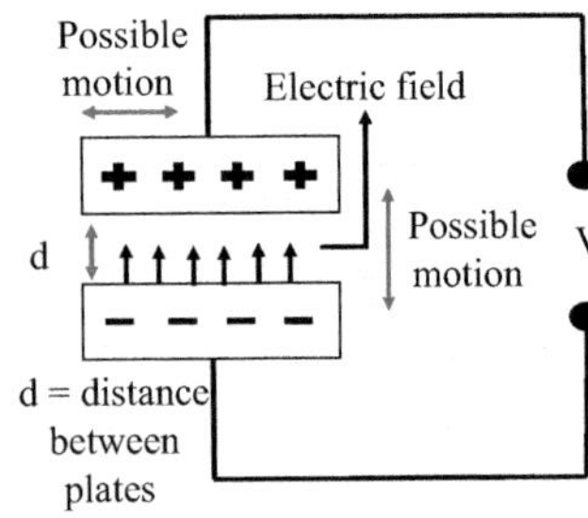

FIGURE 10.9 Electrostatic Transduction Principle Schematic (adapted from Ahmed & Khan, 2021).

brought close to each other and connected galvanically, the charging occurs naturally without the requirement of any external source. Those energy harvesters incorporating material work functions are called work function energy harvesters (WFEH). The main difference between a WFEH and an electrostatic energy harvester is that the WFEH does not require electrets or external power sources.

10.1.2.6 RF Energy Transfer and Generation

In RF energy harvesting (RFEH), electrical energy is harvested from RF signals by a dedicated, ambient, or unknown RF source. Such energy harvesting is a suitable replacement for batteries to avoid limitations like limited lifespan, physical size, or difficulties in replacing batteries. Electromagnetic waves abundantly available in urban environments radiate energy in different frequency bands and can be intercepted at physically inaccessible places making it a unique energy source with diverse applications. Such energy harvesting can be coupled with data transfer in wireless communication setups like wireless power communication (WPC), simultaneous wireless information and power transfer (SWIPT), and their variants (Bi et al., 2015). Radio waves available in the environment can be used as an alternative source of energy for powering small devices requiring low power like sensors, smart tags, WSNs, and other IoT devices (Visser & Vullers, 2013), small devices used for monitoring the environment in smart cities (Liu et al., 2017), and smart tags and sensors required in manufacturing industries (Stoopman et al., 2017). The power density for RF energy varies from 0.2 nW/cm^2 to 1 mW/cm^2 (Yildiz, 2009).

Several approaches can be used to convert the RF signals into DC power, including the dedicated transfer used for high power values or harvesting the ambient energy with low power values (Marian et al., 2012). The power level produced by a circuit that harvests RF energy from a dedicated RF source at a short range, like an RFID chip, can be around 50 nW/cm^2. Contrarily, a system that is used to harvest energy from any ambient RF source achieves power levels of only around 2 nW/cm^2. In either case, the harvested energy can be used to recharge the batteries of an embedded device or a power storage device or directly used to power the circuit or a chip for sensing and data transfer. Some typical ambient sources are bluetooth, wi-fi, global system for mobile communication (GSM)/cellular, frequency modulation (FM)/television (TV)/digital television (DTV), or radio waves. Such a source can be used in applications where the replacement of batteries is cumbersome, expensive, complicated, or impossible (Soyota et al., 2016). A schematic of the ambient RF energy harvesting system is shown in Figure 10.10.

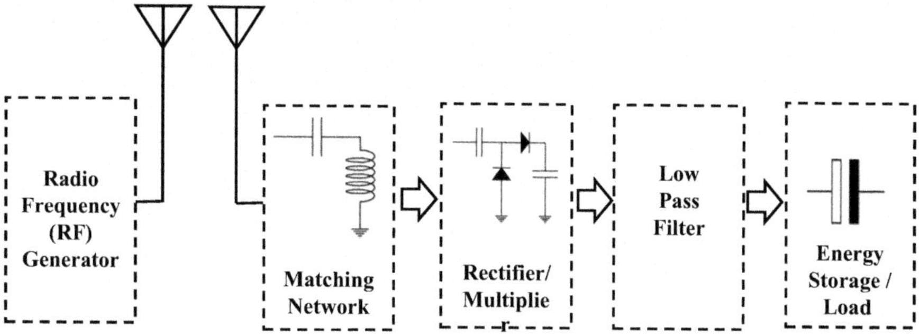

FIGURE 10.10 Ambient RF energy harvesting (modified from Divakaran & Krishna, 2019).

10.1.2.7 Micro-Electro-Mechanical Transducers

Advancements in low-powered electronics and very large-scale integration circuitry have significantly reduced the power consumption of many electronic components. Resultantly, chips consume minimal power creating the opportunity to utilize small-scale energy harvesters as a viable power source. Microelectromechanical systems (MEMS) scale piezoelectric energy harvesters are becoming feasible in this context (Safaei et al., 2019). Wang and Song (2006) proposed a nanoscale energy harvester using zinc oxide nanowire arrays. Micro-scale energy harvesters are useful for self-powered standalone sensors, smart tags, and wireless sensing nodes, which are considered a pivoting component of future industrial setups based on Industry 4.0. Jeon et al. (2005) proposed a MEMS energy harvester based on a cantilever beam having dimensions $100 \times 60 \times 0.48\ \mu m^3$ while producing a power of 1.01 μW at 2.4 V; however, the frequency was very high (13.9 kHz), which is not practical for ambient sources. To reduce the frequency and bring it within the range of frequencies in ambient sources, researchers explored various design modifications, including the use of a high aspect ratio with a tip mass (Fang et al., 2006), zigzag structure design (Karami & Inman, 2011), a fixed-fixed design (Berdy et al., 2012), etc.

10.1.2.8 Hybrid Energy Harvesters

Hybrid energy harvesters scavenge renewable energy from one or more ambient sources using more than one transduction mechanism such as piezoelectric, electromagnetic, pyroelectric, triboelectric, thermoelectric, PV, RF, or work function-based energy harvesting (Liu et al., 2021). An example of such a hybrid piezoelectric, electromagnetic energy harvester is shown in Figure 10.11. A combination of more than one excitation phenomenon can also be used with a single transduction mechanism, such as piezoelectric energy harvesting systems using a combination of base excitation and vortex-induced vibrations (Naseer & Abdelkefi, 2022).

10.1.3 Potential Sources of Energy Harvesting in Modern Industry

Ambient energy sources can be categorized in many ways, including direct sources of energy like electricity, batteries, and heat sources or indirect sources like mechanical

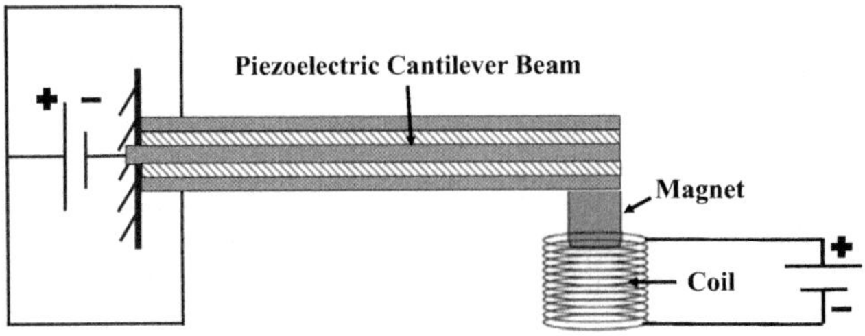

FIGURE 10.11 Hybrid energy harvester (adapted from Ahmad & Khan, 2021).

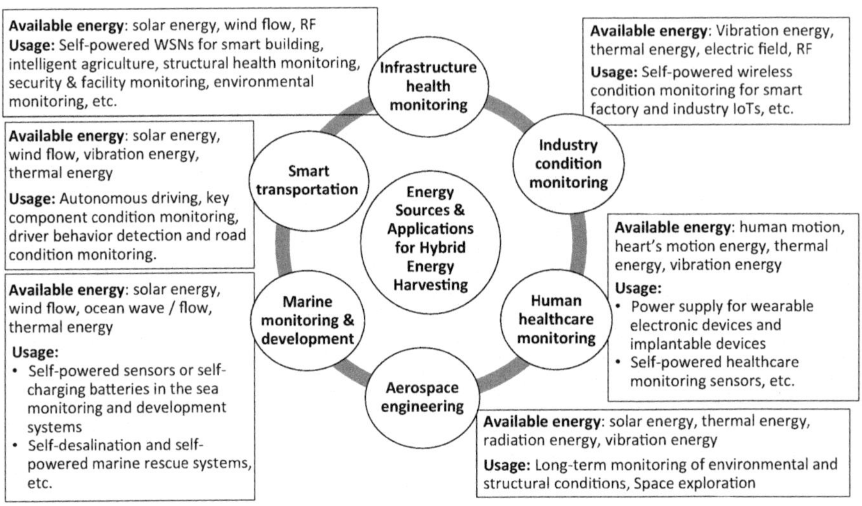

FIGURE 10.12 Energy sources and application prospect for hybrid energy harvesting systems (adapted from Liu et al., 2021).

vibrations, light, heat, sound, pneumatic or hydraulic pressure, wind, radio frequency, or difference in work potential. Similarly, other ways to categorize can be continuous, period, semi-periodic, or occasional sources, etc. Many researchers categorize the ambient sources as per the environment or desired application. A graphical depiction of energy sources and their applications for hybrid energy harvesting is shown in Figure 10.12.

10.2 VIBRATIONAL CONTROL AND ENERGY HARVESTING

Vibration-based systems are either designed to dampen the existing vibrations (called vibrational control) or to harvest energy. Interestingly, both applications have opposite objectives; hence the techniques also vary significantly. In vibrational control, active or passive dampers are used to control the vibrations. Examples of such applications include suspension of automotive to dampen the road bumps for a smooth feel of the ride, wing designs to avoid galloping or flutter-based vibrations, dampers to reduce the transfer of vibrations from machines to base platforms for a better factory environment, and use of shims to control vibrations in the nuts/bolts, etc. On the contrary, in the case of energy harvesting, the objective is to convert the ambient vibrations into a useful form of energy like electricity. Logically, higher amplitudes are desirable in the case of vibrational energy harvesting for higher output power; however, factors like space, size, material specifications, or impact on the functionality of the system itself or the environment limit the extent forcing researchers to find alternate ways for enhancing harvested power. Many researchers investigated the use of nonlinearities like nonlinear spring force, magnetic force, or axial compressive force in the design of vibrational energy harvesting systems. In such a study, Naseer et al. (2019) concluded that nonlinearities might cause hardening or softening behaviors with prominent hysteresis regions, causing an increase in the output levels and broadening of the synchronization region, which is desirable for optimization of the harvested power under given ambient conditions.

10.3 APPLIED PERSPECTIVE – ENERGY HARVESTING IN INDUSTRY 4.0

Despite the well-established energy harvesting potential from mechanical vibrations, the practical application of such energy harvesters is complicated because of the characteristics of each method with specific limitations like source intensity, output instability, dependence on ambient conditions or load, and issues related to finite lifespan considering the mechanical, wear, deformation, and abrasion (Williams et al., 2021) issues. Although initially, some researchers (Cassidy et al., 2011) discussed the methodologies for the use of mechanical energy for large-scale energy production of up to 100 kW (Zuo & Tang, 2013), the majority of the research remained focused on low-power devices like static or motion sensors, wearable devices, smart tags, and cameras. In subsequent paragraphs, it is discussed separately why energy harvesters are more relevant to any future industry and how smart tags and sensors are reshaping the future industrial workspaces, ensuring a safe, self-accounting, secure, and efficient IIoT-based industrial environment.

10.3.1 Why Energy Harvesters Are More Relevant in Future Industry?

Latest trends in IoT, digitalization, big data, and 5G communication are pointing toward exponential growth in the use of WSNs designed to deliver measured data directly to a cloud or server with reliability, authenticity, and consistency for further processing by a decision support system. Such decision support systems can be further augmented by artificial intelligence (AI), making the future smarter and capable of handling flexible production, changing demands, and real-time updating on manufacturing situations (Kanoun et al., 2021). For this purpose, factories must collect and process data consistently from diverse sources, including physical, operational, digital, and human assets. The use of wireless sensing networks has a pivotal role in this regard. Powering such a network of sensors, nodes, and tags through wireless means is still a challenge. In this perspective, batteries not only have a limited lifespan, but there is a hidden cost of battery waste which is being paid in terms of irreparable loss to the environment. In the context of environmental concerns, batteries are made up of some of the most toxic materials, including lead, potassium hydroxide, mercury, sulfuric acid, manganese, steel, carbon graphite, cadmium, zinc, lithium, cobalt, nickel, ammonium chloride, and some rare earth minerals (Sovacool, 2022). Batteries' waste brings these toxic materials back to earth, most likely in hazardous form, polluting the soil and under surface water reservoirs. If batteries are used to power sensing nodes, wireless tags, and other IIoT devices, the use of batteries can increase manifold, causing further damage to already depleting clean natural resources. On the other hand, energy harvesting from ambient sources like light, sound, heat, or mechanical vibrations is emerging as a potential solution; hence, energy harvesting is becoming more and more relevant for any future industrial setup.

10.3.2 Smart Manufacturing, IoT, and Energy Harvesters

In 2005, the first report on the IoT was published by International Telecommunication Union (ITU). Since then, researchers have been exploring new ways and means to connect anything to anyone, anywhere and anytime. Advances in communication speed, modes of communication, and cloud computing platforms, along with compatible data analytic tools, are providing the enabling environment for the realization of the concept of IoT with diverse applications in many industries, including autonomous transportation, smart buildings, efficient energy grid, environmental monitoring, biomedical and healthcare, defence industries, and smart manufacturing (Sanislav et al., 2021). It is predicted that more than 50 billion devices will be connected to the internet in the next couple of years, with around 60% related to IoT, which shows that with this speed, the realization of the once-dreamed IoT is not that far where our physical world becomes a web of interconnected and mutually communicating data nodes that are collecting, processing and sharing useful data with each other in real-time. Such a massive increase in IoT devices will likely impact every aspect of our daily lives. Still, its implications for the manufacturing industry are manifold due to the increased desire to achieve Industry 4.0 goals.

10.3.3 Uses of Energy Harvester-Based Smart Devices in the Industry

Continuous surveillance, AI-based condition monitoring, and smart predictive maintenance are key factors in smart industries. These require continuous and reliable tracking of materials, manpower, and other production resources. Such heavy monitoring of production resources and the environment is only possible through well-deployed WSNs. Here it is important to note that there is no single type of self-powered wireless device, which can fit all monitoring needs because the availability of ambient energy sources varies significantly within the same factory premises. Hence, great deliberation is required before selecting a particular wireless sensor for given applications under certain ambient conditions. Some examples of industrial applications and suitable energy converters are tabulated in Table 10.2 (Kanoun et al., 2021).

TABLE 10.2
Applications of Energy Harvesting Converters in WSNs for Smart Factories

Serial	Converters	Example for Applications
1.	Vibration converters	• Monitoring of conveyor systems • Monitoring of transport systems • Monitoring and retrofitting of machines and tools • Predictive maintenance
2.	Thermoelectric converters	• Thermoelectric conversion from solar energy • Monitoring on pipes and ducts in dark places • Heat dissipating machines and heat pipes • Industrial waste heat recovery • Wearable devices in human-machine interaction
3.	Solar cells	• Warehouse management and localization • Environments with few contaminations • Stationary nodes • Logistics
4.	Hybrid converters	• To support weak sources • Systems with increased power demand • Four systems needing a fallback option (e.g., security)
5.	Wireless power transfer	• Covered/sealed applications • Additional supply for interrupted sources • Rotating and moving parts (e.g., conveyor systems) • Systems with high power demand • Localization of tools and goods

10.3.4 The Safety, Security, Monitoring, and Fail-Safe Handling

Industrial environments are generally vulnerable to extra heat, noise, or mechanical vibrations. In any future IIoT-enabled industry, one can expect sensors and data transfer nodes on various items exposed to potentially dangerous conditions. Powering such sensing devices with traditional batteries poses serious hazards of explosion, necessitating the need for battery-less and maintenance-free IIoT devices. Hence, the fusion of energy harvesting with IoT in an industrial environment is considered the best enabling technology for a sustainable and eco-friendly Industry 4.0 environment (Alegret et al., 2019). From this perspective, some industries, like the petroleum and petrochemical industries, face challenges in implementing the Industry 4.0 concept.

10.3.5 Smart Tags and Standalone Sensors

Although the use of WSNs in the industrial environment has been discussed earlier, smart tags and standalone sensors have applications in diverse fields beyond the industrial environment, like the defense industry, surveillance, wildlife tracking, transportation, supply chain, services sector, and manufacturing industries. Hence, this section is focused exclusively on smart tags and standalone sensors. In general, smart tags are used to collect special information or the location of the tagged items. At the same time, sensors are designed to collect data from an environment, such as temperature, sound, pressure, mechanical vibrations, etc. However, the data is transmitted to a central unit for processing in both cases. Hence, any WSN used in Industry 4.0 environment will essentially comprise thousands of static standalone sensing nodes with limited data processing and transmission capabilities over a short range with limited power requirements (Williams et al., 2021). To achieve effective IIoT, WSNs must be autonomous, perpetually operable, and reliable with little or no maintenance requirements, necessitating powering these sensors from ambient sources using carefully selected energy harvesting techniques. The developments in low-power and lightweight sensors, modern software optimization methods, lightweight communication protocols, low-power radio transceivers, and more reliable energy harvesting techniques, are complementing each other to the possibility of realization of once dreamed self-powered, maintenance-free, perpetually working, and reliable WSN. The taxonomy for industrial energy-aware wireless sensing networks is graphically represented in Figure 10.13.

10.3.6 Use of Energy Harvesters in the Industry – Case Studies

10.3.6.1 Case Study 1: Kinetic EM Energy Harvester for Railway Applications

The latest case study by Hadas et al. (2022) about using energy harvesters in railway applications is included here. Moving trains generate high vibrations in the train structure and the train track. Modern train network requires many sensing nodes integrated to form a reliable railway management and control system. In this study, a kinetic energy-based electromagnetic energy harvester is designed, manufactured, and tested for railway applications. Measured values of regional and express trains are used to test the energy harvester with an integrated resistive load and wireless transmission mechanism. A schematic of the energy harvester is depicted in Figure 10.14 where Z depicts the vertical amplitude of the sleepers. A mechanical resonator is utilized to convert kinetic energy into free vibrations of a mass.

The results showed that peak values of 300 mW and 600 mW were achieved for regional and express trains, respectively. However, mean values were much lower in each case. The authors suggested custom designing the energy harvester for each application due to the diversity in power requirements and inconsistencies in available mechanical vibrations.

10.3.6.2 Case Study 2: Circular Cylinder-Based Wind Energy Harvester with Different Rod-Shaped Attachments

In piezoelectric-based energy harvesters, circular cylinders are typically used to generate vortex-induced vibrations. However, VIV-based energy harvesters have a major problem: they generate useful energy only if the vortex shedding

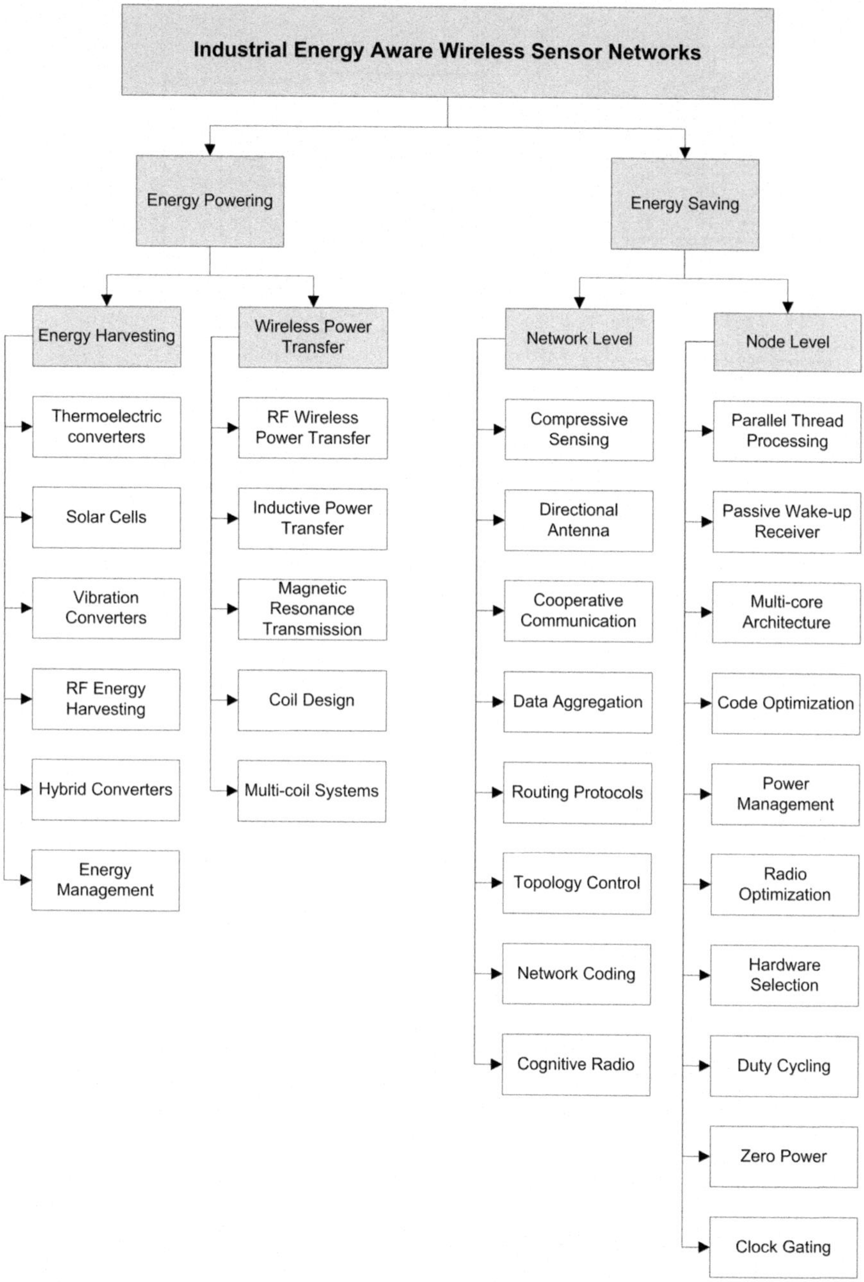

FIGURE 10.13 Taxonomy for industrial energy aware wireless sensing networks (adapted from Kanoun et al., 2021).

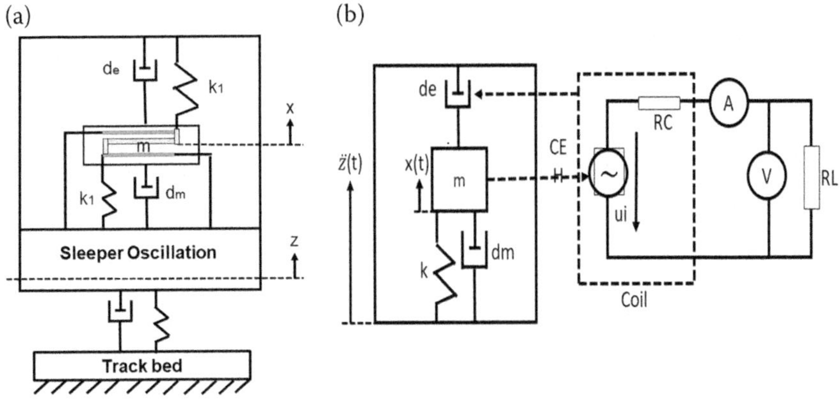

FIGURE 10.14 (a) Schematic of kinetic energy harvesting using train vibrations, and (b) coupling of mechanical and EM models (adapted from Hadas et al., 2022).

frequency is near the system's natural frequency, which is only true within a narrow range of incoming fluid flow. In this experimental case study, Hu et al. (2018) utilized three different rod-shape attachments on the circular cylinder to take advantage of the integrated effects of VIV and galloping excitations. It was found that using three rod-type attachments at the circumference of a circular cylinder at 60 degrees, as shown in Figure 10.15, provided higher power than the

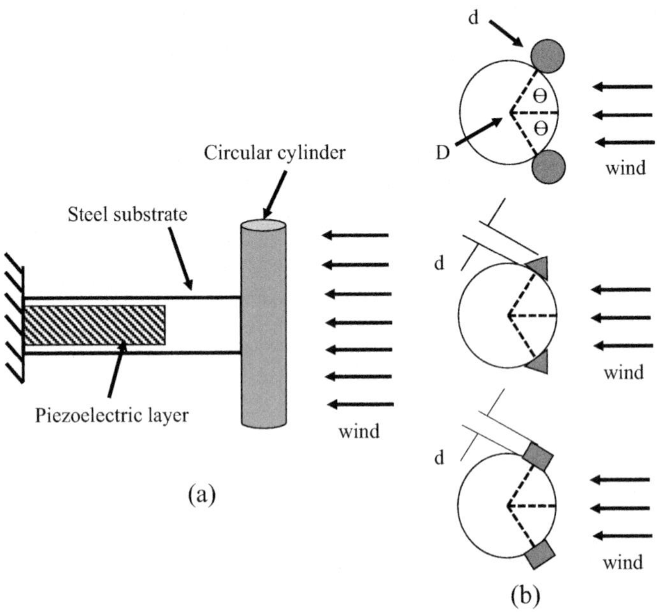

FIGURE 10.15 (a) Schematic of typical VIV-based energy harvester, and (b) top view of using different rod-shape attachments (adapted from Hu et al., 2018).

system with any other arrangement of attachments or without any attachment. The study has provided a sound base for the utilization of the interacting behavior of vortex-induced vibrations and galloping for optimum power output.

10.4 FUTURE TRENDS AND POSSIBILITIES

Industrial environments need high levels of reliability and consistency in power and data transmissions, and that desired level still needs to be achieved in most of the available energy harvesting systems. However, as energy harvesting from ambient sources has become the research hotspot and with the increasing volume of research being published on this subject, it is hoped that overcoming such challenges in the practical implementation of energy harvesting systems in real life, in general, and Industry 4.0, in particular, is not that far. The combination of ambient sources, transduction mechanisms, and excitation phenomena duly supported by nonlinear mathematical tools are likely to provide the required optimization for producing long-life, self-powered, and reliable sensing nodes fulfilling the needs of any future industry. In this regard, advancements in high-speed connectivity with low power requirements are crucial for an early breakthrough. The successful realization of the Industry 4.0 concept is deeply dependent on the successful implementation of energy-harvesting technologies for powering wireless sensing nodes. In fact, applications of energy harvesters and wireless sensing networks are far-reaching, affecting diverse fields like biomedical, wildlife protection, early warning, and environmental safety. Hence, any breakthrough in the development of self-powered, long life, and reliable sensing network is likely to transform human civilization in multiple ways bringing newer possibilities, opportunities, and hazards.

REFERENCES

Aabid, A., Raheman, M. A., Ibrahim, Y. E., Anjum, A., Hrairi, M., Parveez, B., Mohammed Zayan, J. 2021. A systematic review of piezoelectric materials and energy harvesters for industrial applications. *Sensors*, 21(12): 4145.

Ahmad, M. M., Khan, F. U. 2021. Review of vibration-based electromagnetic–piezoelectric hybrid energy harvesters. *International Journal of Energy Research*, 45(4): 5058–5097.

Alegret, R. N., Aragones, R., Oliver, J., Ferrer, C. 2019. Exploring IIoT and energy harvesting boundaries. In *IECON 2019-45th Annual Conference of the IEEE Industrial Electronics Society*, 1: 6732–6736.

Baek, C., Park, H., Yun, J. H., Kim, D. K., Park, K. I. 2017. Lead-free BaTiO3 nanowire arrays-based piezoelectric energy harvester. *MRS Advances*, 2(56): 3415–3420.

Beeby, S. P., Tudor, M. J., White, N. M. 2006. Energy harvesting vibration sources for microsystems applications. *Measurement Science and Technology*, 17(12): R175.

Berdy, D. F., Srisungsitthisunti, P., Jung, B., Xu, X., Rhoads, J. F., Peroulis, D. 2012. Low-frequency meandering piezoelectric vibration energy harvester. *IEEE Transactions on Ultrasonics, Ferroelectrics, and Frequency Control*, 59(5): 846–858.

Bi, S., Ho, C. K., Zhang, R. 2015. Wireless powered communication: Opportunities and challenges. *IEEE Communications Magazine*, 53(4): 117–125.

Cassidy, I. L., Scruggs, J. T., Behrens, S., Gavin, H. P. 2011. Design and experimental characterization of an electromagnetic transducer for large-scale vibratory energy harvesting applications. *Journal of Intelligent Material Systems and Structures*, 22(17): 2009–2024.

Chen, Y., Zhang, Y., Zhang, L., Ding, F., Schmidt, O. G. 2017. Scalable single crystalline PMN-PT nanobelts sculpted from bulk for energy harvesting. *Nano Energy*, 31: 239–246.

da Silva, F. S. T., da Costa, C. A., Crovato, C. D. P., da Rosa Righi, R. 2020. Looking at energy through the lens of Industry 4.0: A systematic literature review of concerns and challenges. *Computers & Industrial Engineering*, 143: 106426.

Díez, P. L., Gabilondo, I., Alarcón, E., Moll, F. 2018. A comprehensive method to taxonomize mechanical energy harvesting technologies. In *2018 IEEE International Symposium on Circuits and Systems (ISCAS)*, 1–5.

Divakaran, S. K., Krishna, D. D. 2019. RF energy harvesting systems: An overview and design issues. *International Journal of RF and Microwave Computer-Aided Engineering*, 29(1): e21633.

Elvin, N., Erturk, A. 2013. Introduction and methods of mechanical energy harvesting. In *Advances in Energy Harvesting Methods*, 3–14, New York, NY: Springer.

Fang, H. B., Liu, J. Q., Xu, Z. Y., Dong, L., Wang, L., Chen, D., Liu, Y. 2006. Fabrication and performance of MEMS-based piezoelectric power generator for vibration energy harvesting. *Microelectronics Journal*, 37(11): 1280–1284.

Hadas, Z., Rubes, O., Ksica, F., Chalupa, J. 2022. Kinetic Electromagnetic Energy Harvester for Railway Applications—Development and Test with Wireless Sensor. *Sensors*, 22(3): 905.

Hu, G., Tse, K. T., Wei, M., Naseer, R., Abdelkefi, A., Kwok, K. C. 2018. Experimental investigation on the efficiency of circular cylinder-based wind energy harvester with different rod-shaped attachments. *Applied Energy*, 226: 682–689.

Jeon, Y. B., Sood, R., Jeong, J. H., Kim, S. G. 2005. MEMS power generator with transverse mode thin film PZT. *Sensors and Actuators A: Physical*, 122(1): 16–22.

Kanoun, O., Khriji, S., Naifar, S., Bradai, S., Bouattour, G., Bouhamed, A., Viehweger, C. 2021. Prospects of Wireless Energy-Aware Sensors for Smart Factories in the Industry 4.0 Era. *Electronics*, 10(23): 2929.

Karami, M. A., Bilgen, O., Inman, D. J., Friswell, M. I. 2011. Experimental and analytical parametric study of single-crystal unimorph beams for vibration energy harvesting. *IEEE Transactions on Ultrasonics, Ferroelectrics, and Frequency Control*, 58(7): 1508–1520.

Kumar, S., Yadav, P. 2022. Energy Harvesting–Based Architecture in IoT: Basics of Energy Harvesting, Key Technology for Enhancing the Life of IoT Devices, Challenges of IoT in Terms of Energy and Power Consumption. In *Energy Harvesting*, 1–18, Chapman and Hall/CRC.

Lee, M. H., Choi, H. I., Jung, Y. G., Lee, S., Song, T. K. 2022. Bipolar cycling effects in BiFeO3–BaTiO3 piezoelectric ceramics. *Current Applied Physics*, 44: 6–11.

Liu, H., Fu, H., Sun, L., Lee, C., Yeatman, E. M. 2021. Hybrid energy harvesting technology: From materials, structural design, system integration to applications. *Renewable and Sustainable Energy Reviews*, 137: 110473.

Liu, M., Lin, R., Zhou, S., Yu, Y., Ishida, A., McGrath, M., Zuo, L. 2018. Design, simulation and experiment of a novel high-efficiency energy harvesting paver. *Applied Energy*, 212: 966–975.

Liu, J., Xiong, K., Fan, P., Zhong, Z. 2017. RF energy harvesting wireless powered sensor networks for smart cities. *IEEE Access*, 5: 9348–9358.

Maamer, B., Boughamoura, A., El-Bab, A. M. F., Francis, L. A., Tounsi, F. 2019. A review on design improvements and techniques for mechanical energy harvesting using piezoelectric and electromagnetic schemes. *Energy Conversion and Management*, 199: 111973.

Marian, V., Allard, B., Vollaire, C., Verdier, J. 2012. Strategy for microwave energy harvesting from ambient field or a feeding source. *IEEE Transactions on Power Electronics*, 27(11): 4481–4491.

Muscat, A., Bhattacharya, S., Zhu, Y. 2022. Electromagnetic vibrational energy harvesters: a review. *Sensors*, 22(15): 5555.

Naseer, R., Dai, H., Abdelkefi, A., Wang, L. 2019. Comparative study of piezoelectric vortex-induced vibration-based energy harvesters with multi-stability characteristics. *Energies*, 13(1): 71.

Naseer, R., Abdelkefi, A. 2022. Nonlinear modeling and efficacy of VIV-based energy harvesters: Monostable and bistable designs. *Mechanical Systems and Signal Processing*, 169: 108775.

Pan, S., Zhang, Z. 2019. Fundamental theories and basic principles of triboelectric effect: A review. *Friction*, 7(1): 2–17.

Pozo, B., Garate, J. I., Araujo, J. Á., Ferreiro, S. 2019. Energy harvesting technologies and equivalent electronic structural models. *Electronics*, 8(5): 486.

Safaei, M., Sodano, H. A., Anton, S. R. 2019. A review of energy harvesting using piezoelectric materials: state-of-the-art a decade later (2008–2018). *Smart Materials and Structures*, 28(11): 113001.

Sanislav, T., Mois, G. D., Zeadally, S., Folea, S. C. 2021. Energy harvesting techniques for internet of things (IoT). *IEEE Access*, 9: 39530–39549.

Sovacool, B. K. 2022. The hidden costs of batteries. *Science*, 377(6605): 478–478.

Soyata, T., Copeland, L., Heinzelman, W. 2016. RF energy harvesting for embedded systems: A survey of tradeoffs and methodology. *IEEE Circuits and Systems Magazine*, 16(1): 22–57.

Stoopman, M., Philips, K., Serdijn, W. A. 2017. An RF-powered DLL-based 2.4-GHz transmitter for autonomous wireless sensor nodes. *IEEE Transactions on Microwave Theory and Techniques*, 65(7): 2399–2408.

Visser, H. J., Vullers, R. J. 2013. RF energy harvesting and transport for wireless sensor network applications: Principles and requirements. *Proceedings of the IEEE*, 101(6): 1410–1423.

Wang, Z. L., Song, J. 2006. Piezoelectric nanogenerators based on zinc oxide nanowire arrays. *Science*, 312(5771): 242–246.

Williams, A. J., Torquato, M. F., Cameron, I. M., Fahmy, A. A., Sienz, J. 2021. Survey of energy harvesting technologies for wireless sensor networks. *IEEE Access*, 9: 77493–77510.

Yildiz, F. 2009. Potential Ambient Energy-Harvesting Sources and Techniques. *Journal of technology Studies*, 35(1): 40–48.

Zhang, R., Olin, H. 2020. Material choices for triboelectric nanogenerators: a critical review. *EcoMat*, 2(4): e12062.

Zhang, L. B., Abdelkefi, A., Dai, H. L., Naseer, R., Wang, L. 2017. Design and experimental analysis of broadband energy harvesting from vortex-induced vibrations. *Journal of Sound and Vibration*, 408: 210–219.

Zhou, Q., Lau, S., Wu, D., Shung, K. K. 2011. Piezoelectric films for high frequency ultrasonic transducers in biomedical applications. *Progress in Materials Science*, 56(2): 139–174.

Zuo, L., Tang, X. 2013. Large-scale vibration energy harvesting. *Journal of Intelligent Material Systems and Structures*, 24(11): 1405–1430.

11 Internet of Things for Manufacturing Industry

Atal Anil Kumar
Department of Engineering, Faculty of Science, Technology and Medicine, University of Luxembourg, Luxembourg

Usman Qamar
Department of Computer and Software Engineering, College of Electrical and Mechanical Engineering, National University of Sciences and Technology, Islamabad, Pakistan

Kanwal Naveed and Uzair Khaleeq uz Zaman
Department of Mechatronics Engineering, College of Electrical and Mechanical Engineering, National University of Sciences and Technology, Islamabad, Pakistan

11.1 WORLD IS CONNECTED

Internet of Things (IoT) is an emerging area. IoT applications are already being leveraged in diverse domains, such as medical services, smart retail, customer service, smart homes, environmental monitoring, and industrial internet. Consequently, there will be a significant increase in spending on the design and development of IoT applications and analytics. Spending in this market is expected to increase substantially. The technology reached 100 billion dollars in market revenue for the first time in 2017, and forecasts suggest that this figure will grow to around 1.6 trillion by 2025 (Santhosh et al., 2020). This increasing market also reflects many jobs currently available in IoT. IoT technologies are the most disruptive technologies of this century. Countries around the globe are naturally strategizing to take the maximum benefit from IoT.

IoT is a new factor of production and has the potential to introduce new sources of growth, changing how work is done and reinforcing the role of people to drive growth in business. The impact of IoT technologies on business is projected to increase labor productivity by up to 40% and enable people to make more efficient use of their time (Kalsoom et al., 2021).

With the rapid advancement in sensing, computing, and embedded technologies, several industries are looking to adopt low-power interconnected devices for monitoring and improving their production line in the context of Industry 4.0 (Galati & Bigliardi, 2019; Frank, et al., 2019). IoT is broadly being used for three main domains of manufacturing: (a) operations which include monitoring,

DOI: 10.1201/9781003327523-13

optimization of performance, human-machine interaction, smart manufacturing, and so on (Lu, 2017; Morimoto, 2013; Xu et al., 2016); (b) production asset management and maintenance which includes quality, efficiency, resource tracking and monitoring (Lin et al., 2018; Boyes et al., 2018; Birkel & Hartmann, 2019); and finally, (c) field service, which includes installation, repair, and maintenance of the industrial equipment (Boyes et al., 2018; Büyüközkan & Göçer, 2018).

This chapter will briefly present the importance of IoT in the manufacturing sector and the benefits, challenges, and opportunities it provides in making the company Industry 4.0 compatible.

11.2 ENABLERS OF IOT

The development of IoT technologies at a rapid pace plays an essential role in the broad adoption of IoT in manufacturing. In general, some core technologies help implement IoT-driven solutions, as shown in Figure 11.1. This section provides a brief description of these technologies.

11.2.1 Radio-Frequency Identification

An RFID system consists of RFID tags and readers which use electromagnetic fields to transfer data, as shown in Figure 11.2. The RFID tags contain information about the objects on which they are fixed, while the RFID readers can read this information without requiring a line of sight and transfer them for tracking the movement of these tags in real-time (Juels, 2006). Such systems are well suited for various manufacturing applications such as supply chain management (Sarac et al., 2010), production scheduling (Huang et al., 2008), tracking of parts, etc.

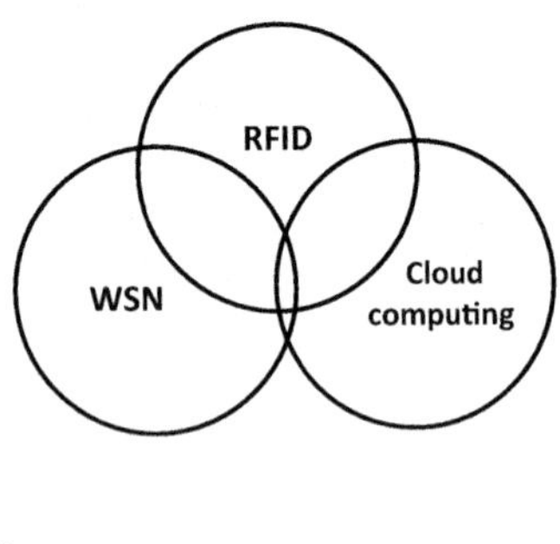

FIGURE 11.1 The key enablers for implementation of IoT in manufacturing.

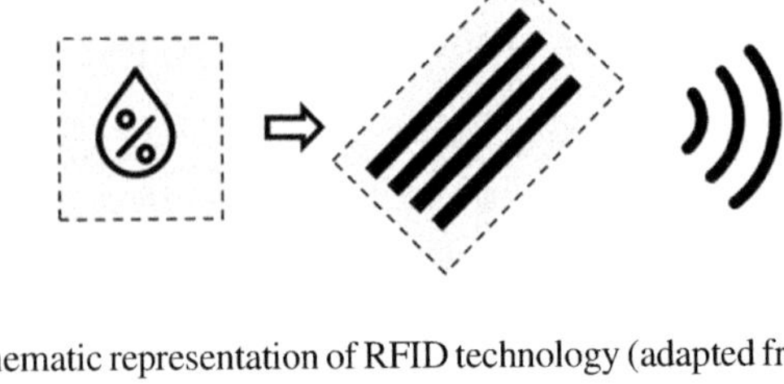

FIGURE 11.2 Schematic representation of RFID technology (adapted from Costa et al., 2021).

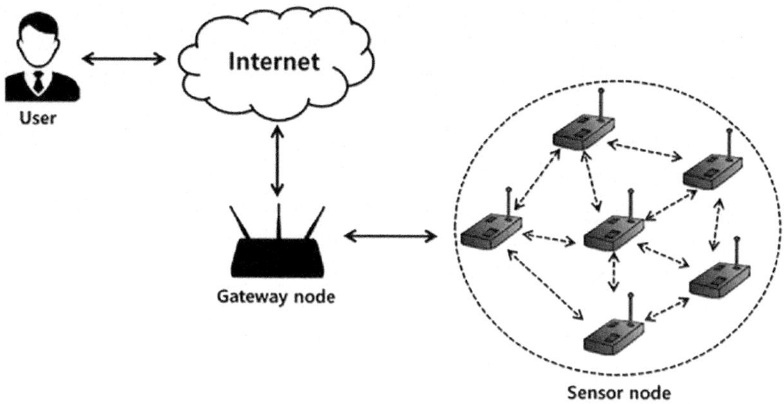

FIGURE 11.3 Schematic representation of WSN with its node (adapted from Yu & Park, 2020).

11.2.2 Wireless Sensor Networks

Wireless sensor networks (WSNs) consist of sensor nodes that are spatially distributed for sensing the environment, conducting computations, and communicating with other nodes (Yick et al., 2008). They operate to ensure the best connectivity by sending their data via multi-hop spreading to the base station. They are primarily deployed in a collective pattern as individual nodes are tiny, energy-constrained devices with limited memory, as shown in Figure 11.3. WSNs are suitable for manufacturing applications requiring flexible configuration and convenient wireless integration (Yang et al., 2018). WSNs and RFID are used in synergy to obtain efficient outcomes (Wang et al., 2013).

11.2.3 Cloud Computing and Big Data

Cloud computing facilitates efficient management of a vast, shared pool of configurable computing resources and can be released with minimal management effort. Cloud computing provides the necessary platform for conveniently handling an important resource generated due to implementing the plethora of sensors in modern manufacturing lines, big data (Mell & Grance, 2009). Big data are generally different from conventional data sets in terms of volume, variety, and velocity (Mayer-Schönberger & Cukier, 2013). The combination of cloud computing and big data is being applied in the entire lifecycle of products, with significant impacts being seen in design innovation, manufacturing intelligence, cost reduction, quality improvement, and customer satisfaction (Li et al., 2015).

11.3 BENEFITS OF IOT

Various benefits are obtained due to the implementation of IoT technologies. IoT sensors will help the company gather all sorts of data from virtually any physical equipment in the production line. Such data can be stored and available for analysis in real-time. Useful information can be extracted from such data using artificial intelligence (AI), machine learning (ML) or deep learning algorithms. This information can help the management improve operational efficiency by making meaningful decisions.

It is also reported that the digitalization of factories significantly decreases the number of product errors and defects. IoT can reduce the rate of accidents caused by human workers. Moreover, IoT systems help make industrial robots more autonomous, interconnected, and cooperative. The robots will be able to achieve a higher degree of interaction with human workers safely.

IoT technologies help the manufacturing sector to make their production line driven by real-time demand with increased product customization. They also allow a higher level of agility and adjustability, shifting the manufacturing industry's focus from product-oriented to customer-oriented.

11.4 APPLICATIONS OF IOT IN MANUFACTURING

IoT is being used in different manufacturing applications. The purpose of this section is to enlist some of them and provide insights about how they can be used further for other applications.

11.4.1 Proactive Maintenance

Proactive maintenance focuses on the continuous monitoring of the machines and equipment. This approach mainly focuses on identifying the root cause of a failure or machine breakdown and proactively scheduling maintenance interventions to correct the reason contributing to the failure. IoT technologies can help collect more data about machines and equipment during operation. This data can then be used to model the machine's behavior for detecting failure. Such implementation can prepare the production process for Industry 4.0 and enable a better engagement of the maintenance department.

11.4.2 Reshoring

IoT helps decentralize factory automation by shifting manufacturing from locations with labor shortages to higher demand and innovation locations.

11.4.3 Safe Human Workplaces

IoT helps maintain the profiles and skill sets of the workers, which can aid the company in distributing tasks considering the capabilities and interests of the workers. This will always ensure that the workers can work on tasks that are interesting for them.

11.4.4 Retrofitting of Legacy Machines for Smart Manufacturing

For several reasons, many existing industries still rely on their old (legacy) machines for producing their products. Most of the legacy machines lack real-time and in-process sensing and control systems. This is foreseen as one of the reasons why small manufacturers cannot compete with other digitally advanced companies. Using IoT technologies to retrofit such legacy machines can potentially make such small companies digitally competent.

11.5 IOT MANUFACTURING POLICIES AND STRATEGIES

The practical implementation of IoT technologies requires policies and strategies formulated by government agencies and industrial organizations. Several steps have been taken worldwide, a few of which are reported here (Yang et al., 2019).

11.5.1 The USA

Advanced manufacturing has been identified as one of the important topics to revitalize US leadership in manufacturing, creating high-quality jobs, and ensuring national security [Advisors on Science and President's Council, and Technology (US), 2011]. The focus of next-generation manufacturing is the effective use and coordination of automation, sensing, networking, data, information, and computation.

Manufacturing USA has established several networked manufacturing innovation institutes such as Advanced Functional Fabrics of America, American Institute for Manufacturing Integrated Photonics (AIM Photonics), America Makes, Advanced Robotics Manufacturing (ARM), Advanced Regenerative Manufacturing Institute (ARMI), and so on.

11.5.2 China

Persistent challenges faced by the manufacturing industry in China resulted in the formulation of the "Made in China 2025" strategy in 2015 to provide a ten-year action plan to transform the manufacturing sector. Ten important areas have been targeted for boosting economic growth: information technology, aviation, railway equipment, power grid, new materials, machinery, robotics, maritime equipment, energy-saving vehicles, and medical devices. To compete globally, manufacturers are being asked to improve their smartness levels by developing unmanned manufacturing systems, implementing IoT in their manufacturing networks, and using industrial cloud platforms and big data analytical tools.

11.5.3 The European Union

The manufacturing sector contributes to 80% of all European Union (EU) exports, and the European Commission continuously strengthens its competitiveness by implementing innovative and disruptive technologies in its manufacturing plants. IoT, big data, AI, additive manufacturing (AM), collaborative robotics, and blockchain technologies have been identified as opportunities for making industries smarter. This has led to the establishment of several European digital innovation hubs to help small, medium, and large companies. These policies are complemented by national initiatives from different European countries, e.g., Germany: Industry 4.0, Smart service world, high-tech Strategy 2020; Netherlands: Smart industry; France: Alliance industry of the future, Industrie du future, Nouvelle France Industrielle, and so on. Additionally, the United Kingdom has laid out its strategic plans for 2013–2050 to contribute toward the rapid global manufacturing revolution.

Apart from these policies, manufacturing giants such as General Electric, Cisco, Intel, AT&T, and IBM founded the "Industrial Internet Consortium (IIC)" in 2014 to help industries adopt IoT systems for their applications to increase their outputs.

11.6 CASE STUDIES

11.6.1 Case Study – Real-Time Visibility for a Firearm and Accessory Manufacturing Company

The results of this case study are from the company SilencerCo, a manufacturer of firearm suppressors. They implemented a real-time visibility system to integrate information from their machine tools to improve decision-making and set benchmarks for future improvement, as shown in Figure 11.4. This system was provided to them by MachineMetrics, an industrial IoT analytics platform, using which the operators

FIGURE 11.4 Implementation of a real-time visibility system by MachineMetrics (MachineMetrics, n.d.).

could understand the machines' functioning elaborately. Following the implementation of its IoT system, the company demonstrated remarkable improvements in both the efficiency and profitability of its production line.

For the year 2019 alone, the company was able to achieve a 5% increase in their overall equipment effectiveness (OEE), was able to eliminate 11.5K hours of unplanned downtime, able to reach a 200% improvement in good part production, achieve an 8% increase in machine utilization, and had 100% increase in the data visibility.

11.6.2 Case Study – Implementation of IoT technology in Volvo Production Plant, Ghent

The automotive company Volvo uses RFID in one of their largest car production factories at Ghent to refill parts on the line. Due to the increasing diversification in customer requirements, several models are assembled in the same production line, which makes the assembly process complex. RFID technology helped the plant track its production assets during the entire production and assembly phase. It also helped them collect data from the shop floor and warehouse in real-time, which they combined with their decision-making models. At Ghent, they also have implemented a large fleet of automated guided vehicles (AGVs) equipped with wireless trackers and sensors.

The decision-makers at the company reported that equipping conveyors, automatic strapping machines, and warehouse vehicles with intelligent sensors helped boost their production. The company adopted a proactive approach to maintenance and collected performance data in real-time to prevent problems before breakdown. The real-time information enabled them to minimize material delivery delays and helped the management make better decisions to improve the responsiveness to the ever-changing market (Bazaru, 2018).

11.7 CHALLENGES AND OPPORTUNITIES IN MANUFACTURING

Implementation of IoT in manufacturing faces several barriers and challenges. Some of the challenges are highlighted in this section to guide the reader understanding them beforehand (Seedao et al., 2021).

11.7.1 High Investment

Implementation of IoT requires a considerable number of sensing, actuating, and data-collecting devices, including cameras, RFID tags, robots, and computers. Several existing infrastructures need to be more flexible, which would require them to be redesigned. In addition to the devices mentioned above, the industry should also need to invest in networking infrastructure with efficient energy management. All these require a very high investment which is only sometimes possible for small-scale industries. On the other hand, bigger companies need to be more explicit about the return on investment (ROI) and are hesitant to take such risks.

11.7.2 Big Data with a Long Analytic Process

IoT technologies need a huge amount of data due to the high number of sensors and interconnected devices (Singh & Bhanot, 2020). Humans cannot analyze such data and require high computational power devices. The analysis of these data can be time-consuming and complex.

11.7.3 Effectiveness of Data

Utilization of big data is also another critical challenge faced by companies. The vast volume of data needs to be processed to identify the essential parameters required for the company. In many cases, companies need to be clearer about how to use the enormous data collected to use it for their benefit effectively.

11.7.4 Cyber Security

The need to ensure data security and prevent cyber-attacks are also considered potential barriers to adopting IoT for manufacturing. Since IoT technologies connect physical equipment with digital operations, a potential security breach can result in excessive damage to both digital data and the production line.

11.7.5 Lack of Skilled Labor

Increasing the level of digital innovation can lead to many manufacturing jobs being obsolete. Implementing innovative IoT technologies requires adequate training to be imparted to the workers, with which improvements can be made. Since the manufacturing industry still relies heavily on labor input, there is a need to train the workers to adapt to work in the presence of such technologies.

11.8 CONCLUSION

As an emerging technology, IoT is widely accepted in the manufacturing industry. However, there are certain barriers (both technical and non-technical) that must be overcome to adopt it entirely by different companies. The extent of its implementation will differ from one sector to another as various industries have different priorities. Despite these challenges, IoT will undoubtedly provide many benefits when combined with other Industry 4.0 technologies elaborated in this book.

REFERENCES

Bazaru, S.K.G. 2018. Implementation of IoT Technologies in Manufacturing Process. Italy.

Birkel, H.S., Hartmann, E. 2019. Impact of IoT Challenges and Risks for SCM. *Supply Chain Management: An International Journal*, 24(1): 39–61.

Boyes, H., Hallaq, B, Cunningham, J., Watson, T. 2018. The Industrial Internet of Things (IIoT): An Analysis Framework. *Computers in Industry*, 101: 1–12.

Büyüközkan, G., Göçer, F. 2018. Digital Supply Chain: Literature Review and a Proposed Framework for Future Research. *Computers in Industry*, 97: 157–177.

Costa, F., Genovesi, S., Borgese, M., Michel, M., Dicandia, F.A., Manara, G. 2021. A Review of RFID Sensors, the New Frontier of Internet of Things. *Sensors*, 21(9): 3138.

Frank, A.G., Dalenogare, L.S., Ayala, N.F. 2019. Industry 4.0 Technologies: Implementation Patterns in Manufacturing Companies. *International Journal of Production Economics*, 210: 15–26.

Galati, F., Bigliardi, B. 2019. Industry 4.0: Emerging Themes and Future Research Avenues Using a Text Mining Approach. *Computers in Industry*, 109: 100–113.

Huang, G.Q., Zhang, Y.F., Jiang, P.Y. 2008. RFID-Based Wireless Manufacturing for Real-Time Management of Job Shop WIP Inventories. *The International Journal of Advanced Manufacturing Technology*, 36(7): 752–764.

Juels, A. 2006. RFID Security and Privacy: A Research Survey. *IEEE Journal on Selected Areas in Communications*, 24(2): 381–394.

Kalsoom, T., Ahmed, S., Rafi-ul-Shan, P.M., Azmat, M., Akhtar, P., Pervez, Z., Imran, M.A., Rehman, M. 2021. Impact of IoT on Manufacturing Industry 4.0: A New Triangular Systematic Review. *Sustainability*, 13(22): 12506.

Li, J., Tao, F., Cheng, Y., Zhao, L. 2015. Big Data in Product Lifecycle Management. *The International Journal of Advanced Manufacturing Technology*, 81(1): 667–684.

Lin, D., Lee, C.K.M., Lau, H., Yang, Y. 2018. Strategic Response to Industry 4.0: An Empirical Investigation on the Chinese Automotive Industry. *Industrial Management & Data Systems*. Emerald Publishing Limited.

Lu, Y. 2017. Industry 4.0: A Survey on Technologies, Applications and Open Research Issues. *Journal of Industrial Information Integration*, 6: 1–10.

MachineMetrics. n.d. "Real-Time Visibility - A Competitive Edge for SilencerCo." https://www.Machinemetrics.Com/Downloads#casestudies.

Mayer-Schönberger, V., Cukier, K. 2013. Big Data: A Revolution That Will Transform How We Live, Work, and Think. *Houghton Mifflin Harcourt.*

Mell, P., Grance, T. 2009. Perspectives on Cloud Computing and Standards. *USA, Nist.*

Morimoto, R. 2013. A Socio-Economic Analysis of Smart Infrastructure Sensor Technology. *Transportation Research Part C: Emerging Technologies*, 31: 18–29.

Of Advisors on Science, President's Council, and Technology (US). 2011. *Report to the President on Ensuring American Leadership in Advanced Manufacturing*. Executive Office of the President, President's Council of Advisors on Science and Technology.

Santhosh, N., Srinivsan, M., Ragupathy, K. 2020. Internet of Things (IoT) in Smart Manufacturing. In *IOP Conference Series: Materials Science and Engineering*, 764: 12025.

Sarac, A., Absi, N., Dauzère-Pérès, S. 2010. A Literature Review on the Impact of RFID Technologies on Supply Chain Management. *International Journal of Production Economics*, 128(1): 77–95.

Seedao, M.R.S., Yenradee, P., Raweewan, M. et al. 2021. Step-by-Step Lean IoT Implementation for SMEs: A Case Study of Full Automation Production Line. *Thammasat University*.

Singh, R., Bhanot, N. 2020. An Integrated DEMATEL-MMDE-ISM Based Approach for Analysing the Barriers of IoT Implementation in the Manufacturing Industry. *International Journal of Production Research*, 58(8): 2454–2476.

Wang, L., Xu, L., Bi, Z., Xu, Y. 2013. Data Cleaning for RFID and WSN Integration. *IEEE Transactions on Industrial Informatics*, 10 (1): 408–418.

Xu, M., Song, C., Ji, Y., Shih, M.-W., Lu, K., Zheng, C., Duan, R., et al. 2016. Toward Engineering a Secure Android Ecosystem: A Survey of Existing Techniques. *ACM Computing Surveys (CSUR)*, 49(2): 1–47.

Yang, C., Shen, W., Wang, X. 2018. The Internet of Things in Manufacturing: Key Issues and Potential Applications. *IEEE Systems, Man, and Cybernetics Magazine*, 4 (1): 6–15.

Yang, H., Kumara, S., Bukkapatnam, S.T.S., Tsung, F. 2019. The Internet of Things for Smart Manufacturing: A Review. *IISE Transactions*, 51(11): 1190–1216.

Yick, J., Mukherjee, B., Ghosal, D. 2008. Wireless Sensor Network Survey. *Computer Networks*, 52(12): 2292–2330.

Yu, S., Park, Y. 2020. SLUA-WSN: Secure and Lightweight Three-Factor-Based User Authentication Protocol for Wireless Sensor Networks. *Sensors*, 20(15): 4143.

12 Product Life Cycle Management and Cloud Manufacturing

Afshan Naseem and Yasir Ahmad
Department of Engineering Management, College of Electrical and Mechanical Engineering, National University of Sciences and Technology, Islamabad, Pakistan

Uzair Khaleeq uz Zaman
Department of Mechatronics Engineering, College of Electrical and Mechanical Engineering, National University of Sciences and Technology, Islamabad, Pakistan

12.1 PRODUCT LIFE CYCLE MANAGEMENT AND CLOUD MANUFACTURING

New manufacturing dynamics and increasing competition compel businesses to expand the information systems, processes, and decision-making techniques to manage production. The essential aspects of winning over the competition in the manufacturing business include applying the concepts of product life cycle management (PLM) and cloud manufacturing (CM) (Santos et al., 2018; Fisher et al., 2018). PLM is described by the National Institute of Standards and Technology (NIST) as "a vision or a business strategy for creating, sharing, and managing information about a product, process, people, and services within and across the extended and networked enterprise covering the entire life cycle spectrum of the product" (Rachuri et al., 2006). PLM not only assimilates data, people, processes, and business systems but also acts as the product information backbone for businesses and extended enterprises (Saaksvuori & Immonen, 2004). The users and system developers should comprehensively grasp PLM's concept, components, functionalities, scope, and relative positioning within the businesses to execute PLM systems effectively (Ameri & Dutta, 2005).

Moreover, CM has generated considerable research interest, and currently, several definitions exist. CM system is developed for many business situations via different cloud forms such as private, community, and public clouds (Tao et al., 2011). CM also presents numerous ways to avoid the obstacles that hinder sustainable manufacturing. With the help of CM, data may be remotely processed without extra corporate expertise, which is especially advantageous for small and

DOI: 10.1201/9781003327523-14

medium enterprises (SME) since they might not have the capabilities for data analytics inside the company (Fisher et al., 2018). To provide customized products, CM can facilitate and help manufacturers manage diverse resources and capabilities (Yang et al., 2017). Furthermore, to meet client needs, manufacturing processes must become agile and adaptable to enable effective customization promptly. A possible interface for clients to upload their customized demands in the production process is provided by CM, which encourages improved communication between consumers and firms (Cerdas, 2017). The subsequent paragraphs highlight the importance and use of PLM and CM in the modern manufacturing industry.

12.1.1 Sustainable Production Using PLM

The term 'life cycle' commonly specifies the complete set of phases that could be considered independent stages, followed by a product, from the cradle to the grave (Stark, 2011). United Nations Environmental Program (UNEP) defines life cycle management (LCM) as "the application of life cycle thinking to modern business practice, to manage the total life cycle of an organization's products and services toward more sustainable consumption and production" (Jensen and Remmen, 2006) and is defined from diverse perspectives in the literature (Westkämper et al., 2000). Product life cycle is defined mainly by three phases: (1) beginning of life (BOL), (2) middle of life (MOL), and (3) end of life (EOL) (Kiritsis et al., 2003). BOL encompasses product designing and production; MOL encompasses good consumption, service, and maintenance; and EOL encompasses when items are dismantled, recycled, remanufactured, reused, or disposed of. Traditionally, LCM has been viewed as a component of PLM, a holistic business approach emphasizing software solutions and a commercial viewpoint. However, PLM is viewed as a collection of tools and methods that transformed from a group of engineering-focused tools into all-encompassing solutions. (Corallo et al., 2013). Figure 12.1 shows the various product life cycle stages. PLM may handle each of these stages concurrently.

The sustainability paradigm is implemented and interpreted differently in many industries. Sustainable manufacturing can be defined as the ability to utilize natural resources for manufacturing in an intelligent manner by developing products and solutions that can meet economic, environmental, and social goals while improving

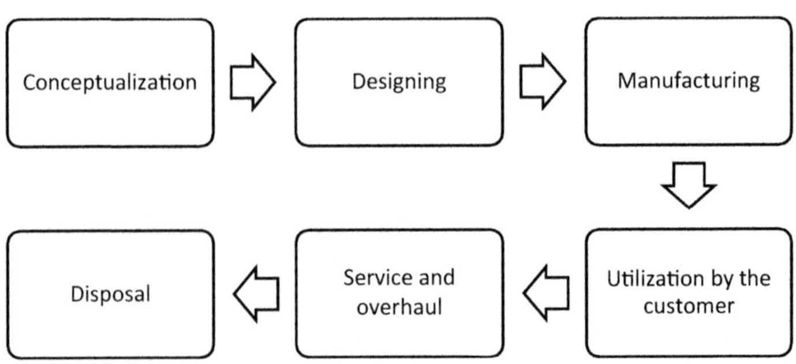

FIGURE 12.1 Generic product life cycle (adapted from Singh et al., 2019).

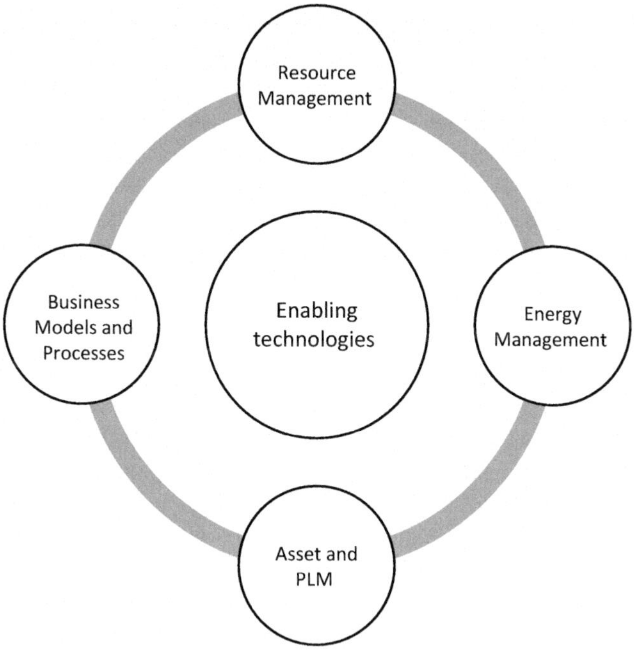

FIGURE 12.2 Research clusters in sustainable manufacturing (adapted from Garetti & Taisch, 2012).

the standard of living for the people. (Garetti & Taisch, 2012). The research clusters in sustainable manufacturing are depicted in Figure 12.2. The critical research areas in sustainable manufacturing may be clustered to build a holistic picture. Figure 12.2 presents five predominant research clusters considered essential to understanding the overall spectrum of research. Enabling technologies cluster includes issues related to new production processes, advanced manufacturing technology, and information and communication technologies (ICTs). The resource and energy management clusters encompass the issues related to the scarcity of resources and energy-efficient manufacturing. Asset and PLM cluster is concerned with sustainable life cycle management. Also, business models and processes cluster cover new ways to organize sustainable businesses.

LCM focuses on the methodical integration of sustainable product development into business strategies, planning, product development, buying decisions, and communication initiatives. LCM is an adaptive integrated management framework of concepts, strategies, and procedures that incorporates environmental, social, and economic elements of goods, processes, and organizations instead of being a single instrument or methodology. Various frameworks, such as the total LCM framework, are based on viable system modeling operating at strategic and operational layers (Herrmann et al., 2007). As a result of the requirement for the strategic shift from the present manufacturing paradigm to manufacturing sustainability, enterprises are currently trying to accomplish the year 2030 sustainability targets (Jamwal et al., 2021; Machado et al., 2020; Malek & Desai, 2020; Sharma et al., 2020). There has been an effort to integrate sustainability knowledge with PLM techniques to assist designers in considering all three sustainability aspects (Hauschild et al., 2020).

Sustainability is defined as "meeting the needs of the present without compromising the ability of future generations to meet their own needs" (Brundtland, 1987). It is considered a difficult issue at present in manufacturing. It will be critical for future generations and industries as natural resources are not immeasurable and cannot fulfill future generations' demands. Sustainable manufacturing refers to creating manufactured products that use processes that minimize the negative environmental impacts, use natural resources effectively and conserve energy, are safe for communities, employees, and consumers, and are economically sound (Hauschild et al., 2020). Some studies have reported that modern production methods using sustainable technologies minimize labor inputs, lower overall energy usage, and improve condition-based maintenance forecasts (Wu et al., 2017; Zhang et al., 2017). Moreover, given the difficulties in implementing green manufacturing, it is essential to distinguish between the terms 'green' and 'sustainable' (Byggeth & Hochschorner, 2006). Both terminologies, in many situations, still need to be explained, and the picture becomes more complicated when there are numerous kinds of product returns and recovery.

Sustainability design goals can mitigate the complications through a systematic problem description. Decision support mechanisms include using cases of target breakdowns in the subsystems, systematic reduction of solution space, and assistance in design activities to achieve sustainability design targets. Engineering techniques can be implemented to streamline the process of information retrieval and decision support, like product data management (PDM)/PLM or computer-aided design/engineering (CAD/CAE) (Buchert et al., 2017). Regarding the quantity of energy consumed globally and CO_2 emissions, the building and automotive domains are considered the most important (Främling et al., 2013). Moreover, since the advent of the circular economy movement, final landfill disposal has become a last resort for most products (Kloepffer, 2008).

A significant issue for sustainable PLM is to measure the impact of manufacturing processes on the environment precisely. Life cycle assessment or life cycle analysis (LCA) is considered a methodological framework to estimate and assess the environmental effects of the life cycle of a product (Rebitzer et al., 2004). Closed loop life cycle management (CL2M) is an approach that focuses on tracking and managing system information. It offers chances to lessen life cycle operations' inefficiencies and increase competitiveness in the Internet of Things (IoT) era (Kiritsis, 2011). The overarching goal of CL2M is to be capable of continuously enhancing product design, production, use, and disposal handling, resulting in greater quality, fewer breakdowns, less need for repair, and continuous maintenance of an operation at the highest degree of energy and resource efficiency. Due to evolving technology, CL2M enables product information collection and multi-organizational usage.

Intelligent products have a managerial consequence in which it is challenging to either expand or modify information systems modeled on demand and supply, including enterprise resource planning (ERP) systems, to meet the demands of sustainable PLM. The interoperability among devices and other information systems can only be achieved if it is technically simple and financially viable enough to meet the requirements of sustainable PLM with multiple entities involved in the manufacturing processes. The adoption and evolution of sustainable PLM may take longer if processes' loosely coupled integration is not considered.

12.1.2 Synergetic Production Through CM

CM is one of the promising manufacturing paradigms that evolved from advanced manufacturing concepts, including agile manufacturing, application service provider (ASP) and manufacturing grid (MGrid). The term first emerged in 2010, and till now, there are numerous definitions of CM. However, none of them are currently standardized (Fisher, et al. 2018; Liu et al., 2018; Tao et al., 2011). Today, the importance of CM in the manufacturing world has been progressively realized by practitioners, and both academic research and industrial implementation of CM, are witnessing swift developments. CM is an intelligent manufacturing paradigm with eight prominent and representative technologies, including Cyber-Physical systems (CPS), the IoT, cloud computing, sensors, big data, additive manufacturing (AM), energy saving, and holograms.

The following four core CM operations necessitate knowledge assistance (Tao et al., 2011):

- Manufacturing resource and capability assessment, connection, and virtualization
- Cloud services, match and search, accumulation, and structure
- Optimal distribution and organizing
- Enterprise system management.

Lack of expert knowledge in an organization and availability of access to quantitative environmental effect data of manufacturing processes are the two obstacles cited by businesses (Bey et al., 2013). Moreover, traditional automation consists of sophisticated process control based on data gathered from an individual process plant's pressure, temperature, flow rate, and level sensor measurements, for instance. CM helps make more knowledgeable decisions about remote processing of plant operations by collecting and analyzing data from various sources of plants. The fact that CM is a centralized data source for multiple sites makes the system more accessible (Guo 2016; Helo et al., 2019).

The CM platform, which supports many tenants, can pair up various cloud users. Local manufacturers should connect and improve communication to discover current ineffective waste treatment techniques and local, sustainable alternatives. For example, a non-renewable source may be replaced by an alternate feed source for a procedure that CM finds in the waste of another nearby company. Due to its position at the top of the European Union (EU) waste hierarchy, this would be the preferable method of treatment (Fisher, et al. 2018; Tao, et al. 2018). A typical CM architecture is shown in Figure 12.3. The bottom section encapsulated the services to be managed through a centralized platform. The middle section makes the backbone of this architecture concerned with managing the CM services. The top section of the architecture includes the application layers, which act as a portal for interface with the system for cloud users.

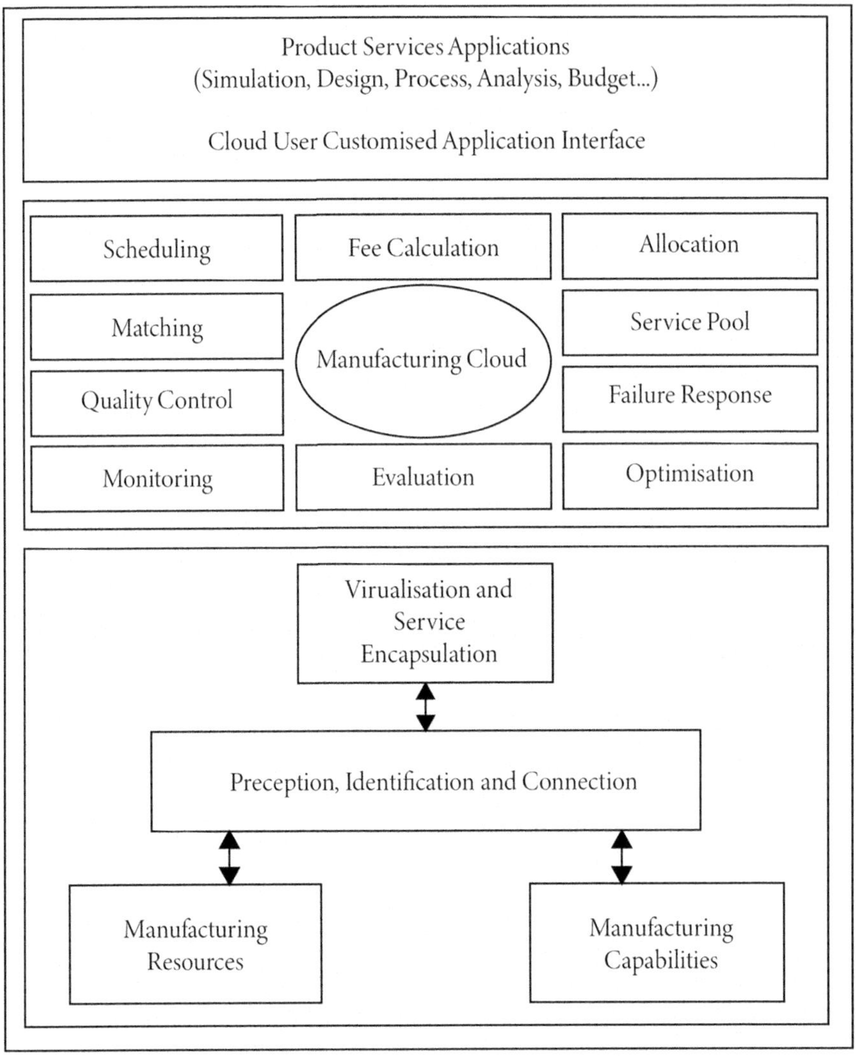

FIGURE 12.3 Cloud manufacturing architecture (Tao et al., 2011).

According to Zhang et al (2014), each manufacturing technique or model has a specific emphasis and has contributed significantly to the growth of the manufacturing industries. These two elements are as follows: structure-oriented concepts, which emphasize the establishment of businesses, and technology-oriented concepts, which emphasize the use of technologies. The exchange of information throughout the supply chain is a crucial component of several cloud manufacturing techniques. 'Servitization,' which refers to the process by which production gradually switches from the old product-oriented type to the service-oriented kind, is a component of the CM idea. All manufacturing tools and resources are offered as services and the features of the new manufacturing period are agility, information exchange, and networking with a temporary collaboration (Ford et al., 2012).

Further highlighting the distinctions between cloud computing and cloud-based manufacturing, Tao et al. (2011) investigated the concept, architecture, and standard features of cloud-based computing and manufacturing and how the concepts relate to each other. The relation of CM with cloud computing can be described through three perceptions: resource, application, and technology. CM may be referred to as the "manufacturing version of cloud computing". Ren et al. (2015) expanded the architecture into six layers: user interface (UI), toolkit, middleware, virtual pool, and resource perception. The researchers explored the critical demands of cloud-based manufacturing and developed a CM platform acting as a prototype. CM is a framework for providing global, practical, on-demand network access to a collection of configurable manufacturing resources swiftly employed and distributed with less involvement from management or service providers (Xu, 2012).

12.1.2.1 Case Study – CM in Siemens and BMW

CM was implemented as the manufacturing-as-a-service model by Ivanov (2021) and Mourtzis (2022). Two examples illustrate this concept. Firstly, the cloud-based manufacturing platform of Siemens and MindSphere utilizes Industry 4.0 principles by using advanced analytics for digitally managed interconnected machines and systems across globally dispersed physical factories. With advanced analytics and artificial intelligence (AI), MindSphere drives IoT solutions from the edge to the cloud using data from linked goods, facilities, and systems to optimize processes, produce best quality products, and introduce novel business models. Secondly, the open manufacturing platform (OMP) developed by BMW and Microsoft embodies key Industry 4.0 concepts of cross-enterprise collaboration of machines and HR. Supporting collaboration and data transparency in Industry 4.0 networks is the primary goal of OMP. The OMP's goal is also to make manufacturing more intelligent by utilizing open standards and data analytics to solve practical issues as efficiently as possible while maximizing the use of resources (Ivanov et al., 2022). Since there are several service providers and required functions in systems, managing and matching various desired tasks to a particular service provider is a sensitive topic in the CM world (Aghamohammadzadeh & Valilai, 2020; Liu et al., 2019a).

CM uses a share-to-gain mindset rather than a conventional compete-to-win approach using the industrial IoT and services. A diverse network of machines enables a broader range of manufacturing resources centered on exploiting enterprises' capabilities. It might allow immediate communication across numerous manufacturing facilities that are spread out geographically, thereby optimizing a network's value chain.

12.1.3 Expanding PLM and CM Horizons with Industry 4.0

Industry 4.0 has a significant impact on the manufacturing industry. The concept is based on developing intelligent factories, innovative products, and smart services entrenched in the IoT. The basic framework of Industry 4.0, to boost the development of the manufacturing industry by utilizing the power of communications technology and innovative inventions, was first published by Kagermann and Wahlster (2013). CM is recognized as one of the main pillars for achieving smart manufacturing (SM) (Xu, 2012) in Industry 4.0. Adding emergent techniques like IoT, AI, data analytics, and digital delivery services are impacting SM practice in the Industry 4.0 period (Jamwal et al., 2021; Machado et al., 2020). The product life cycle is getting shorter, and in new manufacturing setups, particular emphasis is given to managing it. It has only recently become clear that a variety of IoT and industrial internet technologies and methodologies can give considerable and innovative solutions to manage the product life cycle information, including providing access and real-time insights in the data of several PLM-related activities.

The term 'cloud-based design and manufacturing (CBDM)' refers to a service-oriented approach to product development in which service users can alter the configuration of goods or services as well as manufacturing systems using infrastructure-as-a-service (IaaS), platform-as-a-service (PaaS), hardware-as-a-service (HaaS), and software-as-a-service (SaaS), in response to swiftly shifting consumer demands. On-demand self-service, constant access to networked data, quick scalability, resource pooling, and virtualization are the defining features of CBDM. Cloud deployment models come from private, public, and hybrid clouds.

Moreover, three computing layers—the edge, the fog, and the cloud—can further encapsulate the modern computational landscape, or 'technology stack,' which facilitates vertical and horizontal integration and CPS intelligence. According to philosophy, this is the Fourth Industrial Revolution of the concentrated "biological mind," which has evolved into the dispersed cyber-physical "digital mind" (Aazam et al., 2018; Lee, 2008; Monostori et al., 2016; Morgan et al., 2021; Pisching et al., 2018). Knowledge and data are gathered, saved, and exchanged along the supply chain, a feature of CM. Limited data sharing may occur in the early phases of a partnership between cloud users, but data sharing could rise with trust as the relationship develops. Recently, the idea of 'Social Manufacturing,' in which producers and consumers collaborate to produce physical goods, has been getting attention (Hamalainen & Karjalainen, 2017).

Manufacturing organizations are beginning to utilize cloud computing to produce goods within their customers' price ranges, boost productivity, and facilitate a smoother work inflow. Cloud computing has emerged as one of the industry's primary enablers, revolutionizing its business models (Pathak & Bhatt, 2021). Access to business-critical data and analytics will help organizations adopt the cloud, which will help them stay competitive and be essential to their survival.

Many scholars (Lee, 2008; Grieves, 2005; Stark, 2011) believe PLM as a management idea supported by ICT in the form of PLM systems. Geographical information systems (GIS) can be used in the planning of sustainable manufacturing; by mapping the same factor across time and space. Therefore, an overview of changes is developed, which makes it easier and more accurate to anticipate future developments and make sound planning strategies in urban localities (Terzi et al., 2010). By implementing GIS, 'Territory' at the macro level adds to the structure of PLM. It makes it able to provide a proper catalog of the surroundings, including environmental impact on natural ecosystems, transportation, community demographics, public safety, utilities, services, and accessibility. A geographically sizable technical system's environmental effects may traverse international, regional, and even continental boundaries, impacting various civilizations or groups of people with various perceptions of environmental change (Vadoudi et al., 2014).

Product developers' and service providers' tasks have increased further than production and servitization, including maintenance, reuse, and recycling until the end of the value chain. So, today, managing a product's entire life cycle is essential. Product life cycle technologies and management systems also need information flow management along the product value chain. PLM has been shown to boost after-sales services, increase revenues, improve product delivery timeliness, and increase organizational competitiveness (Singh & Misra, 2018). For manufacturers to remain competitive, PLM is needed now more than ever as post-production tasks become even more crucial.

Four aspects have influenced PLM systems' institutionalization in a manufacturing firm: PLM system implementation risks, successful PLM system implementation, PLM institutionalization barriers, and individual perception of PLM systems' implementation and institutionalization (Singh et al., 2022). Many optimizations and evaluation tools that have been projected in the relevance of PLM include technique for order of preference by similarity to ideal solution (TOPSIS) (Rostamzadeh et al., 2018), effects analysis (Tomasic et al., 2017), Kano model (Chen & Chuang, 2008), analytic hierarchy process (AHP), etc. AHP is a frequent evaluation approach and a common evaluation indicator for the designing stage, which includes appearance, performance, developmental cycle, and environmental friendliness (Chen et al., 2011). Moreover, typical optimization approaches are shown by numerous meta-heuristic algorithms (Wang et al., 2018). Common evaluation measures included in manufacturing stages are makespan, price, energy, and resource utilization (Li et al., 2017). The quality of service (QoS), which measures the degree to which a consumer receives personalized services, is a crucial indicator for assessing the effectiveness of the service stage (Tao et al., 2009).

12.2 MAINTAINING CONTROL OF PRODUCTION IN INDUSTRY 4.0

A substantial breakthrough in the industrial environment is anticipated, with a rise in the challenges in production setup and goods. Organizations must address contemporary difficulties such as shorter product life cycles, increasingly customized goods, and global competition. For the enhancements of manufacturing competitiveness and meeting the market challenges, advancements in information technology are considered mandatory. Therefore, computer science, communication technologies, and information developments assist in integrating and interconnecting smart appliances, machines, manufacturing programs, and parts. Industries are developing great intelligence, and the gap between the actual and virtual worlds is lessening. This transformation is converting the manufacturing sector into what is referred to as Industry 4.0, which will result in increased efficiency, precision, and economic advantages (Schützer et al., 2019). The fact that manufacturing firms have started using GIS in their PLM structures to attain sustainability is not surprising, particularly given that

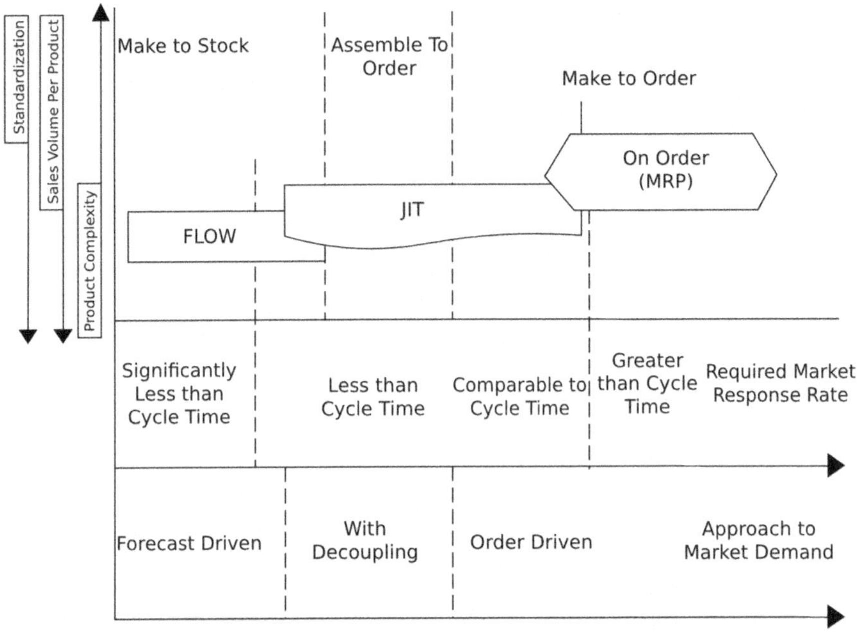

FIGURE 12.4 Coherence diagram using several manufacturing setup methodologies (Grassi et. al, 2020).

much of the data that organizations typically use includes significant spatial components (estimates range between 50% and 85%) (Azaz, 2011).

In Figure 12.4, a coherence diagram shows various manufacturing processes appropriately placed for the product's customization amount and the needed marketplace reaction time. The market requires short reaction times in the left part of Figure 12.4. There are various production approaches that should be appropriately positioned for product customization and market response time. As it is known that the market needs short response times, emphasizing the requirement to retain completed product stockpiles and to carry out forecast-based production planning, is necessary. Here, if customization is increasing, the just-in-time (JIT) technique is applied to keep a constant and well-maintained production flow across the entire setup (Grassi et al., 2020).

12.2.1 Product Life Cycle Management Frameworks and Methods

Manufacturing has a crucial role in both developing and developed countries. PLM idea is explained as a product-centric-life cycle-oriented business model which is supported through ICT (Terzi et al., 2010). The objective is to achieve the required performance and sustainability for the product and related services. According to Cao et al. (2009), PLM advanced steadily, beginning from the improvement of fundamental product-centric IT tools (e.g., CAD) during the 1970s to broad product-centric IT applications [e.g., computer-aided quality (CAQ)] during the 1980s. Later on, since the year 2000, improving tools for product data management

(PDM) in the 1990s has helped evolve PLM. The goals include driving innovation, accelerating product development time, improving quality, reducing costs, visualizing the information of the product, and lessening the communication barriers. PLM has become an advanced business approach designed to include; managerial features (e.g., integrated approach), technological features (e.g., product information backbone), and collaborative features (e.g., integrating people, process, and data) as unique components (Corallo et al., 2013).

PLM broadens the spectrum of stakeholders to cooperate across the full product life cycle, made possible by the advancement of the Internet and IT. It entails many product lifetime stages, including research and development, manufacture, supply, user services, maintenance, and recycling. PLM is considered a strategic way to improve the enterprise's product competitiveness. The automotive and aerospace sectors were the pioneers in opting for the PLM, as their products are somewhat complicated and require a long life (Grieves, 2005). However, for some industries like textile and apparel, there needs to be more research on PLM to address the challenges of sustainability, traceability, and transparency in the sector and inter-industry collaborations (Conlon, 2020). Most businesses are eager to use PLM since it can provide highly effective advances in the manufacturing industries. Moreover, PLM implementation errors are scant, and institutionalizing PLM is more complex (Singh & Misra, 2018). In addition, new manufacturing dynamics and increasing competition force companies to enhance their information systems, decision-making techniques, and processes to manage production. One essential prospect is seeking help from PLM (Santos et al., 2018).

The 'lean' idea is the predecessor of PLM (Grieves, 2005). PLM has a lengthy acceptance cycle due to holistic mapping and review of several organizational processes (Schuh et al., 2008). PLM implementation typically coincides with goals for organizational reform, and to support change, collaboration is an essential requirement to materialize the PLM notion.

Besides role definitions and computerizing existing processes, PLM deployment necessitates a fundamental rethinking of business processes to take advantage of technological advancements and address the problems and opportunities faced by the industry (Schuh et al., 2008). PLM is an idea that is focused on products and aids in managing the entire product portfolio across each product's life cycle stages (Ameri & Dutta 2005; Singh & Misra, 2018; Singh et al., 2022). Practically, PLM implementation might be complex for some businesses (Batenburg et al., 2006; Silventoinen et al., 2011) since the business processes are not well defined and the desired transformation is not carried out to the desired scale. Capability maturity models also help make the PLM vision more manageable (Silventoinen et al., 2011) and facilitate establishing a shared vision to prioritize process improvement plans in firms (Swarr, 2011).

12.2.1.1 Digital Twin in Product Life Cycle Management

By adopting a generic competitive process framework developed by Casadesus-Masanell and Ricart (2010), a detailed digital twin's application structure and procedure are used to examine digital twin perspectives in PLM stages. With the adoption of this framework, companies may concentrate on essential PLM features for product enrichment. Integration of digital twin web (DTW) in engineering PLM is a paradigm transformation which may help firms to build up better procedures to manage every product life cycle level, beginning with idea generation and progressing to designing, analysis, certification, production, operation, maintenance, and, eventually, discarding. Moreover, PLM integrates data, processes, people, and technology to provide product information support, as shown in Figure 12.5 (Qi & Tao, 2018). A DTW may simulate hundreds of procedures and alterations for every stage during the product's life cycle. Users may undergo various 'what if' testing situations for materials, designing, production factors, operational circumstances, and logistics, among other things. Additionally, the impacts of the changes on the various life cycle stages may be evaluated. For instance, a few of the benefits of DTW include extensive collection and processed data storage through the production level and prompt information utilization through production errors or challenges and part defect to highlight crucial manufacturing phases. Additionally, a client may offer to customize the requirements. Repair procedures can be scheduled depending on the awareness of the complete product operation history during the product life cycle resulting in high machine availability, significantly fewer downtimes, and fast operation time.

There are four stages in the digital twin life cycle: (1) creation, (2) manufacture, (3) operation, and (4) discarding. This technique of digital twin life cycle is analogous to a typical product life cycle. Broadly speaking, a product is any digital or physical good created through a value-adding process and is then introduced to the market to meet customers' demands. Contextually, digital twin is responsible for reducing the gap between the virtual and real worlds, as well as the advantages that might result from close contact in production environments under Industry 4.0. Furthermore, while comparing digital twin lifespan and product life cycle, it is considered that digital twin is a complicated setup and, as a result, visualizing the implementation phases toward a fully operational procedure should be done in a methodical way (Schützer et al., 2019).

Furthermore, digital twin technologies and frameworks improve the design phase in a dynamic, responsive, and all-encompassing manner. It significantly impacts the manufacturing front, with an ample variety of creative and inventive research aimed at making the production process more efficient, dependable, and adaptive. In logistics, it employs real-time tracking and other technologies to make processes more efficient. As the industry moves towards industrial robots in applications such as smart warehouses, digital twins are being used in the improvement of warehouse safety and efficiency. In the usage phase, digital twin's capabilities include forecasting and designing next-generation goods, product enhancement, and assisting industrial asset maintenance. Operations, reconfigurations, and maintenance procedures may be enhanced by employing data and analytics by sensors integrated with smart goods and equipment. Finally, Govindan and Soleimani's (2014) term 'reverse logistics' strives to limit adverse effects on humans and the environment by prioritizing disposal, remaining lifespan forecast, smart recycling, and material recovery (Lim et al., 2020a).

12.2.1.2 Product Life Cycle Improvement Through Supply Chain

In contrast to the typical automation pyramid, the product life cycle of Industry 4.0 will have a decentralized flow of information. This will improve level-to-level

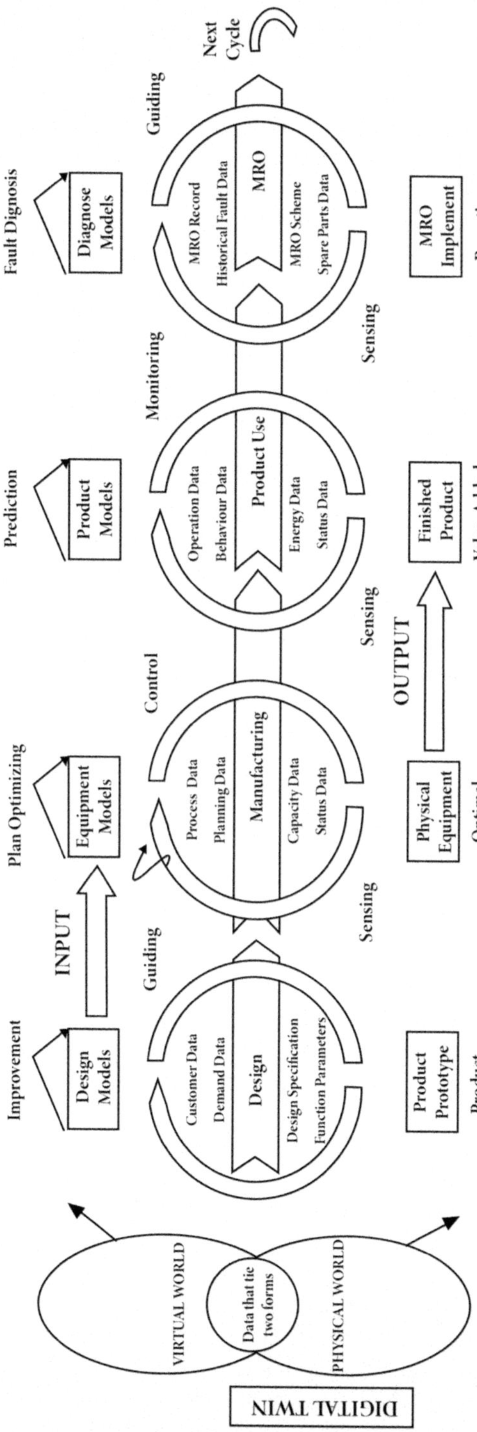

FIGURE 12.5 Digital twin and product life cycle (Qi & Tao, 2018).

communication and increase the visibility of many product life cycle phases, improving automation by making the performance more dynamic. Also, increasing system complexity will bring many security risks (Chhetri et al., 2018). Moreover, Industry 4.0 may accelerate the product life cycle and supply chain (20% to 50% less time to market), increase flexibility (30% to 50% less machine downtime), and improve precision (up to 85% forecasting accuracy) with a 3%–5% improvement in productivity. Nevertheless, the integration of components will impact the complex product life cycle supply chain by bringing different security concerns to the manufacturing system's confidentiality, availability, and integrity (Chhetri et al., 2018).

12.2.1.3 Blockchain-Empowered Product Life Cycle Management

A decentralized network based on user-generated content allows all to engage in the full product's life cycle. It challenges the conventional production paradigm, and the innovative model steadily moves the importance from producers to customers by focusing on product design and manufacturing. Furthermore, because of social resources and major manufacturing network distribution, prosumers (people who play both the role of provider and consumer) can participate more effectively in product designing, manufacturing procedures, and low-carbon methods, thereby expanding the innovation periphery and saving energy indefinitely, as shown in Figure 12.6 (Leng et al., 2020). Recent advances in edge computing and fog computing provide a new impetus to reconsider blockchain applications in manufacturing systems (Heinrichs, 2013). Blockchain uses a decentralized peer-to-peer (P2P) communication mode to process information between machines efficiently; thus, it significantly enhances process flexibility and social sustainability (Venkatesh et al., 2020).

12.2.1.4 Big Data and Product Life Cycle

Advancements in information technologies are propelling the production sector into the era of big data. Data mining and analysis are becoming increasingly important in manufacturing business management. Big data may give methodical direction for associated manufacturing processes by gathering and analyzing a wide range of data generated throughout the product's lifetime. Moreover, it can assist company managers in resolving operational and decision-making issues. The usefulness of manufacturing big data may be thoroughly investigated to improve manufacturing efficiency. Now, big data drives SM on three levels: association, forecasting, and control. Its purpose is to derive a new value from several data sets' association and arithmetical properties (Leng et al., 2020).

Big data in manufacturing refers to a high degree of structured, semi-structured, and unstructured information created over the product's lifetime. Manufacturing's rising digitization is generating opportunities for SM. Manufacturers might identify production process bottlenecks, understand the reasons and consequences of problems, and develop remedies using big data analytics based on cloud computing. As a result, manufacturing processes can be upgraded to increase production efficiency by making manufacturing leaner and increasing competitiveness. The essential knowledge is derived through manufacturing big data, which is used to improve product designing, production, MRO (Maintenance, Repair, and Overhaul), etc. This can also assist manufacturers in transitioning to SM (Tao et al., 2018).

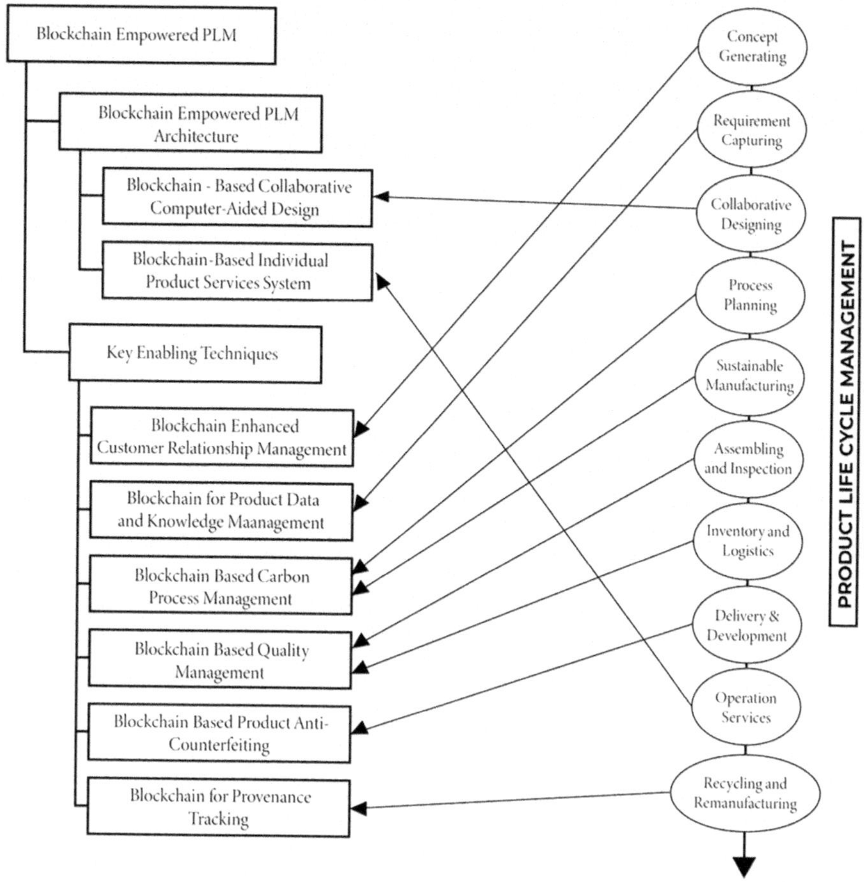

FIGURE 12.6 Blockchain and product life cycle management (Leng, 2020).

12.2.1.5 AI Applications for PLM

AI application for PLM in advanced manufacturing is scarce compared to other disciplines (Zhong et al., 2017). Manufacturers are cautious about embracing AI in PLM because of the high necessity for quality, dependability, precision, and cost-effectiveness in production, which is true in the case of SMEs (Tao et al., 2018). However, the benefits of AI to PLM are too many to count. First, a complicated industrial process can be simplified toward a lesser labor-intensive state because AI can substitute people for hazardous and repetitive tasks. Second, a costly manufacturing process can become more affordable since a proven AI solution can last long. Third, because many AI algorithms are attuned to big industrial data, manufacturing decision-making is moving more and more toward data-driven approaches. Manufacturers are better prepared to respond to changes in the industry's environment thanks to their improved capacity for handling massive data volumes (Lu, 1990; Wuest et al., 2016).

AI and machine learning (ML) are being incorporated in other verticals of industries, such as aerospace, so that they can strive toward technological advances. The primary feature is the concurrent engineering capabilities like collaborations between interdepartmental operations to reduce the non-value-added activities during the product development phases (Ghahramani, 2015).

While using big data in manufacturing, data sources, data processing, and applications are used. The data sources are comprised of manufacturing resources, information systems, and the Internet. Then data is collected, analyzed, and visualized. It further leads to smart design, planning, production, MRO, as shown in Figure 12.7 (Qi & Tao, 2018).

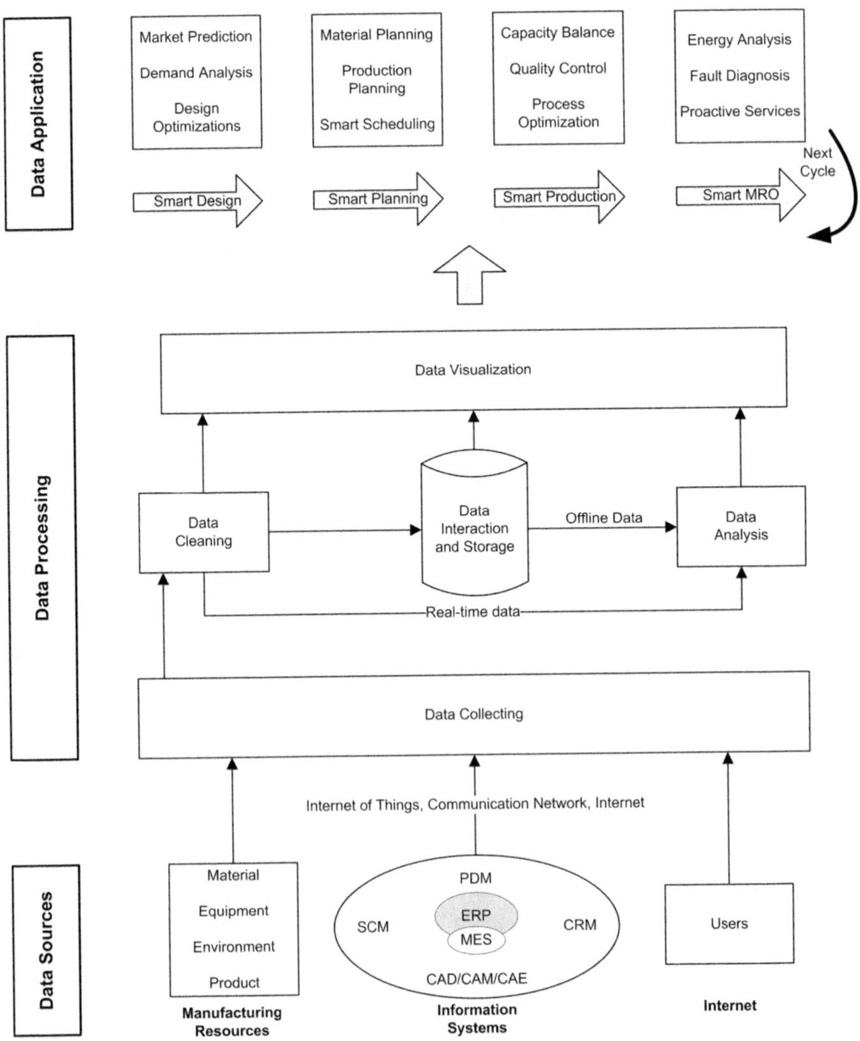

FIGURE 12.7 Sources, processing, and applications of big data in manufacturing (Qi & Tao, 2018).

12.2.2 Case Study – Upgradation of Haier Via Industry 4.0

Haier is a Chinese company that globally supplies home appliances and consumer electronics. The supply includes a wide variety of appliances, for example, refrigerators, laundry machines, microwave ovens, air conditioners, laptops, and mobile phones. One of the plants of Haier, i.e., household appliance manufacturing, is considered among the intelligent industries. The upgradation of Industry 4.0 was completed in 2017, in which integrated architecture was implemented to some extent for effective PLM. Some improvements were: traceability of bills of materials, integration of management information systems (MIS), data accessibility, customization, less response time, low defected parts, minimization of the delivery period, the viability of privacy protection, and reduction in disposal expenditure.

12.3 CM: A DATA-DRIVEN SYNERGETIC APPROACH

12.3.1 CM Platforms and Possibilities

CM leverages cloud computing, big data, service-oriented technologies (SOTs), IoT, and big data to enable large-scale sharing and on-call usage of ubiquitously dispersed production capabilities and resources along with free circulation and transactions. CM applications included the ones built on cloud computing (e.g., cloud-based software applications given by the operators), thus covering the concepts of cloud computing-based manufacturing (Coullon & Noyé, 2018). However, the opposite is not valid i.e., cloud computing-based manufacturing cannot be understood to be cloud manufacturing. Cloud computing is merely one of the numerous supporting technologies for cloud manufacturing from a technical standpoint. Resultantly, cloud-computing manufacturing might be viewed as a subset of CM.

Xu (2012) proposed two strategies for implementing cloud computing in the manufacturing sector, including direct cloud computing implementation and the manufacturing version of cloud computing. He emphasized that cloud computing plays a significant role in manufacturing by offering solutions that may be customized.

Most production capabilities and resources supplied through multiple service suppliers are generally encased and virtualized as basic manufacturing cloud services in a cloud manufacturing system. Following that, functionally comparable MCSs are assigned to the same candidate cloud manufacturing system set. While a service demander submitted his manufacturing job to the setup and asked for manufacturing services, a succession of cloud manufacturing systems from various candidate cloud manufacturing system sets are picked and combined into a virtual manufacturing system (Liu & Zhang, 2017).

Multiple users may simultaneously make service requests in a CM service-oriented paradigm. The users submit their required tasks to a CM platform. Because manufacturing virtualized services are operated and managed centrally in a CM environment, several manufacturing tasks can be carried out simultaneously. In order to improve the performance of a cloud-manufacturing platform, it is critical to schedule manufacturing tasks optimally in CM. One of the key aspects of cloud manufacturing technology that requires effective scheduling algorithms is the distribution of resources (Ghomi et al., 2019).

The recent example modernizes and converts conventional manufacturing setup into the SM object (SMO) that intelligently enables manufacturing facilities to be shared and disseminated. SMOs within the CM setup can automatically sense, develop a connection, respond, interact, and collaborate. Consequently, production capabilities and resources can be encased in a variety of services that understand flexible-to-access, easy to deploy, and simple to invoke through advanced technologies, including virtualization, service-based, and cloud computing methods. CM can provide reliable, safe, sustainable, on-demand, and good quality facilitation for the complete manufacturing lifecycles because of its smart and consolidated management and practices. CM is currently attracting much interest due to its various benefits. Various SMOs having shared capacities may be intelligently sensed via a larger internet connection. The deployment of public, private, communal, and hybrid cloud with ubiquitous access to diverse consumers might enable the wider Internet. IoT technologies enable the entire manufacturing procedure life cycle that is virtualized into multiple cloud-based services, allowing manufacturing facilities and potentials to be shared (Zhong et al., 2016).

A complete CM system necessitates many crucial technological advancements, such as using embedded sensors for real-time resource monitoring (Lindström et al., 2014) and expanding cloud services to manage massive supplier networks (Hosono & Shimomura, 2013). CM is also considered a network manufacturing method that offers consumers a range of on-demand manufacturing services based on their demands. The network is used in this manner to arrange the online production facilities.

Manufacturing service suppliers integrate various manufacturing options for the entire product life in cloud platforms as a kind of service via technologies like perception and virtualization. These resources may include hardware, software, data, knowledge, and mode. Manufacturing service requesters send manufacturing needs to the cloud platform for every level or varied granularity manufacturing service needs of the product's entire life cycle, discovering or seeking manufacturing services already there on the cloud platform. The role changes dynamically when any user operates as a manufacturing service supplier or a service requester. In a CM system, a person is a manufacturing service provider if they supply a manufacturing service. A person is referred to as the manufacturing service requester when they acquire a manufacturing service. The cloud platform operators are primarily responsible for ensuring the efficient administration and operation of the cloud platform and coordinating the connection between manufacturing service requesters and manufacturing service suppliers. The operators can give resource users a flexible and dynamic service based on their application needs. The basis of the cloud manufacturing operation is to distribute manufacturing services between supply and demand, and its operation is primarily to realize manufacturing as a service process.

The primary goal of CM is comprehensive resource sharing and highly efficient collaborative business. Manufacturing facilities and potentials in a value chain or among various value chains are accumulated in cloud platforms as manufacturing cloud services, allowing users to get manufacturing services on-demand and realize personalized, scalable, adaptable, and highly configured production and services (Liu et al., 2019a).

Security and hybrid manufacturing are examples of cyber-physical applications. The information technology section includes IoT and cloud computing, and the component design complexity, precision, and dependability are all associated with AM/3D printing. Moreover, manufacturing service activities include design, testing, production, and scheduling, while the design is an example of resource-shared services. Simulation and optimization are included in the modeling area, and customer needs and quick and adaptable operations are all part of the agile-manufacturing category.

Therefore, a comprehensive CM system is made up of three types of stakeholders: operators, providers, and customers. The mutual collaboration among them ensures the long-term operation of a CM system, as shown in Figure 12.8 (Liu et al., 2018).

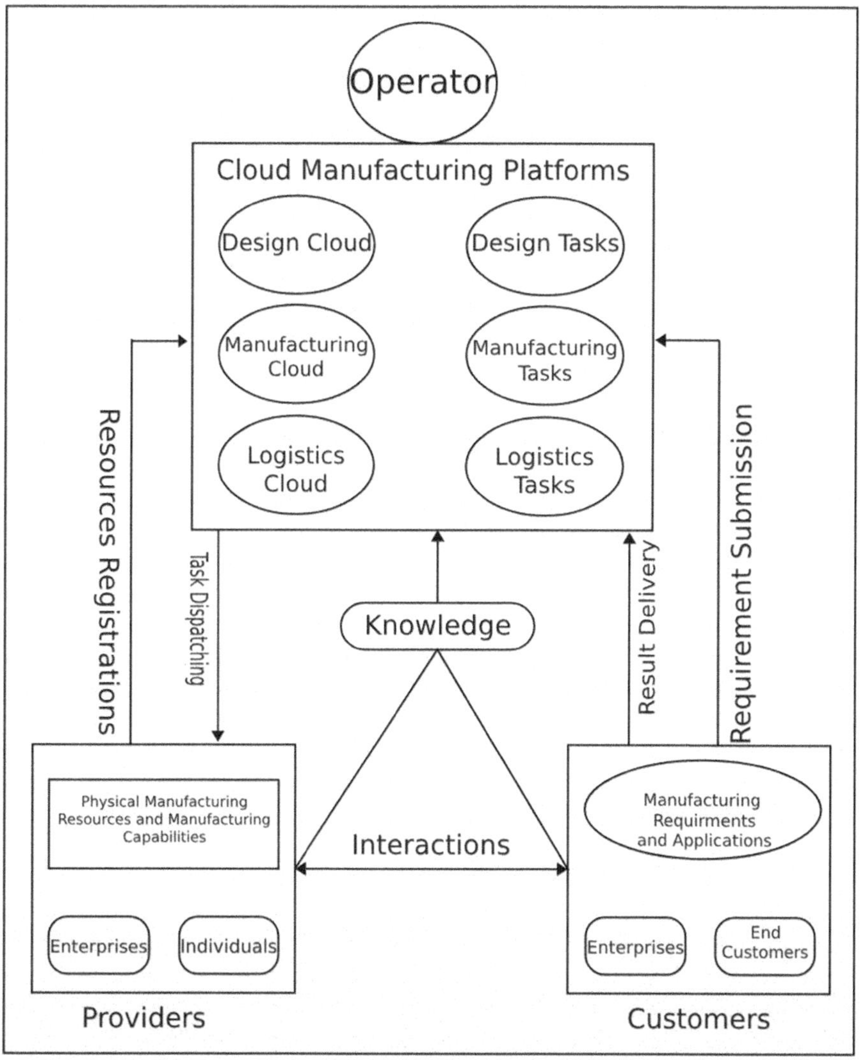

FIGURE 12.8 Operation model of cloud manufacturing (Liu et al., 2018).

12.3.1.1 Characteristics of CM

There are a few characteristics of CM, including flexibility and scalability, on-demand, knowledge-intensive, manufacturing service, and multi-tenancy. 'Adaptability and scalability' refer to real-time data observation for production setup optimization. At the same time, 'on-demand' is involved in the customization of the products and making amendments per customers' requirements. Moreover, 'knowledge intensive' shows the utilization of data for process conditions' optimization, while 'manufacturing service' presents the outsourcing or subcontracting of a few supply chain components. Last, 'multi-tenancy' refers to coordination with numerous other manufacturers' excess flows.

12.3.1.2 Application Platforms of CM

The CM platform's structure is shown in Figure 12.9. It consists of three primary modules: database, intelligent evaluation and optimization, and decision-making. Both suppliers and clients can use the latter as a beneficial tool to advocate the best production strategy and provide the best manufacturing options, as shown in Figure 12.9 (Simeone et al., 2021). Furthermore, the report defines the following CM application modes: public clouds, private clouds, local clouds, and hybrid clouds. Each application mode relates to a unique application service platform. Public, private, and local cloud are now the most popular CM service platforms. Hence, the current state of the following three categories of practiced service platforms shall be examined.

12.3.1.2.1 Public CM Service Platform

Several organizations take part in this platform. Organizations may fuse all production facilities from their area, the country, and the world. In the United States, a company called MFG, typically uses a public CM service platform. Similarly, another company, COM, has the world's biggest manufacturing capabilities trade platform, and is devoted to providing worldwide manufacturing partners with a rapid and efficacious trade network. Moreover, the German CREMA platform creates methodologies to model, configure, execute, and monitor manufacturing procedures, offering end-to-end assistance to CM via developing real-world manufacturing setups and validating CM in real-world manufacturing scenarios. 'Chinese Aerospace Cloud Network' and 'Manufacturing Cloud' are open internet-industry-cloud service platforms that provide the majority of manufacturing firms and technical professionals with massive, novel, and comprehensive industrial cloud resources, software, and services.

12.3.1.2.2 Private CM Service Platform

This platform can generally be utilized by big businesses or intra-group networks that focus on integrating manufacturing facilities and potentials among the working cluster. Microsoft, Amazon, Google, IBM, and Ali Baba are 'private cloud' manufacturing service platforms. Integrating such resources inside the organization enhances usage efficiency, reduces costs, and expands product design capabilities and innovation.

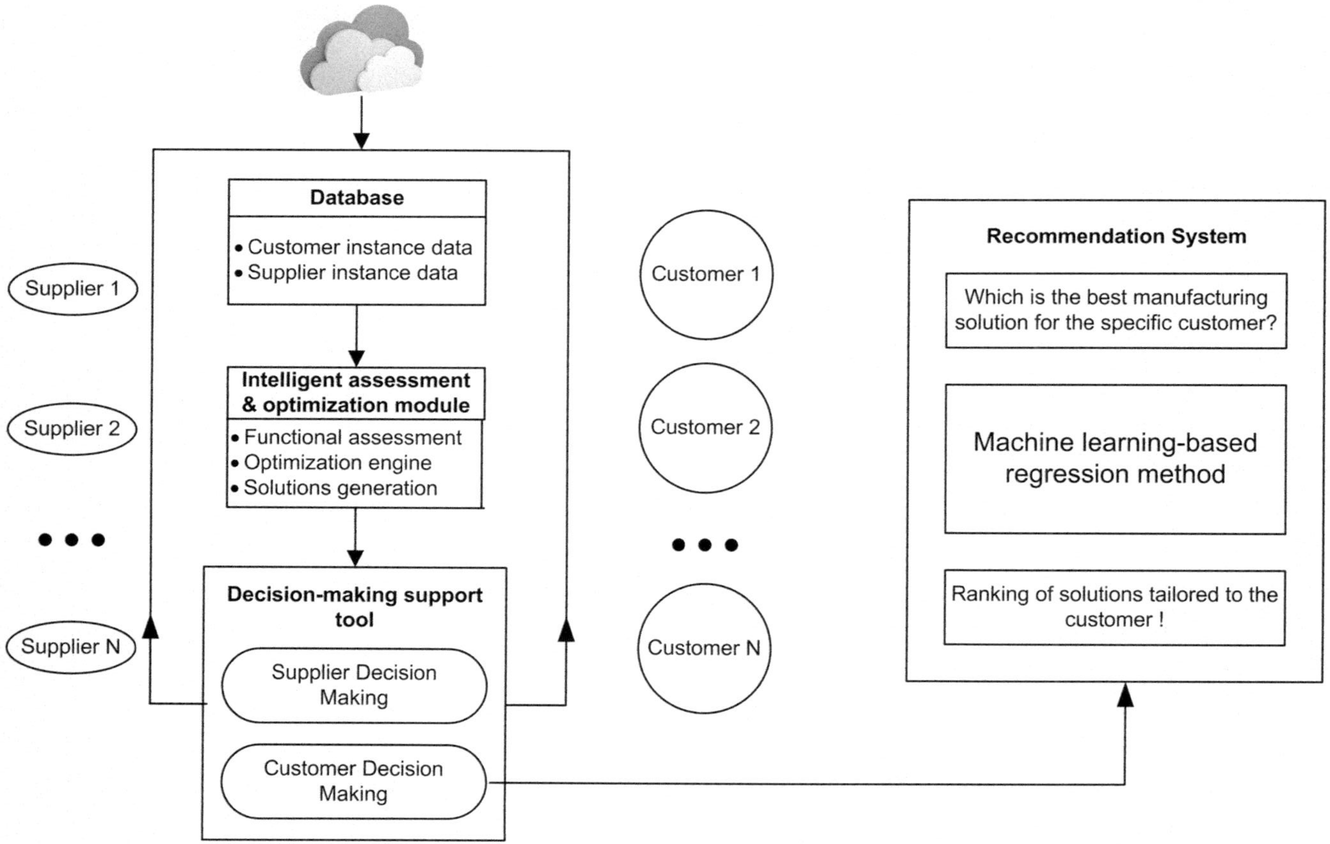

FIGURE 12.9 Cloud manufacturing platform scheme (Simeone et al., 2021).

12.3.1.2.3 Regional CM Service Platform

It is an extensive manufacturing service setup based upon the local manufacturing sector centered inside major firms and connected by the manufacturing sector. Currently, this platform is not extensively employed because there is a requirement for integrating manufacturing enterprises across the region and implementing a substantial amount of engineering. China's CAXA cloud is a general local CM service platform. It consists of 19 regionally processed resource-sharing service platforms that may first accomplish the effectively shared and optimized manufacturing and processing facilities in the area and support regional manufacturing growth.

After analyzing the CM service platform's application status, it has been observed that 'public' and 'private' cloud manufacturing service platforms account for the majority of the CM service platform's application, whereas the 'regional cloud' and 'hybrid cloud' manufacturing service platforms are less developed having limited practical successes. Moreover, there are support systems which are provided by CM service. Such systems are required to manage operations in a cloud-based manufacturing platform. Services provided by CM involved the following:

- Supporting data security and multi-agent
- Manufacturing process performance optimization
- Remote real-time collaboration method along with integrated communication
- Incident recovery in operations
- Data melding and transformation
- Access management for many consumers
- Customizing platform functionality

12.3.1.3 Challenges

CM mainly offers service aid to managing tasks and transactions in SMEs. It calls for a virtual manufacturing setup with an upgraded intelligent matching engine to maintain the supply-demand balance. CM manages service resources in the following stage to allow users to submit the manufacturing service needs. It is connected to a series of complicated manufacturing procedures comprising project match and automatic transactions management. CM technology was developed in response to the manufacturing industry's demand to become more cohesive and linked. Predictably, there are particular challenges at the cutting edge of the CM ideas, and the future progression of vital technologies may face hurdles. CM is a paradigm that blends the application of emerging computing technology along with sophisticated manufacturing models. The manufacturing models are becoming more closely tied to network-related technologies, which are frequently subjected to cyber assaults and the resulting security needs. In the digital world, there are broad chances of cyber-attacks, which span the community, financial, and industrial realms. Engineers need to be made aware of the existing and future hazards of these cyber-attacks as they are unable to detect the reason when attacked accurately. Considering, understanding, and resolving the current manufacturing shortcomings should be the initial steps to avoid, detect, and mitigate cyber-attacks. Most manufacturing vulnerabilities are associated with production control, system design, manufacturing cyber security, and quality control research weaknesses. The lack of such expertise posed significant hurdles to the widespread CM Implementation and adoption. This gap can be addressed by investigating CM security in manufacturing systems.

On the one hand, product and service complexity is expanding, while on the other, globalization is increasing. Manufacturers must compete in a worldwide market; thus, they must build, plan, and manage their facilities in various situations. Both service and system complexity are rising, yet low-cost production is still sought to preserve profitability. These complications and problems, like globalization, increased energy usage, and highly complicated information technology, have encouraged businesses to adopt cloud-based software.

12.3.1.4 Potential

Per the state-of-the-art, cloud technology is inexpensive, has a layered design, is adaptable, and has a fast reaction time. The advantages make large-scale network systems like agile manufacturing more secure, interoperable, and performant. Cloud technology's advantages in advanced agile manufacturing applications include parallel processing, shared services, broad access, and flexible network resource (Sturm et al., 2017). One of the essential requirements for CM adoption is the ability to access production resources and realize intelligent insight (Haleem & Javaid, 2019). CM is gaining traction as an integrated technology and service-based manufacturing paradigm that can transform the recent manufacturing sector into highly inventive, connected, and collaborative (Lim et al., 2020b).

12.3.2 Industrial Internet of Things in CM

Data, information, and knowledge (DIK) and how to use them properly to service a firm's business and product development and create value for the customer (Schuh et al., 2008) are at the core of PLM. Industrial internet platforms can handle data and information flows in real-time for PLM and support DIK conversions throughout the product life cycle (Menon et al., 2019).

Multiple platforms can be used to conduct an in-depth analysis of industrial Internet platforms from the perspective of PLM that represent various platform domains and have distinctive characteristics as platforms, especially considering their capacity to deal with various problems. Furthermore, five platforms have been chosen that give analysis to emphasize the influence on the management of DIK within different PLM stages via industrial Internet as a technology enabler (Menon et al., 2019).

The new technology has enabled us to connect humans, machines, systems, logistics, and products, thus, delivering innovative products with smart engineering. The new PLM needs to be an end-to-end digitalized product life cycle. The concept of PLM digital twin (and thread) is expected to be "a virtual representation of a physical object or system across its life cycle, using real-time data to enable understanding, learning, and reasoning" (Hubert, 2018).

Furthermore, CM is an idea and framework based on cloud computing, IoT, and other innovative mechanics in informative manufacturing scenarios. In the integrated network, intelligence, and service-oriented platform, CM may be able to promote the manufacturing sector, raising information on agile manufacturing to the expected degree. IoT and cloud computing can comprehend intelligence and machine to machine (M2M) connections, which include man-to-man, man-to-machine, and machine-to-machine, as well as on-call consumption and resource collaboration. A critical obstruction for advanced manufacturing processes is the realization of transition and conversion from production-based manufacturing into service-based manufacturing.

CM combined with IoT, and cloud computing also helps improve agile manufacturing practices and efficiency, reduces cost, and enhances security and service-based manufacturing, ultimately creating advanced agile manufacturing. Indeed, CM makes use of the relationship between cloud computing and IoT to integrate the manufacturing resources into the agile manufacturing process. The reciprocal and shared impacts between IoT for the SM means and potential, internet of services (IoS) based upon cloud computing, and Internet of users for utilization across the product life cycle define the interaction between CM, IoT, and cloud computing. CM-cloud computing focuses on selecting on-time computations services via continual and extremely dependable practices in a dispersed production setting.

12.3.2.1 Maturity Model

The company's strategies, organizational structure, technological development, and operational processes are impacted by the transformation process under Industry 4.0. Therefore, senior management should support proactive projects and investments, mostly to enhance market competitiveness and tackle complex processes. In this context, a maturity model can evaluate and transform the company's business goal and scope and understand the internal and external resources needed to implement Industry 4.0 strategies.

12.3.2.2 Case Study – Twining of CM and IoT in Huaiji Dengyun

Intelligent manufacturing is the need of today. In this context, the twining of CM and IoT plays a significant role. Radio Frequency Identification (RFID) is one IoT technology implemented on the manufacturing front. Vast records can be developed using this technology. A Chinese SME, Huaiji Dengyun Auto-parts (Holding) Co. Ltd, is a real-life automotive manufacturer. Huaiji specializes in producing engine valves. The company uses top advanced technologies, including CM, big data, and IoT. While considering the quantities and types, Huaiji is a giant Chinese engine valve producer. They have the capacity to produce 31.5 million valves annually. Moreover, they can manufacture above 8000 product categories utilized in aquatic, automobile, and other engines. They have four manufacturing shop floors and numerous production lines set on every shop floor. Huaiji employed RFID technology so that logistics operations and CM management could be facilitated to make decisions for advanced productions. The shopfloors have 1,000 machines and 55 buffers. There is an RFID reader on each machine and a buffer for identification. CM shopfloor logistics management supported by RFID-enabled big data is implemented here, and four hundred logistics operators are employed. All of them carry RFID mobile readers as well as RFID staff cards. After initial deployment, real-time logistic statistics are created and gathered to support the logistics management. Huaiji opted for the CM to fully utilize RFID logistics data, which is collected for advanced decision-making, including production planning and scheduling. Numerous services are created in a private cloud to utilize real-time data of RFID to support shopfloor management. A few important services are production planning, scheduling, data capture, visibility, and traceability (Zhong et al., 2016).

12.4 COMPREHENSIVE APPROACH TO ACHIEVE SUSTAINABILITY

Methods for sustainability claim to have a holistic view. They must oversee the product life cycle management and the flow of information through each stage of the life cycle. Social responsibility is impacted by sustainability; however, how it is achieved depends on how it is put into practice, i.e., by optimizing the resource consumption throughout the product lifespan and maintaining the quality of goods and services as sustainable as possible. Information is a strong foundation for process optimization and product quality.

In addition to being environmentally beneficial, a green product should require minimal waste and sustainable upkeep. Additionally, because it might be reused and have a third age after its initial usage, even though this thing is useless in the first world, it may still be helpful to someone else. Thus, a green or sustainable PLM strategy has a mission to supply products that satisfy customer needs taking advantage of the company's innovation, quality, and sustainable production system considering all the life cycle impacts. The main objective is to promote sustainability through the use of green products and processes by facilitating the sharing of data, information, and knowledge of all phases of the product life cycle among all the driving forces involved (internal and external) (Vila et al., 2015).

12.4.1 Future Trends

Since the service-based issues in CM are non-polynomial (NP)-hard-natured, a substantial amount of research has been done to use meta-heuristic algorithms to identify optimal or nearly optimal solutions (Hayyolalam et al., 2019; Zhou & Yao, 2017a; Zhou & Yao, 2017b). However, difficulties in this field include slow convergence and local optimum traps creating more issues. By enabling service matcher agents to communicate with one another in a decentralized setting, blockchain is a solution that can address the manufacturing-related problems outlined before (Aghamohammadzadeh & Valilai, 2020).

Presently, SM is a popular research area on a global scale, and many nations share the goal of implementing SM. Excellent prospects exist in this situation for the continued development of CM. Unfortunately, there exist several difficulties with the idea, the technology, and the practical use of CM. Furthermore, CM's advancement is hampered by the absence of a standardized definition (Liu et al., 2019b).

The rapid progress of nascent information technologies, including cloud computing, the IoT, big data, and AI, is changing advanced manufacturing paradigms and impacting every facet of PLM (Tao et al., 2018). The developed countries keenly promote the coming industrial revolution through national manufacturing plans, for example, Industry 4.0 in Germany, Industrial Internet in the USA,

Industry 2050 Strategy in the UK, Manufacturing Innovation 3.0 in South Korea, Society 5.0 in Japan, and Made in China 2025 Strategy in China (Zhou, 2015).

For CM to attain process resilience, big data technologies will be required. In order to identify trends, patterns, and optimized processes, manufacturers collect and analyze vast volumes of data all through the production process (Waller & Fawcett, 2013). Process manufacturing generally gathers many data from process control systems, but it is only sometimes used, frequently only for the after-the-fact study of catastrophes. The volume and variety of data created in manufacturing systems will expand due to CM's reliance on sophisticated sensing and pervasive communication technology (Bao et al., 2012).

12.4.2 Conceptual Framework

Pursuing improved product quality, cost-effectiveness and innovation has put the manufacturing industry under increased demand. The manufacturers are required to swiftly reciprocate market transformations by adjusting their approaches to lessen the product life cycle, mainly the product designing and manufacturing stage. Increased demand for customized and personalized items requires manufacturers to prepare and use more adaptable, responsive, and speedy manufacturing and service processes. Manufacturers need more collaboration efforts using CM to control the end-to-end life cycle of products globally. Manufacturers must improve the manufacturing processes in response to the heightened competition for low cost, improved quality, lessened lead times and systems intelligence using ML and AI.

The broad contours of a system are shown in Figure 12.10. Manufacturers must enhance their information monitoring, tracking, and management systems in

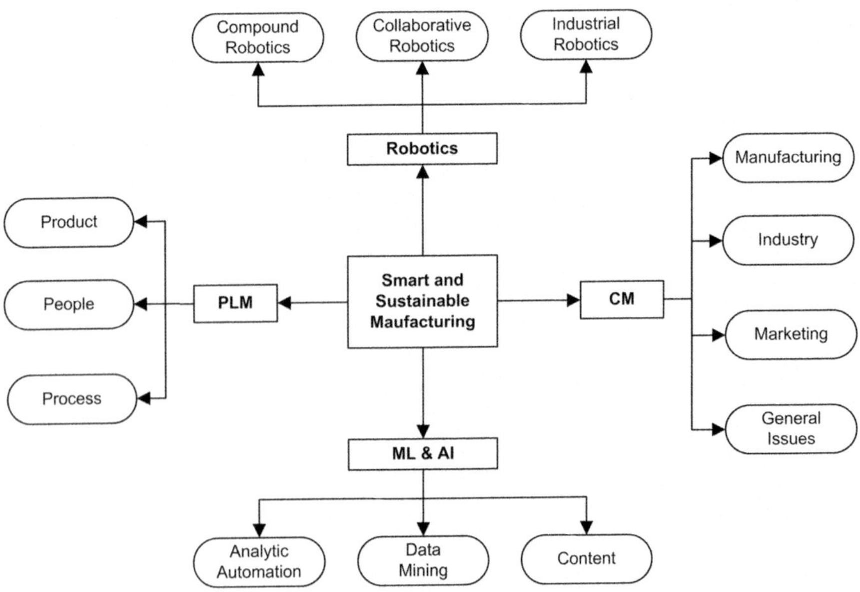

FIGURE 12.10 Conceptual framework.

response to high client expectations for service quality. The manufacturers need to set up a more responsive online service platform giving remote product repair, maintenance, and upgradation, incorporating all the stakeholders. Manufacturers must analyze, mine, and integrate a range of high-dimensional, unlabeled, unstructured, and non-standardized data using tools and techniques of ML due to the large number of big industrial data that has emerged. To foster the budding demands of manufacturing industry, academia and industry need to devote immense efforts to developing new information technologies (Tao et al., 2015), in which AI plays a vital role. Moving from Industry 4.0 to Industry 5.0 requires better data analytics and intelligent systems to face uncertain or risky manufacturing scenarios with customized products according to the customers' needs.

REFERENCES

Aazam, M., Zeadally, S., Harras, K. A. 2018. Deploying fog computing in industrial internet of things and industry 4.0. *IEEE Transactions on Industrial Informatics*, 14(10): 4674–4682.

Aghamohammadzadeh, E., Valilai, O. F. 2020. A novel cloud manufacturing service composition platform enabled by Blockchain technology. *International Journal of Production Research*: 58(17): 5280–5298.

Ameri, F., Dutta, D. 2005. Product lifecycle management: Closing the knowledge loops. *Computer-Aided Design and Applications*, 2(5): 577–590.

Azaz, L. 2011. The use of geographic information systems (GIS) in business. *International Conference on Humanities, Geography and Economics*: 299–303.

Bao, Y., Ren, L., Zhang, L., Zhang, X., Luo, Y. 2012. Massive sensor data management framework in cloud manufacturing based on Hadoop. *IEEE 10th International conference on industrial informatics*: 397–401.

Batenburg, R., Helms, R. W., Versendaal, J. 2006. PLM roadmap: Stepwise PLM implementation based on the concepts of maturity and alignment. *International Journal of Product Lifecycle Management*, 1(4): 333–351.

Bey, N., Hauschild, M. Z., McAloone, T. C. 2013. Drivers and barriers for implementation of environmental strategies in manufacturing companies. *CIRP AnnalsI*, 62(1): 43–46.

Brundtland, H. G. 1987. Report of the World Commission on Environment and Development: Our Common Future. Oxford University Press.

Buchert, T., Pförtner, A., Stark, R. 2017. Target-driven sustainable product development. *Sustainable Manufacturing*: 129–146.

Byggeth, S., Hochschorner, E. 2006. Handling trade-offs in ecodesign tools for sustainable product development and procurement. *Journal of cleaner production*, 14(15-16): 1420–1430.

Cao, H., Folan, P., Mascolo, J., Browne, J. 2009. RFID in product lifecycle management: A case in the automotive industry. *International Journal of Computer Integrated Manufacturing*, 22(7): 616–637.

Cerdas, F., Juraschek, M., Thiede, S., Herrmann, C. 2017. Life Cycle Assessment of 3D Printed Products in a Distributed Manufacturing System. *Journal of Industrial Ecology*, 21(S1): S80–S93.

Chen, C.-C., Chuang, M.-C. 2008. Integrating the Kano model into a robust design approach to enhance customer satisfaction with product design. *International Journal of Production Economics*, 114(2): 667–681.

Chen, C.-W., Chan, C.-L., Cheng, C.-Y. 2011. Using AHP for determining priority in seamless strategy: A case study of the click-and-mortar bookstore. *International Journal of Electronic Business Management*, 9(2): 95.

Chhetri, S. R., Faezi, S., Rashid, N., Al Faruque, M. A. 2018. Al Faruque. 2018. Manufacturing supply chain and product lifecycle security in the era of industry 4.0. *Journal of Hardware and Systems Security*, 2(1): 51–68.

Conlon, J. 2020. From PLM 1.0 to PLM 2.0: The evolving role of product lifecycle management (PLM) in the textile and apparel industries. *Journal of Fashion Marketing and Management: An International Journal*, 24(4): 533–553.

Corallo, A., Latino, M. E., Lazoi, M., Lettera, S., Marra, M., Verardi, S. 2013. Defining product lifecycle management: A journey across features, definitions, and concepts. *ISRN Industrial Engineering*, 2013: 170812.

Coullon, H., Noyé, J. 2018. Reconsidering the relationship between cloud computing and cloud manufacturing. In *Service Orientation in Holonic and Multi-Agent Manufacturing*: 217–228.

Fisher, O., Watson, N., Porcu, L., Bacon, D., Rigley, M., Gomes, R. L. 2018. Cloud manufacturing as a sustainable process manufacturing route. *Journal of Manufacturing Systems*, 47: 53–68.

Ford, A. L., Williams, J. A., Spencer, M., McCammon, C., Khoury, N., Sampson, T. R., Panagos, P., Lee, J. M. 2012. Reducing door-to-needle times using Toyota's lean manufacturing principles and value stream analysis. *Stroke*, 43(12): 3395–3398.

Främling, K., Holmström, J., Loukkola, J., Nyman, J., Kaustell, A. 2013. Sustanable PLM through Intelligent Products. *Engineering Applications of Artificial Intelligence*, 26(2): 789–799.

Garetti, M., Taisch, M. 2012. Sustainable manufacturing: Trends and research challenges. *Production Planning & Control*, 23(2–3):83–104.

Ghahramani, Z. 2015. Probabilistic machine learning and artificial intelligence. *Nature*, 521(7553): 452–459.

Ghomi, E. J., Rahmani, A. M., Qader, N. N. 2019. Cloud manufacturing: Challenges, recent advances, open research issues, and future trends. *The International Journal of Advanced Manufacturing Technology*, 102(9–12): 3613–3639.

Grassi, A., Guizzi, G., Santillo, L. C., Vespoli, S. 2020. A semi-heterarchical production control architecture for industry 4.0-based manufacturing systems. *Manufacturing Letters*, 24: 43–46.

Grieves, M. 2005. *Product Lifecycle Management: Driving the Next Generation of Lean Thinking: Driving the Next Generation of Lean Thinking* (1st edition). McGraw Hill.

Guo, L. 2016. A system design method for cloud manufacturing application system. *The International Journal of Advanced Manufacturing Technology*, 84(1): 275–289.

Haleem, A., Javaid, M. 2019. Additive manufacturing applications in industry 4.0: A review. *Journal of Industrial Integration and Management*, 4(04): 1930001.

Hamalainen, M., Karjalainen, J. 2017. Social manufacturing: When the maker movement meets interfirm production networks. *Business Horizons*, 60(6): 795–805.

Hauschild, M. Z., Kara, S., Røpke, I. 2020. Absolute sustainability: Challenges to life cycle engineering. *CIRP Annals*, 69(2): 533–553.

Hayyolalam, V., Pourghebleh, B., Pourhaji Kazem, A. A., Ghaffari, A. 2019. Exploring the state-of-the-art service composition approaches in cloud manufacturing systems to enhance upcoming techniques. *The International Journal of Advanced Manufacturing Technology*, 105(1): 471–498.

Heinrichs, H. 2013. Sharing economy: A potential new pathway to sustainability. *GAIA-Ecological Perspectives for Science and Society*, 22(4): 228–231.

Helo, P., Phuong, D., Hao, Y. 2019. Cloud manufacturing – Scheduling as a service for sheet metal manufacturing. *Computers & Operations Research*, 110: 208–219.

Herrmann, C., Bergmann, L., Thiede, S., Halubek, P. 2007. Total life cycle management-an integrated approach towards sustainability. *3rd International Conference on Life Cycle Management, Zurich*.

Hosono, S., Shimomura, Y. 2013. Towards establishing mass customization methods for cloud-compliant services. *In The Philosopher's Stone for Sustainability*: 447–452.

Hubert, P. 2018. Learning from system engineering to deploy product lifecycle management. *IFAC-PapersOnLine*, 51(11): 1592–1597.

Ivanov, D. 2021. Digital supply chain management and technology to enhance resilience by building and using end-to-end visibility during the COVID-19 pandemic. *IEEE Transactions on Engineering Management*: 1–11.

Ivanov, D., Dolgui, A., Sokolov, B. 2022. Cloud supply chain: Integrating Industry 4.0 and digital platforms in the supply chain-as-a-service. *Transportation Research Part E: Logistics and Transportation Review*, 160: 102676.

Jamwal, A., Agrawal, R., Sharma, M., Dangayach, G. S., Gupta, S. 2021. Application of optimization techniques in metal cutting operations: A bibliometric analysis. *Materials Today: Proceedings*, 38: 365–370.

Jensen, A. A., Remmen, A. 2006. "Background Report for a UNEP Guide to Life Cycle Management: A bridge to sustainable products." https://wedocs.unep.org/20.500.11822/33072.

Kagermann, H., Wahlster, W. 2013. Recommendations for implementing the strategic initiative INDUSTRIE 4.0 (Industrie 4.0 Working Group) [Working Paper]. *National Academy of Science and Engineering*.

Kiritsis, D. 2011. Closed-loop PLM for intelligent products in the era of the Internet of things. *Computer-Aided Design*, 43(5): 479–501.

Kiritsis, D., Bufardi, A., Xirouchakis, P. 2003. Research issues on product lifecycle management and information tracking using smart embedded systems. *Advanced Engineering Informatics*, 17(3): 189–202.

Kloepffer, W. 2008. Life cycle sustainability assessment of products. *The International Journal of Life Cycle Assessment*, 13(2): 89.

Lee, E. A. 2008. Cyber Physical Systems: Design Challenges. *11th IEEE International Symposium on Object and Component-Oriented Real-Time Distributed Computing (ISORC)*: 363–369.

Leng, J., Ruan, G., Jiang, P., Xu, K., Liu, Q., Zhou, X., Liu, C. 2020. Blockchain-empowered sustainable manufacturing and product lifecycle management in industry 4.0: A survey. *Renewable and Sustainable Energy Reviews*, 132: 110112.

Li, X., Peng, Z., Du, B., Guo, J., Xu, W., Zhuang, K. 2017. Hybrid artificial bee colony algorithm with a rescheduling strategy for solving flexible job shop scheduling problems. *Computers & Industrial Engineering*, 113: 10–26.

Lim, K. Y. H., Zheng, P., & Chen, C. H. 2020b. A state-of-the-art survey of Digital Twin: techniques, engineering product lifecycle management and business innovation perspectives. *Journal of Intelligent Manufacturing*, 31(6): 1313–1337.

Lim, M. K., Xiong, W., Lei, Z. 2020a. Theory, supporting technology and application analysis of cloud manufacturing: a systematic and comprehensive literature review. *Industrial Management & Data Systems*, 120(8): 1585–1614.

Lindström, J., Löfstrand, M., Reed, S., Alzghoul, A. 2014. Use of cloud services in functional products: Availability implications. *Procedia CIRP*, 16: 368–372.

Liu, B., Zhang, Z. 2017. QoS-aware service composition for cloud manufacturing based on the optimal construction of synergistic elementary service groups. *The International Journal of Advanced Manufacturing Technology*, 88(9-12): 2757–2771.

Liu, Y., Wang, L., Wang, X. V. 2018. Cloud manufacturing: Latest advancements and future trends. *Procedia Manufacturing*, 25: 62–73.

Liu, Y., Wang, L., Wang, X. V., Xu, X., Jiang, P. 2019a. Cloud manufacturing: Key issues and future perspectives. *International Journal of Computer Integrated Manufacturing*, 32(9): 858–874.

Liu, Y., Wang, L., Wang, X. V., Xu, X., Zhang, L. 2019b. Scheduling in cloud manufacturing: State-of-the-art and research challenges. *International Journal of Production Research*, 57(15-16): 4854–4879.

Machado, C. G., Winroth, M. P., Ribeiro da Silva, E. H. D. 2020. Sustainable manufacturing in Industry 4.0: An emerging research agenda. *International Journal of Production Research*, 58(5): 1462–1484.

Malek, J., Desai, T. N. 2020. A systematic literature review to map literature focus of sustainable manufacturing. *Journal of Cleaner Production*, 256: 120345.

Menon, K., Kärkkäinen, H., Wuest, T., Gupta, J. P. 2019. Industrial internet platforms: A conceptual evaluation from a product lifecycle management perspective. *Proceedings of the Institution of Mechanical Engineers, Part B: Journal of Engineering Manufacture*, 233(5): 1390–1401.

Monostori, L., Kádár, B., Bauernhansl, T., Kondoh, S., Kumara, S., Reinhart, G., Sauer, O., Schuh, G, Sihn, W., Ueda, K. 2016. Cyber-physical systems in manufacturing. *CIRP Annals*, 65(2): 621–641.

Morgan, J., Halton, M., Qiao, Y., Breslin, J. G. 2021. Industry 4.0 smart reconfigurable manufacturing machines. *Journal of Manufacturing Systems*, 59: 481–506.

Mourtzis, D. 2022. The mass personalization of global networks. In *Design and Operation of Production Networks for Mass Personalization in the Era of Cloud Technology*: 79–116.

Pathak, A., Bhatt, M. G. 2021. Synergetic manufacturing systems anchored by cloud computing: A classified review of trends and perspective. *Materials Today: Proceedings*, 46: 212–216.

Pisching, M. A., Pessoa, M. A., Junqueira, F., dos Santos Filho, D. J., Miyagi, P. E. 2018. An architecture based on RAMI 4.0 to discover equipment to process operations required by products. *Computers & Industrial Engineering*, 125: 574–591.

Qi, Q., Tao, F. 2018. Digital twin and big data towards smart manufacturing and industry 4.0: 360 degree comparison. *IEEE Access*, 6: 3585–3593.

Rachuri, S., Foufou, S., Kemmerer, S., Rachuri, S. 2006. Analysis of Standards for Lifecycle Management of Systems for US Army: A preliminary investigation. US Department of Commerce. *National Institute of Standards and Technology*.

Rebitzer, G., Ekvall, T., Frischknecht, R., Hunkeler, D., Norris, G., Rydberg, T., Schimdt, W. P., Suh, S., Weidema, B. P. Pennington, D. W. 2004. Life cycle assessment: Part 1: Framework, goal and scope definition, inventory analysis, and applications. *Environment international*, 30(5): 701–720.

Ren, L., Zhang, L., Tao, F., Zhao, C., Chai, X., Zhao, X. 2015. Cloud manufacturing: From concept to practice. *Enterprise Information Systems*, 9(2): 186–209.

Rostamzadeh, R., Ghorabaee, M. K., Govindan, K., Esmaeili, A., Nobar, H. B. K. 2018. Evaluation of sustainable supply chain risk management using an integrated fuzzy TOPSIS- CRITIC approach. *Journal of Cleaner Production*, 175: 651–669.

Saaksvuori, A., Immonen, A. 2004. Integration of the PLM system with other applications. In A. Saaksvuori & A. Immonen (Eds.), *Product Lifecycle Management*: 57–73.

Santos, K. C. P. D., Freitas Rocha Loures, E. D., Junior, O. C., Santos, E. A. P. 2018. Product lifecycle management maturity models in Industry 4.0. In *IFIP International Conference on Product Lifecycle Management*: 659–669.

Schuh, G., Rozenfeld, H., Assmus, D., Zancul, E. 2008. Process oriented framework to support PLM implementation. *Computers in Industry*, 59(2): 210–218.

Schützer, K., de Andrade Bertazzi, J., Sallati, C., Anderl, R., Zancul, E. 2019. Contribution to the development of a Digital Twin based on product lifecycle to support the manufacturing process. *Procedia CIRP*, 84: 82–87.

Sharma, R., Jabbour, C. J. C., de Sousa Jabbour, A. B. L. 2020. Sustainable manufacturing and industry 4.0: What we know and what we don't. *Journal of Enterprise Information Management*, 34(1): 230–266.

Silventoinen, A., Pels, H. J., Kärkkäinen, H., Lampela, H. 2011. Towards future PLM maturity assessment dimensions. *Proceedings of PLM11 8th International Conference on Product Lifecycle Management*: 1–14.

Simeone, A., Zeng, Y., Caggiano, A. 2021. Intelligent decision-making support system for manufacturing solution recommendation in a cloud framework. *The International Journal of Advanced Manufacturing Technology*, 112(3): 1035–1050.

Singh, S., Misra, S. C. 2018. Identification of barriers to PLM institutionalization in large manufacturing organizations: A case study. *Business Process Management Journal*, 25(6): 1335–1356.

Singh, S., Misra, S. C., Kumar, S. 2019. Critical barriers to PLM institutionalization in manufacturing organizations. *IEEE Transactions on Engineering Management*, 68(5): 1436–1448.

Singh, S., Misra, S. C., Kumar, S. 2022. What does it take for your organization to institutionalize product lifecycle management? *IEEE Engineering Management Review*, 50(1):132–137.

Stark, J. 2011. Traditional pre-PLM environment. *In Product Lifecycle Management*: 295–309.

Sturm, L. D., Williams, C. B., Camelio, J. A., White, J., Parker, R. 2017. Cyber-physical vulnerabilities in additive manufacturing systems: A case study attack on the. STL file with human subjects. *Journal of Manufacturing Systems*, 44: 154–164.

Swarr, T. 2011. A capability framework for managing social and environmental concerns. *The International Journal of Life Cycle Assessment*, 16(7): 593–595.

Tao, F., Hu, Y., Zhao, D., Zhou, Z., Zhang, H., Lei, Z. 2009. Study on manufacturing grid resource service QoS modeling and evaluation. *The International Journal of Advanced Manufacturing Technology*, 41(9): 1034–1042.

Tao, F., Zhang, L., Venkatesh, V. C., Luo, Y., Cheng, Y. 2011. Cloud manufacturing: A computing and service-oriented manufacturing model. *Proceedings of the Institution of Mechanical Engineers, Part B: Journal of Engineering Manufacture*, 225(10): 1969–197.

Tao, F., Zhang, L., Liu, Y., Cheng, Y., Wang, L., Xu, X. 2015. Manufacturing service management in cloud manufacturing: Overview and future research directions. *Journal of Manufacturing Science and Engineering*, 137: 4.

Tao, F., Zhang, M., Liu, Y., Nee, A. Y. 2018. Digital twin driven prognostics and health management for complex equipment. *CIRP Annals*, 67(1): 169–172.

Terzi, S., Bouras, A., Dutta, D., Garetti, M., Kiritsis, D. 2010. Product lifecycle management—From its history to its new role. *International Journal of Product Lifecycle Management*, 4(4): 360.

Tomasic, I., Andersson, A., Funk, P. 2017. Mixed-effect models for the analysis and optimization of sheet-metal assembly processes. *IEEE Transactions on Industrial Informatics*, 13(5): 2194–2202.

Vadoudi, K., Troussier, N., Zhu, T. W. 2014. Toward sustainable manufacturing through PLM, GIS and LCA interaction. *2014 International Conference on Engineering, Technology and Innovation (ICE)*: 1–7.

Vila, C., Abellán-Nebot, J. V., Albiñana, J. C., Hernández, G. 2015. An Approach to Sustainable Product Lifecycle Management (Green PLM). *Procedia Engineering*, 132: 585–592.

Waller, M. A., Fawcett, S. E. 2013. Data science, predictive analytics, and big data: A revolution that will transform supply chain design and management. *Journal of Business Logistics*, 34(2): 77–84.

Wang, L., Guo, S., Li, X., Du, B., Xu, W. 2018. Distributed manufacturing resource selection strategy in cloud manufacturing. *The International Journal of Advanced Manufacturing Technology*, 94(9): 3375–3388.

Westkämper, E., Alting, L. Arndt, G. 2000. Life Cycle Management and Assessment: Approaches and Visions Towards Sustainable Manufacturing (keynote paper). *CIRP Annals*, 49(2): 501–526.

Wu, D., Jennings, C., Terpenny, J., Gao, R. X., Kumara, S. 2017. A Comparative Study on Machine Learning Algorithms for Smart Manufacturing: Tool Wear Prediction Using Random Forests. *Journal of Manufacturing Science and Engineering*, 139(7): 071018 (9 pages).

Wuest, T., Weimer, D., Irgens, C., Thoben, K. D. 2016. Machine learning in manufacturing: Advantages, challenges, and applications. *Production & Manufacturing Research*, 4(1): 23–45.

Xu, X. 2012. From cloud computing to cloud manufacturing. *Robotics and Computer-Integrated Manufacturing*, 28(1): 75–86.

Yang, C., Lan, S., Shen, W., Huang, G. Q., Wang, X., Lin, T. 2017. Towards product customization and personalization in IoT-enabled cloud manufacturing. *Cluster Computing*, 20(2): 1717–1730.

Zhang, L., Luo, Y., Tao, F., Li, B. H., Ren, L., Zhang, X., Guo, H., Cheng, Y., Hu, A., Liu, Y. 2014. Cloud manufacturing: A new manufacturing paradigm. *Enterprise Information Systems*, 8(2): 167–187.

Zhang, Y., Ren, S., Liu, Y., Si, S. 2017. A big data analytics architecture for cleaner manufacturing and maintenance processes of complex products. *Journal of Cleaner Production*, 142: 626–641.

Zhong, R. Y., Lan, S., Xu, C., Dai, Q., Huang, G. Q. 2016. Visualization of RFID-enabled shopfloor logistics big data in cloud manufacturing. *The International Journal of Advanced Manufacturing Technology*, 84(1-4): 5–16.

Zhong, R. Y., Xu, X., Klotz, E., Newman, S. T. 2017. Intelligent Manufacturing in the Context of Industry 4.0: A Review. *Engineering*, 3(5): 616–630.

Zhou, J. 2015. Intelligent mannfacturing-main direction of Made in China 2025. *China Mechanical Engineering*, 26(17): 2273.

Zhou, J., Yao, X. 2017a. Hybrid teaching–learning-based optimization of correlation-aware service composition in cloud manufacturing. *The International Journal of Advanced Manufacturing Technology*, 91(9): 3515–3533.

Zhou, J., Yao, X. 2017b. A hybrid approach combining modified artificial bee colony and cuckoo search algorithms for multi-objective cloud manufacturing service composition. *International Journal of Production Research*, 55(16): 4765–4784.

Part III

Operations in Industry 4.0

13 Material Supply and Logistics 4.0

Taiba Zahid
NUST Business School, National University of Sciences and Technology, Islamabad, Pakistan

Mathias Kuhn and Thorsten Schmidt
Technische Universitat Dresden, Dresden, Germany

13.1 INTRODUCTION

It is argued that the heart of digitalization and the Industry 4.0 era is its contribution to production planning, material supply, and logistics systems. Apart from automation in product manufacturing, the Fourth Industrial Revolution demands decentralization and automation at both the production and logistics levels with robust decision support systems. This requirement is due to the digitization of production systems during the change in Industry 4.0. The path leads away from manual planning toward the use of computer systems, which are part of computer-aided designs. Computer-aided technologies take on more and more tasks in product design, manufacturing, quality assurance, and production planning and control systems (PPS). The user is thereby relieved or supported in their work. In recent years, the term "Industry 4.0" has increasingly established itself as a catchphrase for the increasing networking of different systems within a factory or a production system. According to the Federal Ministry of Education and Research in Germany, "Industry 4.0" is an umbrella term for the use of cyber-physical production systems (CPPS) and the networking of machines, employees, and processes, as well as their interactions with one another. This trend toward the increased use of cyber-physical systems (CPS) will continue in the context of globalization, which further creates the opportunity to better respond to the customer's individualization requirements and the increasing variety of products. Moreover, future challenges in the industrial environment require mastering many variants (Bischoff et al., 2015, p. 46). This trend in production goes hand in hand with shorter product life cycles and smaller batch sizes while at the same time increasing demands on logistical target fulfillment. A quick delivery time, in particular, plays a decisive role for the customer when placing an order or making a purchase decision. The shortening of the lead time for construction, work preparation, and prototype production of the products in demand lead to an increase in fuzzy, missing process parameters and disturbance variables, thereby increasing the requirements to order processing in PPS. Stochastically influenced process times are inherent in the process due to the fluctuating individual performance of people and will therefore continue to

DOI: 10.1201/9781003327523-16

exist in the future. Representative examples of the characteristics mentioned include the final assembly of customer-specific machine tools or printing machines, in which several projects with individual objectives also compete for resources, further increasing the problem's complexity. To handle such a situation, central manufacturing execution systems (MES) with integrated advanced planning systems (APS) and decentralized simple priority rules are currently used. MES with APS is only suitable to a limited extent since permanent replanning is necessary due to the imprecise process parameters and disturbance variables. In addition, the required database and feedback quality are often inadequate (Pessl et al., 2013). The dynamic production environment under consideration cannot predict the mode of action of simple priority rules on the logistical target values (Tay & Ho, 2008). Hence, an application leads to good results at best by chance. Because of the fuzzy process parameters and disturbance variables, permanent replanning is, therefore, necessary.

Under the Industry 4.0 vision, promising approaches are being researched to solve the PPS problem. Especially the complete autonomy through self-configuration and self-control of the PPS functionalities should enable the compensation of disruptive influences while adhering to or improving the logistical target values, thereby generating the demand for CPSs. The permanent automatic data acquisition and availability required for such systems and the implementation of a decentralized distributed intelligence is only partially feasible and will remain the subject of research and development for the next few years. On the one hand, the technological implementation and permissibility of a continuous digital recording of human work that are unclear, could be more clearer. On the other hand, interfaces for application programming of machines are not freely accessible.

In general, a material supply system, considering corporate target variables, should control and organize the (production) networks, including various sub-areas of a production system, such as shipping, sales, production, and assembly. It was initially planned in the 1970s, beginning with the MRP-I and MRP-II (Manufacturing Resource/Requirements Planning) concepts, which attempted to plan the resources for a company's production with increasing accuracy and cyclical processing (Wiendahl, 2014). Feedback loops were built in to make possible changes to the plan to ensure the consistency of the individual goals of the MRP-I / II concept. In practice, however, this consistency is rarely achieved, which means that the planning quality of existing concepts has been criticized (Wiendahl, 2014). Over the years, there have been further developments in the field of material supply and enterprise resource planning ERP systems, away from central planning and toward the use of decentralized systems. A PPS is divided into two stages, i.e., the planning of production processes before execution and control of the same during execution. In current PPS practices, special attention must be paid to the interface between a digital PPS model and real production. Malfunctions such as machine failure mean they are not so heavily burdened and can respond more robustly to changing production systems. Therefore, briefly, to solve the material supply and execution problem in the industries under the aspects of the development toward the vision of Industry 4.0, the central requirements for a semi-autonomous system are:

- The best possible compensation of disturbance variables and stochastically influenced process parameters regarding selected target variables through self-control at the order and resource level.
- High robustness of self-control with high logistical goal fulfillment at the same time.
- Short calculation times for the central configuration of the self-control itself, as well as when using this for decentralized sequence formation and resource allocation.
- Use of practical and easy-to-collect information in the production system as a data basis for the configuration and application of self-control.
- Processing and utilization of implicit knowledge from simulation and production through machine learning for the continuous improvement of self-control.
- Differentiated optimization of project-specific (e.g., high individual adherence to deadlines) and production system-side goals (e.g., increased utilization, low inventory) to enable customer and company-oriented process control.

Therefore, this chapter focuses on the material requirement, supply, and logistics systems currently practiced in the industry and the efforts being made for their transformation in the Fourth Industrial Revolution era.

13.2 DEVELOPMENTS: PRODUCTION PLANNING AND MATERIAL SUPPLY SYSTEMS

Production planning and material supply and control are the core of every modern industry. A well-functioning industry cannot exist without efficient chain management within the system. It is an essential building block to keep the company's three major goals (i.e., cost, time, and quality) in line. This whole supply and production control system is divided into two components: the planning and execution stage. Both components were designed on pre-established data and modules applicable in any industry. However, Industry 4.0 promotes a real-time interface between a digital system and real-time production. Figure 13.1 represents the steps

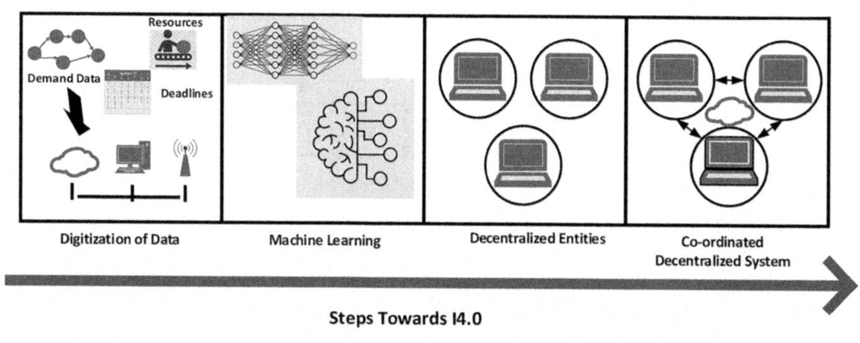

FIGURE 13.1 Industry 4.0 implementation on decentralized systems.

toward decentralized autonomous logistics systems. Moreover, it is challenging to distinguish precisely between the real system and the digital production image as it may vary from system to system. In the following paragraphs, the traditional PPS model from Aachen is intended to provide a more detailed insight into the functions and tasks of a supply and control system.

13.2.1 AACHEN MODEL

Research Institute for Rationalization in Aachen developed one of the best-known models for a PPS. It effectively supports users in practice in the selection and introduction, reorganization, development, and harmonization of the PPS system's concept processes. The main perspectives discussed in the model are:

- Technology (information technology)
- Human (human-oriented)
- Organization (business administration)

The model provides four reference views at various levels of manufacturing systems in an industry (see Figure 13.2). These four views form the basis for understanding and fundamentally developing a PPS concept. The Aachen model helps to see a holistic connection between the PPS. That is why the different views are linked to one another. Also, to get a better insight into the tasks of a PPS, the "task view" of the Aachen model can be used.

In the course of steadily advancing globalization, the development of new sales markets and goods, and the growing international competition between industrial companies, it has become necessary to expand the original task model to include network and cross-sectional tasks. The "network tasks" should summarize all planning and coordinating tasks that originate from the networking between individual

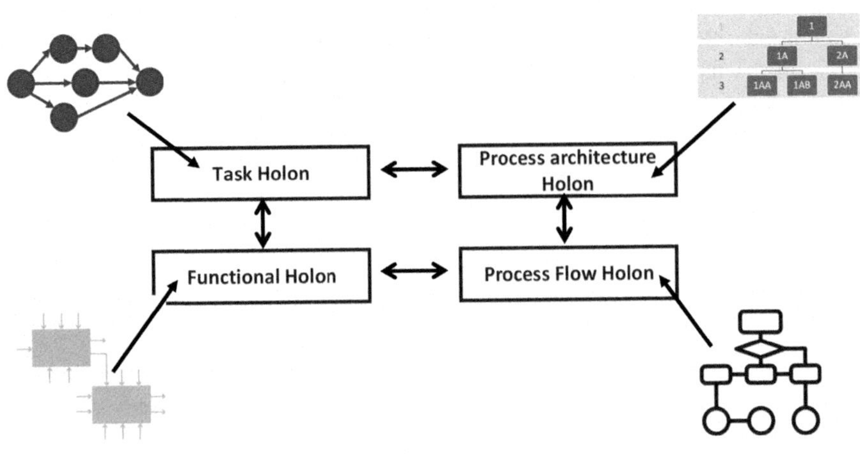

FIGURE 13.2 Aachen model reference views.

companies in a manufacturing network. The "cross-sectional tasks" contain connecting elements from the network and core tasks and assume a coordinating character.

13.3 OPERATIONS IN PPS

The first task in a continuous model is production scheduling. Depending on the manufacturing company, this can take a few weeks to several years and includes the planning or prognosis of primary products to be manufactured for a customer as required or in stock. To overcome this issue, sales and planning forecasting methods such as the moving average or exponential smoothing of higher orders are used to estimate a probable production quantity in the subsequent production periods. In addition, a net primary requirement is determined from existing customer orders compared to the stock levels. This step results in a preliminary production plan, which is included in the rough resource planning compared with the available resources, materials, employees, etc., and checked for feasibility.

The second task is production requirement planning, which is to be settled in the medium-term time horizon and can range from months to weeks or individual days, depending on the manufacturing company. Here the preliminary production plan from the production program planning is checked more closely for its feasibility. In doing so, e.g., utilizing ABC-XYZ analysis, gross secondary and below net secondary needs assessments are carried out to estimate which resources still have to be ordered and which are still in stock. It is also determined which resources can be procured in-house (e.g., semi-finished products) or through external procurement (e.g., highly frequented standard parts). Then the lead time scheduling for creating a network plan, which chronologically structures the various dependencies in the product structure with no capacity restrictions is considered here. As a result, essential dates for individual production orders are the output determined using forward or backward scheduling. Moreover, here it is checked whether an order from the status "today" is feasible and on schedule. After that, the determination of capacity requirements, such as machines, personnel, etc., are planned in detail and coordinated with the existing capacities. Should there be any discrepancies here, for example, due to over- or under-loading, adjustments must be made (e.g., overtime by employees or outsourcing of production orders).

This is further done for the individual work steps of an order, such as setting up, turning, grinding, and individual machines and employees. The operations planned for a capacity/resource in a planning period are in a queue. To decide which work steps are to be done next is usually determined using a sequence planning set and with the help of priority rules such as first-in-first-out (FIFO) or shortest processing time (SPT). The last step in a material execution system is order approval and subsequent monitoring. When the order is released, it is checked whether all planned capacities/resources are available. Suppose this is the case according to the order release procedure, e.g., load-oriented order release. Then, the materials are made available, and all necessary documents for the order are created, such as material slips, route cards, wage cards, etc.

13.4 MATERIAL SUPPLY AND REQUIREMENT PLANNING: BACKGROUND

According to Wiendahl (2014, p. 290), material requirements planning is defined as:

> The material requirements planning has the task of determining the material requirement according to type, quantity, and date (differentiated according to storage or requirement location) for each product unit. The term "material" is used to summarize the material types; raw material, material, semi-finished product, auxiliary material, operating material, part, and group.

Material requirements planning falls into the core tasks of a logistics network, especially production requirements planning. In the intralogistics area of production, the definitions for "material provision" cannot be found in the literature. This definition includes the statements by Lödding (2016, p. 305) as well, which signifies that the "order release" is the point that also triggers the material provision and starts a production order. The "material provision" is part of the order release in a production system assigned to the production control sub-area. Its task is to serve the production processes' individual secondary and tertiary requirements in a timely and quantitative manner.

Figure 13.3 shows the interface between the MES system and real production. This interface is intended to be a boundary between the digital and real parts of the production system. The material staging exceeds this limit since this is the first time that raw material must be transported from a warehouse to the corresponding work center for a production order. The transport system or the logistics employees receive this order from a corresponding production interface. In addition to the material provision, other digital/real interfaces, such as sequence planning or capacity control, can also be recorded. There are three classification features used for order approval in the industry. The first classification feature is a criterion based on which a decision is made about a release. Four characteristics are decisive here:

- No criterion/immediate order approval: An order is released immediately after its creation and, therefore, cannot directly influence the target values of production (inventory, lead time, etc.). It is impossible to differentiate

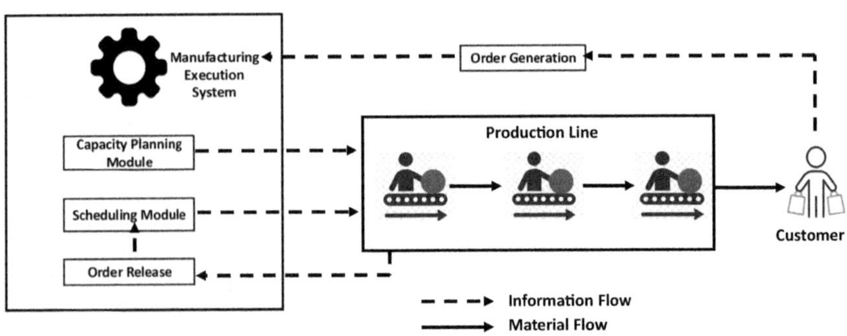

FIGURE 13.3 Traditional "PUSH" manufacturing execution system.

between urgent and non-urgent orders, and it is primarily used in make-to-stock (MTS) production.
- Plan start date: There should be an approval when the planned start date of the order is reached. This is very widespread in most PPS systems.
- Inventory of the production or the work system: The actual issue determines the actual receipt, and release happens when the production stock falls below a plan value, i.e., stock limit.
- Load on the work systems: This refers to the control of a load of individual work systems (e.g., machines) via the order release, delay of orders, which run through overloaded work systems, premature release of orders that run through underloaded work systems, and a special form of the order release regulating the order.

Another classification feature is the level of detail in which the order is released, such as a low level of detail (entire order is released) or a high level of detail (individual work processes of an order are released). For instance, a central inventory control system such as CONWIP (constant work in process) or a decentralized inventory control (for individual work systems/machines) like POLCA (paired-cell overlapping loops of cards) uses different levels of order details. Central inventory control is required with a low level of detail, and decentralized inventory control is governed with a high level of detail. Hybrid inventory control can do both, i.e., the release of entire orders and the release of individual work processes of an order.

The last classification feature of the order release is the trigger logic, which is between the periodic order approval and an event-oriented order release. The periodic order release decides on the release of new orders at fixed times. No orders are released between the times, and the large fluctuations in inventory in production are a disadvantage, as several orders are usually released. The low cost of implementation is advantageous as the event-oriented order release decides after the occurrence of certain events whether a new order should be released. Possible events can be the generation of an order, reaching the planned start date, or other digital signals from a production control.

13.5 ESTABLISHED APPROACHES IN PPS

In a scientific article by Bertolini et al. (2015), a comparison of two hybrid PPS systems was undertaken and compared with a classic "push" system. A "hybrid system" combines a "push" and "pull" control. As examples of hybrid production planning and control (HPPC), strategies such as CONWIP, Workload Control, POLCA, and other similar ones are listed in the literature and discussed in later sections.

13.5.1 CONWIP

The CONWIP production system was introduced in the 1990s (Framinan et al. 2003). The basic idea behind CONWIP is to keep the inventory of a production line or work system constant. An order for a production line or a workstation is released when the stock has fallen below a limit. Such as scenario is usually communicated

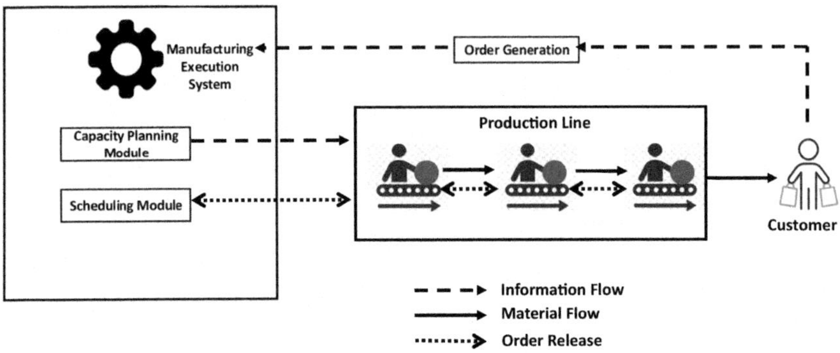

FIGURE 13.4 Constant work in process (CONWIP).

through CONWIP cards wherein the job with the highest priority in the job list is released. Here, the control parameters of this procedure are the number of CONWIP cards, and they control the stock of the production lines and the length of the advance horizon, which defines the maximum period of time with which an order can be released before its actual start date. The constant work-in-process (WIP) for a CONWIP production system is shown in Figure 13.4.

13.5.2 deCONWIP – Decentralized CONWIP

The deCONWIP control is formally like the CONWIP control already presented in Section 13.5.1. Orders in deCONWIP are generated via a central PPS system, and approval is given for each work process. Here, one workstation queries the inventory of the subsequent workstation. Suppose the inventory is below a certain limit, a message is given against the flow of material, and the processing of a work process is released.

13.5.3 LOOR Approach

The LOOR strategy, also known in German as "load-oriented order release," was first introduced by Bechte (1984) and can be operated periodically and via events. According to LOOR, an order is only released when it has been ensured that no stock limit has been exceeded on any subsequent workstation to process the order. An inventory account is assigned to each workstation in this case. When an order is released, it is posted with its order time for the first workstation. For all subsequent work systems of the order, the respective processing time of the work system is booked into the inventory account. The LOOR strategy is depicted in Figure 13.5.

13.6 HYBRID MATERIAL, RESOURCE, AND PRODUCTION PLANNING TECHNIQUES

In literature, two major technologies are used for material and resource planning. One is the simulation-based optimization (SBO) or service-oriented architecture (SoA), and the second is the use of agent systems.

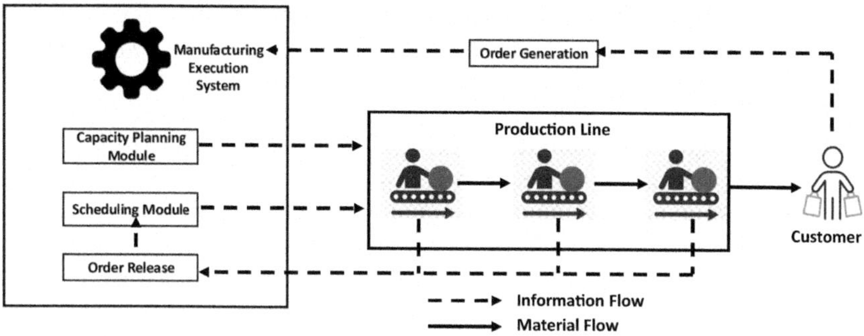

FIGURE 13.5 Load-oriented order release system (LOOR).

13.6.1 Simulation-Based Methodologies

After the work of Kück et al. (2016), a series of publications about production planning and control using SBO technology emerged. Simulation platforms describe a concept for the data-driven method of an SBO, which allows the simulation model to change during the simulation. With the use of heuristic methods, parts of the production plan, which are no longer applicable due to changes in resource availability or order fluctuations, are recalculated, taking real-time information into account. A data exchange platform is used to transfer planning decisions from the SBO framework back to an MES. The approach has been extended to a small supply chain, which includes a supplier, a production facility with processes and machines, and an end customer. The production time of the machines is given with a triangular distribution. The core of the SBO lies in the exchange between a control algorithm (e.g., genetic algorithm, swarm optimization, simulated annealing, etc.) and the supply chain simulation model. The control algorithm generates solutions for variables such as safety stocks or order quantities and allows them to be checked and evaluated by the simulation model. This procedure is repeated iteratively until a stop criterion is reached. The ERP, MES, and production data acquisition (PDA) play an important role in the production supply chain. Orders from customers are sent to the ERP system, which forwards production orders to the MES. The MES serves as an interface between the SBO method and the manufacturing system. Optimization of the simulation model is triggered by an adaptation function that periodically and, based on events, causes the model to be adapted to real production. Moreover, a three-level structure is used to apply SBO at full scale, including the top "decisional level" followed by the "execution control level" and the "physical execution level." In addition, a shared data repository is used to store all information that the MES supplies centrally.

The workers should be offered an assistance system for the existing modules to work in real time. Similarly, data should also be generated through simulations that the system can use. The approach is based on CPS, which aims to transmit data automatically to a decentralized MES. This further allows simulations of the material flow to be run synchronized with the machines. Two datasets are required for the data exchange, which can map the resources, products, and processes and, at the same time, establish a connection between products and processes. In an advanced study,

Luo et al. (2017) proposed a synchronized production and logistic system (SPL) intended to release synergies in decision-making with execution and thus bring about an improvement in the overall performance of a supply system. SPL further applies, in particular, to joint decision-making and the simultaneous optimization of production plans, internal transport, and warehouse planning, which is implemented in a framework with layers. Moreover, production resources equipped with intelligent devices such as RFID tags or portable computers are used in the "environment layer." These resources send data in real-time to the "data service layer" that forwards these to the "decision layer." The subsequent hierarchy is also presented as a complex flexible manufacturing system (CFMS), which, in addition to the physical execution, is also to be simulated as a digital image of reality. A CFMS is a system that combines multi-purpose tools and a material transport system that can exchange parts to be transported between the CFMS. The current configuration of the CFMS in practice is used to build a model of reality, and a simulator evaluates the current listing of the CFMS, such as cycle times, degree of tool utilization, etc. If a similar system configuration has occurred in the past, outcomes can be realized via a knowledge base data module.

13.6.2 Agent-Based Systems

The use of Agent-based systems to control production processes was developed a few decades ago and is experiencing a renaissance again with the help of real-time data. Agents are autonomously acting software, which are characterized by their autonomy and robustness against errors in the system and the adaptation to dynamic systems. The cooperation of several agents is known as a multi-agent system (MAS). MAS is characterized by the decentralized, parallel execution of tasks and work in a dynamic system toward a specific goal. MES based on a holonic system aims to achieve improved coordination between centralized planning and decentralized control decisions in a product-centric production system. The aim is to ensure the production plan's stability while simultaneously reducing the throughput time by creating material buffers. These decisions should be simulated based on agents. At the tactical level of planning, decisions are made about the quantity of each product in a product family within the planning horizon. Moreover, these quantities have to be divided into different lots, and a decision is made later related to which batches will be produced.

13.6.3 Special Techniques: Use of Artificial Intelligence and Smart Systems

Research by Xu (2012) proposed the creation of a smart cloud-based production logistics synchronization system (S-CPLSS), which is service-oriented and has the idea of ubiquitous computing (ubiquity of computer-aided information processing) that solves production logistical problems. The S-CPLSS framework consists of five layers. The physical resource layer represents all the production resources required for the execution of transport tasks, e.g., department stores, forklifts, employees, and transport containers. Building on the physical resource layer, the physical resources are mapped into smart objects in the smart object layer. Here, the cloud manufacturing (CM) resource management layer is responsible for registering the resources in the cloud and their correct definition and virtualization. The cloud management core service layer provides important service functions for data management, task planning, process planning, etc. The cloud management application layer further represents the interface between the system and the user. Based on a model test in China (Xu, 2012), the application of S-CPLSS increased the throughput of the warehouse, decreased the working hours, and improved the utilization rate of the forklifts.

Azouz and Pierreval (2018) presented an intelligent smart Kanban system based on a neural network (NN) and SBO. The goal was to add or remove Kanban cards from the system, depending on the system status. Real-time information was used in this case, where each input layer neuron represented a system variable, such as the number of parts at each production station, the number of finished products, etc. In addition, the input layer included information about the previous ordering behavior of customers. An optimization module helped to determine the optimal weightings within the NN. The output layer, as a result, decided which production stations got new Kanban types, which had to be removed, or if nothing had to be done. The NN consistently achieved the best results in terms of production costs.

Therefore, from the preceding paragraphs, it can be concluded that the central requisite for the execution of Industry 4.0 in the material requirement, production planning, supply chain, and logistics requires intelligent and autonomous decision support systems. In the next section, decision support systems in the context of Logistics 4.0 are discussed.

13.7 DECISION SUPPORT SYSTEMS

To automate various workflows like inventory and bookkeeping, the first information systems were developed in the 1960s. Due to the information technology (IT) practitioners need to understand the management processes, most support systems need improvement. The term "decision support system (DSS)" first appeared in an article by Gorry and Scott Morton in 1971. After this early work, DSS's research area remained theoretical and experimental for more than a decade. These new systems included personal decision support systems, group support systems, negotiation support systems, intelligent DSS, executive information systems and business intelligence, and data warehouses and knowledge management-based DSS. Modern digitization pushes researchers to create support systems that are human interactive. Based on a well-known classification (Power, 2002), subsequent paragraphs provide brief details on types of DSSs.

13.7.1 Communication Driven

Communication-driven DSS places particular emphasis on collaboration, communication, and joint decisions. These systems allow decision-makers to coordinate their activities and share information. The software in communication-driven DSS must enable agents to communicate with each other, facilitate the exchange of information, promote collaboration in group coordination, and support group decision-making tasks. Technologies in communication-driven DSS include client-server and the web.

13.7.2 Data-Driven DSS

Data-driven DSS focuses on the manipulation and application of data available within the organization and external data to support decisions. These systems are mostly used by product/service suppliers, employees, and managers. The first step in operating such a system is accessing data using retrieval tools. A data-driven DSS is then implemented by manipulating the data with the help of special computer-aided tools. Computer-based databases with a "query system" are examples of data-driven DSS. Compared to communication-driven DSS, much newer literature on data-driven DSS focuses more on scheduling.

13.7.3 Document-Driven DSS

Document-driven DSS deal with the use of unstructured documents. Document-driven DSS are newer and more prevalent than other types of DSS. The primary documents used by these systems are oral documents (e.g., transcripts of conversations), written documents (e.g., written reports), and video documents (e.g., television advertisements). Most companies store large numbers of documents, most of which are not used by managers for decision-making. IT tries to present the information in the documents in a usable format to support decision-making processes. Document-driven DSS is set up via client/server systems or the web, and search engines are powerful tools to help decision-making in Document-Driven DSSs.

13.7.4 Knowledge-Driven DSS

Knowledge-driven DSS specializes in solving problems through the application of human-computer systems. It is commonly used to advise managers and contain all the subject matter knowledge, problem-understanding, and problem-solving skills needed to recommend appropriate actions to decision-makers.

13.7.5 Model-Driven DSS

Model-driven DSS deals with the application and handling of quantitative models. In general, mathematical models are used by model-driven DSS to support decisions. Statistical and analytical tools are used at the first level of implementing a model-driven DSS. What-if relationships and sensitivity analysis are also commonly used. Moreover, model-driven DSS uses the parameters and data (not very large databases) provided by the decision-makers to analyze the possible decisions and different options. The model-driven DSS is the most used in scheduling.

13.8 CENTRALIZED VS. DISTRIBUTED DSS IN SUPPLY CHAINS

A supply chain is collectively formed by numerous entities working in a specific sequence to deliver a final product to end users. A single lead controls the entire chain to reach desired objectives in centralized decision-making systems. However, this traditional approach has numerous limitations in the era of globalization. As skills, labor costs, raw materials, geography, natural resources, etc., vary, running an entire chain by a single organization is not desirable. Hence, multiple autonomous entities work together for an efficient supply chain leading to data privacy and security concerns. As a result, limited information is shared between the entities. Moreover, although the output of one entity affects the other, autonomous entities have their objectives and constraints. This necessity further leads to a distributed or decentralized decision-making system. In the era of Logistics 4.0, decentralized and well-coordinated systems are essential. Since material supply and logistics require real-time information and coordination between real-time data and execution systems, most research focuses on data and/or knowledge-driven DSSs. Some of the methodologies include agent-based systems and the use of simulation-based intelligence, which has been discussed in earlier sections. However, using a communication-driven system such as chatbots is still in its fancy and will be discussed in the next sections.

13.9 COMMUNICATION-DRIVEN SYSTEMS: CHATBOTS

Chatbots offer a way to implement a conversational application between a human and an application using a defined data source. Chatbots are often used in web-based customer service and customer service areas to answer general questions because they can simulate human communication and interact intelligently through a dialogue with a human. This type of interaction works best when the questions are related to a specific topic. The use of chatbots for material supply and production execution systems is still at a theoretical stage.

13.9.1 Classification

Chatbots can be classified based on various parameters: the knowledge domain, the service offered, the goals, the method of processing input and generating responses, human help, and the creation method. Classification based on knowledge domain considers the knowledge a chatbot has access to or the amount of data it has been trained on. The classification based on the service offered takes into account the emotional closeness of the chatbot to the user, the level of intimate interaction that takes place, and the task that the chatbot performs. Interpersonal chatbots are in the field of communication and offer services such as restaurant bookings, flight bookings, and frequently asked questions (FAQ) bots. Moreover, chatbots are not companions of the user but instead receive information and pass it on to the user. They can have a personality, can be friendly, and are likely to remember information about the user. The classification according to the goals considers the primary goal that chatbots should achieve. The next sections discuss the potential application of chatbots with reference to material supply and production planning systems.

13.10 POTENTIAL USE OF COMMUNICATION DRIVEN SYSTEMS IN LOGISTICS 4.0

13.10.1 Chatbots in MES

MES is designed to interact with people in factories to enable smoother production management. The purpose of MES is to support human decision-making in activities related to manufacturing operations management. MES achieves this by making production data accessible in real time. Chatbots can be used as conversational information systems for information filtering and processing. However, chatbots are an emerging technology whose application in companies still needs to be sufficiently explored. Chatbots use artificial intelligence (AI) or decision trees while combining chatbots with MES for human-machine interaction can further support the production workforce by promoting flexibility and creativity. MES can be augmented with a chatbot prediction system to improve user experience and support human decision-making. Since, in current times of manufacturing change, a strategy to improve customer responsiveness and reduce time to market is fundamental, examining order management practices proactively by studying previous inputs is a great advantage. Furthermore, MES supports control measures that are carried out in real-time. Live access to production status can improve workflows for the workforce. For this purpose, an MES system with additional intelligence can be developed in a manufacturing company. MES can be augmented with a chatbot prediction system to improve user experience, order management, and support human decision-making. In the industrial environment, a chatbot for MES can be made "intelligent" using machine learning techniques to learn from the users' repeated requests.

13.10.2 CPS and Chatbots

The current development in production aims at a socio-CPS in which people, machines, materials, technologies, and the environment work together. The processes in the companies are integrated into a platform called Industry 4.0. The production is more like a living organism, where the planned dynamics in the logistics and production elements of the system are exchanged by dynamically moving, fully adaptable autonomous agents. Industry 4.0, or CPPS, is a new concept gradually being adopted by manufacturing companies to increase overall efficiency and sustainability while addressing the need for highly customized products with shorter lifecycles and emerging product service systems. Thanks to advances in industrial automation, information and communication technologies (ICT), and control and management models, manufacturing systems and plants have evolved into much more active entities within the broader, intensely collaborative, and more intelligent production environment that characterizes the Industry 4.0 scenario.

13.10.3 Operator 4.0 and Chatbots

To address the issue of communication and achieve a balanced and symbiotic interaction between humans and machines, the concept of "Operator 4.0" was developed. Operator 4.0 aims to create a socially sustainable work environment in future factories, where intelligent and skilled operators will not only perform "collaborative work" with robots but also provide "machine-assisted work." Depending on requirements, this should develop toward "human-automation symbiosis work systems" with the help of human CPS, advanced human-machine interaction technologies, and adaptive automation. In addition, the increasing burden of cognitive tasks and the growing shortage of highly skilled workers require new intelligent interactions between Industry 4.0/CPPS and Operator 4.0. Voice-controlled chatbots give the user intuitive access to a wealth of information and knowledge. Also, humans will play an important role in the holonic manufacturing systems of the future.

A significant problem with emerging systems is predicting the behavior of such complex systems. In the future factories, most factories will be complex systems and employ agents. In such scenarios, it is challenging for people to complete tasks without information from the system. Current production assumes communication and coordination are more uncomplicated as the systems are predictable and manually operated. However, new emerging manufacturing systems will work with agents. Human positions do not entirely replace robotic systems; instead, humans carry out activities that work together with other elements of the system (agents) to achieve their common goal. Here, agents are kept emergent, which makes it difficult for people to work in such systems. However, it is necessary to find ways to fully implement the human in new production systems so that they can communicate with the other agents to increase the activities' effectiveness. Also, chatbot systems have proven to be able to handle communication between humans and different agents. They can be based on specific mechanisms and provide a clear presentation of information. Simultaneously, it is also imperative to find ways to fully implement the human into new production systems so that they can communicate with the other agents to increase the effectiveness of activities.

13.11 CASE STUDY I: DECENTRALIZED SUPPLY CHAIN SYSTEM

13.11.1 Introduction and Motivation

The study proposed by Ghasemi et al. (2017) represents an effort toward implementing Industry 4.0 concepts in the supply chain through decentralized decision-making. Managing a supply chain with an ability to adapt, be resilient, and be responsive requires coordination among all its players, from supplier to end user. Concepts of Industry 4.0 aim to develop a collaborative planning structure amongst suppliers, producers, and distributors. The challenge is to develop a structure that can benefit these autonomous players with their own target values and constraints. In the past, such a problem was solved with centralized decision-making for the complete supply chain. However, this is not possible due to data privacy between manufacturers and distributors. Only limited information regarding demand and supply is usually shared; hence, it is impossible to solve the problem with a centralized decision-making planning system optimally. With this motivation, the authors proposed a decentralized supply chain system to optimally solve material supply problems for both players (i.e., manufacturer and distributor).

13.11.2 Methodology

The approach was implemented in a hardboard industry with a multi-echelon and multi-product problem. After the producer purchases the raw material, hardboards can be prepared and stored in various sizes in a warehouse. The final products are sent to distributors, who are responsible for delivering them to end-users by their own means. As per the size of the factory and warehouses, the producer and distributors' capacity is known. Moreover, both players work independently and are responsible for their transportation and deliveries with a limited available capacity of transportation vehicles. Although the methodology was applied to a hardboard industry, it can be applied to any similar multi-echelon problem with a similar logistics network, such as paper, floor, marble, and other industries. Multi-integer mathematical problem was proposed for the problem, but instead of one centralized model, separate models were created for the manufacturer and distributor with their constraints. A general model included capacity constraints for all logistics network players to minimize the logistics network's overall costs. A coordination mechanism was developed where the information and data sharing were kept minimal between manufacturer and distributor, and where both players try to achieve optimal cost for themselves, disregarding the interests of the other. The coordination mechanism starts with distributor input, who generates a demand and delivery schedule for the manufacturer after analyzing its customer requirements. The manufacturer uses this quantity and prepares a production plan with minimal cost with optimization iterations. However, if it is impossible to generate a solution due to the capacity constraints of the manufacturer, the updated quantity and delivery time are informed to the distributor. If it is acceptable, the distributor produces a new plan and sends it back to the manufacturer. If not, then the problem becomes infeasible. The algorithm stops when there are no production shortages for any product, and both parties have developed a decentralized optimal plan with minimal information sharing. Figure 13.6 depicts the process flow for the coordination mechanism.

13.11.3 Conclusion

The results computed via GAMS software showed that compared to a traditional logistics system for a multi-echelon supply chain, a decentralized system with a

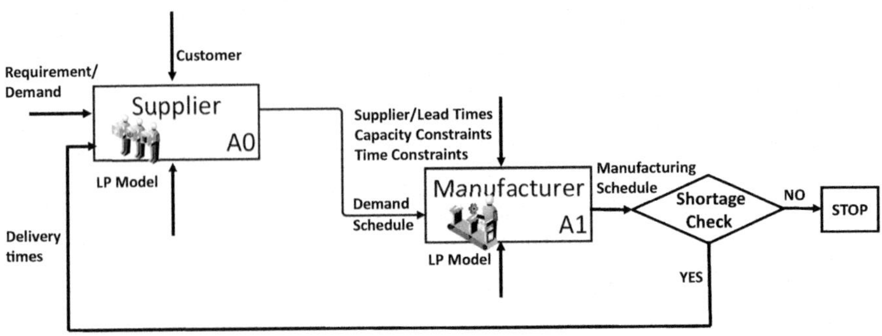

FIGURE 13.6 Coordination mechanism for a two-echelon decentralized supply chain.

coordination mechanism resulted in lower costs and computation time. It was observed that the worst results would be obtained if the manufacturing capability was increased without matching it with the storage capacity of the distributor. Sensitivity analysis was also carried out to observe various parameters in the system. On observing one of the major cost parameters, i.e., transportation costs, it was deduced that transportation costs at different links impact manufacturers and distributors differently. For instance, while increasing the transportation cost from raw material to the supplier, the manufacturer's total costs were increased compared to the distributor's cost. However, less variation was seen in the increase of transportation costs from manufacturer to distributor. The number of iterations used as a stopping criterion didn't significantly affect objective value, perhaps due to the size of the problem. The case study provides an interesting viewpoint of a decentralized planning and execution system for a supply chain built on Industry 4.0 principles.

13.12 CASE STUDY II: LOGISTICS 4.0

13.12.1 Introduction and Background

The case study was conducted by Hermann et al. (2019), which provides a framework to transform logistics processes to have potential advantages with the application of Industry 4.0. These potential advantages can resolve industrial challenges due to changing customer demands, technology, and competition. With the evolution of the Industry 4.0 concept, several theoretical research and case studies have been conducted on product design. However, the motivation behind this study was the need for sufficient research on business processes, material flow, and logistics management concerning Industry 4.0. The study focused mainly on discovering process design frameworks that can be applied to industrial transformation. Moreover, the framework was implied in the automobile industry and analyzed to support future practitioners in the design transformation process. The aim was to create a smart factory with CPS, such as intelligent machines, connected with the help of the internet of things (IoT) and internet of services (IoS). These CPS can create a digital twin of the physical structure. Furthermore, the smart factory makes use of these systems to perform decentralized distributed decision-making.

13.12.2 Findings

The authors used action design research to conduct the study, which is a step-wise transformation process theory. Before the transformation, the current process is analyzed thoroughly, and target values are defined, followed by detailed planning and execution. The case study was performed in the automobile industry. It started with the procurement and delivery of raw materials, which are then brought on a need basis on the assembly lines. Instead of delivering assembled cars, car manufacturers provide complete parts worldwide where cars are assembled, thereby reducing costs. After thoroughly analyzing the system of case study, the following findings were suggested by the authors:

- The information flow was very critical between the logistics chains. A well-communicated system will support decentralized decision-making for implementing Industry 4.0 and reduce overall costs. For instance, the arrival of material should be communicated earlier than its actual arrival. Based on this information, the independent inventory management system and shop floors should schedule their territories for a well-planned system.
- Although the list of intelligent automated systems may be endless, every system needs to be evaluated with its potential economic benefits before application.
- To transform traditional logistics to Logistics 4.0, it is necessary to define current processes in the form of goals and deliverables rather than deterministic sequential tasks. With such a definition, perception changes as it is tried to reach the target values by making independent decisions on task planning to improve goals.

REFERENCES

Azouz, N., Pierreval, H. 2018. Adaptive smart card-based pull control systems in context-aware manufacturing systems: Training a neural network through multi-objective simulation optimization. *Applied Soft Computing*, ISSN: 15684946.

Bechte, W. 1984. Control of throughput time through load-oriented order release in workshop production. *Zul.: Hanover, Univ., Faculty of Mechanical Engineering, dissertation, 1980. Düsseldorf: VDI-Verl. Progress reports of the VDI magazine series*, ISBN 3181470023

Bertolini, M., Romagnoli, G., Zammori, F. 2015. Simulation of two hybrid production planning and control systems: A comparative analysis. *International Conference on Industrial Engineering and Systems Management (IESM)*: 388–397.

Bischoff, J, Hegmanns, T., Henke, M., Hompel, M., Döbbeler, F., Fuss, E. 2015. Developing the potential of using 'Industry 4.0' in medium-sized companies. *Mühlheim*.

Framinan, J.M., Gonzalez, F. M., Ruiz-Usano, R. 2003. The CONWIP production control system: Review and research issues. *Production Planning & Control*, 14(3): 255–265.

Ghasemi, P., Damghani, K. K., Hafezolkotob, A., Raisii, S. 2017. A decentralized supply chain planning model: a case study of hard board industry. *International Journal of Advanced Manufacturing Technology*, 93: 3813–3836.

Hermann, M., Bucker, I., Otto, B. 2019. Industrie 4.0 process transformation: Findings from a case study in automotive logistics. *Journal of Manufacturing Technology Management*, 31(5): 935–953.

Kück, M., Ehm, J., Freitag, M., Frazzon E.M., Pimentel R. 2016. A data driven simulation-based optimisation approach for adaptive scheduling and control of dynamic manufacturing systems. *Advanced Materials Research*, 1140: 449–456.

Lödding, H., 2016. *Process of Production Control*. Berlin, Heidelberg: Springer Berlin Heidelberg. ISBN 978-3-662-48458-6.

Luo, H., Wang, K., Kong, X. T. R., Lu, S., Qu, T. 2017. Synchronized production and logistics via ubiquitous computing technology. *Robotics and Computer Integrated Manufacturing*, 45: 99–115.

Pessl, E., Gabriel, M., Hanusch, S., Neumann, C., Ortner, W., Ropin, H. 2013. Digital production. Study on the status, obstacles and requirements of Austrian small and medium-sized manufacturing companies and analysis of the software manufacturers of MES systems. Kapfenberg. *Institute of Industrial Management/Industrial Economics*, FH JOANNEUM.

Power, D. J. 2002. *Decision support systems: Concepts and resources for managers.* Greenwood Publishing Group, *ISBN: 1-56720-497-X.*

Tay, J. C., Ho, N. B. 2008. Evolving dispatching rules using genetic programming for solving multi-objective flexible job-shop problems. *Computers & Industrial Engineering*, 54(3): 453–473.

Wiendahl, H.P. 2014. *Business organization for engineers. Production planning and control (PPS).* Carl Hanser Verlag GmbH & Co. KG. *ISBN 978-3-446-44053-1.*

Xu, X. 2012. From cloud computing to cloud manufacturing. *Robotics and Computer Integrated Manufacturing*, 28(1): 75–86.

14 Workforce Engagement Within Decentrally Controlled Production Systems

Julia Schwemmer and Thorsten Schmidt
Technische Universitat Dresden, Dresden, Germany

14.1 INTRODUCTION: MOTIVATION FOR THE INVESTIGATION

The Fourth Industrial Revolution concept was introduced at the Hannover Fair in 2011 by Kagermann, Lukas, and Wahlster (Kagermann et al., 2011). Supported by the German government since 2012 (German Federal Ministry of Education and Research, 2017), it involves far-reaching production changes, which entail technological, systematic, and organizational modifications. Consequently, new challenges arise to achieve the desired flexibility of production systems based on self-controlling resource allocation and sequencing. The awakening demand for innovative systems and solution approaches covers the technical areas and the inquisitiveness for new planning and control mechanisms. Even the rostering and scheduling procedures of personnel do not remain unaffected by these innovations, especially regarding the workforce directly involved in production. The role of humans will change in cyber-physical production systems (CPPS), but they will remain a critical factor in a directing and controlling role (Ganschar et al., 2013). Nevertheless, human labor, in general, is a large cost factor for companies. Therefore, the worker is part of the optimization investigations in production systems (Van den Bergh et al., 2013), particularly in high-wage countries. Mainly economic considerations trigger the dynamic, growing research area of personnel rostering and scheduling (Van den Bergh et al., 2013). An efficient workforce roster results in the lowest possible labor costs while all requirements of production are covered. Thus, the aim is to match capacity supply and demand perfectly.

Usually, the demand for workers is derived from the baseline production schedule at the first step of creating a workforce roster. Based on this knowledge, and therefore on a subordinated optimization level, the available employees are mainly assigned to rigid shifts to ensure the necessary capacity. Considering CPPS, primarily decentrally controlled production systems (DCPSs), this classic approach to workforce planning is no longer appropriate. Due to the changing world of work in Industry 4.0 and the missing baseline schedules in DCPS, the attendance roster

 DOI: 10.1201/9781003327523-17

for workers will be challenging (Schwemmer et al., 2022, 2020), as described in Section 14.2 in detail.

For a comprehensive investigation of this topic, employees' perspectives must be included in the considerations. Employee-centered approaches have gained importance, such as higher self-determination of individual working times for a better work-life balance (German Federal Institute for Occupational Safety and Health, 2019). This aspect is discussed in Section 14.3, together with the advantages of time-flexible working conditions. Section 14.4 further deals with how to model and solve this fundamental conflict. Finally, the chapter ends with a summary in Section 14.5.

14.2 DILEMMA OF WORKFORCE ROSTERING AND SCHEDULING IN INDUSTRY 4.0

Industry 4.0 results in a challenge for workforce rostering that is manifested in time conflict of a planning and scheduling dilemma (Schwemmer et al., 2022, 2020), as summarized in Figure 14.1.

Since decentralized production control (DPC) is one of the core parts within the paradigm of Industry 4.0, short reaction time to individual requests of customers and unforeseen process disturbances ensure a demand-oriented production with a high degree of flexibility within DCPS. Due to the DPC, handling customer-specific orders, small batches, and related process disturbances are faced with minimal planning effort. The autonomous communication of resources and orders enables real-time control with decisions made decentrally and distributed at the lowest shop floor level. The decisions are made as they are due. On the shop-floor level, there are no long lead times for decisions on allocating resources and the sequence of operations. Consequently, there is no (detailed) baseline production schedule any longer, and neither is there any exact information available in advance about

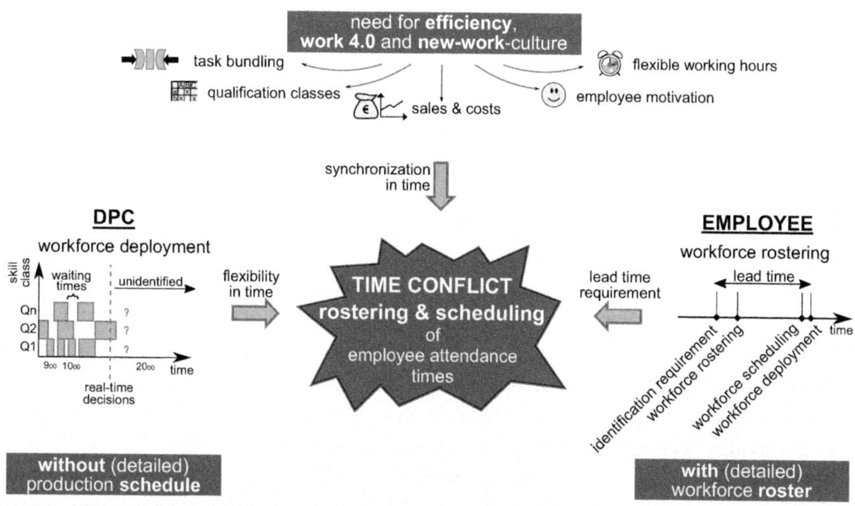

FIGURE 14.1 Dilemma of workforce rostering and scheduling in DCPS.

scheduling decisions, such as which operation on which machine and at which time. Such a scenario leads to severe implications for the workforce roster and the workforce scheduling.

Moreover, in future CPPS scenarios, the workforce will still play a significant role (Neumann et al., 2021; Pinzone et al., 2020; Ganschar et al., 2013). However, since humans are not machines, they cannot be scheduled in the same way. Also, they are not permanently part of the production system, thereby requiring to be scheduled and provided individually with sufficient lead-time via a staff roster. Therefore, for the personnel planning process, many specific regulatory rules need to be observed, such as regulations of a country, collective labor agreements, company guidelines, and employment contracts. In general, these regulations protect the employees and ensure fair working conditions. Together with administrative and individual organizational lead times, the rostering lead process consumes several days and even weeks (Bauer, 2015).

Due to unavailable baseline production schedules, the required operating times of the employee and, therefore, the concrete required attendance times are not known in advance. Based on the baseline schedule, the synchronization of requirement and supply is no longer ensured to project an efficient workforce roster. Consequently, the 'classic' rostering does not work anymore, and the workforce roster cannot be derived based on the necessary information from the baseline schedule (Schwemmer et al., 2022, 2020).

Two further aspects intensify the conflict between DPC and workforce rostering (see Figure 14.1): the changing nature of production workers' tasks and changing work attitude of the employees. The former refers to the technological changes of smart factories, which come hand in hand with the modification of manual activities and tasks of humans. Even if these change, they will only be eliminated partially (Ganschar et al., 2013; Pinzone et al., 2020; Neumann et al., 2021). The deserted factory is a utopian vision not to be striven for (Ganschar et al., 2013). Generally, there still will be human operators in the production systems (Ganschar et al., 2013; Pinzone et al., 2020; Neumann et al., 2021). In large part of future scenarios, the human will particularly assume a coordinating, controlling, and directing role (Ganschar et al., 2013; Pinzone et al., 2020), which will be the focus of this chapter. Increasing automation will replace simple activities that are usually characterized by a high level of working time continuity and reduce the continuity of operating times and the duration of work assignments. Due to the high-tech production system, the tasks of humans will become more complex and require higher qualifications, further leading to an increased workforce specialization (Ganschar et al., 2013). The consequences of such a future work scenario are a high level of expertise required, high specialization with several specific qualification classes, and less consistent periods of working time per qualification class (Schwemmer et al., 2022, 2020). For instance, a real-world application field, with regard to the changing manual tasks, is the production of medical implants (examples of manual tasks are visual inspection, insertion, removal of parts, performing non-standard production steps, etc.).

The latter aspect, intensifying the conflict, is the competition for highly qualified and motivated employees in combination with the increasing trend of employee-centric working conditions, which is discussed in the subsequent section in more detail.

14.3 WORK 4.0, TIME-FLEXIBLE WORKING MODELS, AND INDIVIDUAL WORKING HOURS

In office workplaces, the possibility of adapting working hours to individual needs has existed for years. However, in production workplaces, rigid shift regimes still prevail. From a production planning perspective, this is very convenient. These strict attendance times ensure continuous coverage of the plant capacity during the daily production routine.

Moreover, in the above-described context of the changing world of production, a constant supply of employees may work but will no longer be appropriate or, to be more precise, efficient. The less consistent periods of working time per qualification class would result in frequent and possibly long idle times for the workers. Given some employees' high degree of specialization and associated labor costs, these less consistent periods should not be the goal of a workforce management system. It should provide workers exactly when the production system requires them—which, however, requires a certain flexibility of the workforce resource regarding corporate concerns. This time-related synchronization of labor with actual demand is not a new consideration in the research environment. Near the year 2000, research approaches were striving for synchronization within the production system using flexible working time organization (e.g., see Hung, 1999; Grabot & Letouzey, 2000). However, the above notion does not seem to have been established yet. It may be that the changing requirements from Industry 4.0 will more strongly highlight the advantages of time-flexible working within production for companies and push the development of these methods for implementation.

Furthermore, individual workers usually do not prefer rigid shift times (German Federal Institute for Occupational Safety and Health, 2019) and employee-centered working conditions are gaining importance (Eurofound, 2016) more than payment or career opportunities (German Federal Institute for Occupational Safety and Health, 2019).

For example, a German study in 2017 surveyed 8,700 employees on the importance of working time conditions. Two-thirds of the employees considered it essential to influence the start and end of their working times (Brauner et al., 2018). In the year 2007, in a similar study with about 3,000 employees aged between 25 and 39, 65% (without kids) and 92% (with kids) considered work-life balance more important than the salary (Stutzer, 2012). Also, 59% of employees without kids and 78% with kids were willing to change their employer for a better work-life balance (Stutzer, 2012).

Buzzwords 'Work 4.0' (German Federal Ministry of Labour and Social Affairs, 2017; German Federal Institute for Occupational Safety and Health, 2019) and 'New Work' (Bergmann, 2019) are the terms that are mainly connected with this trend. The keyword 'Work 4.0' not only includes the changes in everyday working life brought by the new technologies (e.g., digitalization with a higher degree of integration and cooperation) but also discusses the associated opportunity of flexible work arrangements regarding time and location (German Federal Ministry of Labour and Social Affairs, 2017; German Federal Institute for Occupational Safety and Health, 2019). Moreover, the keyword 'New Work' (Bergmann, 2019) focuses primarily on self-fulfillment, which is mainly based on a high sense of purpose and autonomy in own work and freedom for creativity and self-determination to gain personal development.

In the context of the discussed issue, flexible working time arrangements based on the principle of work-life balance (WLB) stand out in particular. WLB is the compatibility of time in terms of the working day with private and family interests. For this purpose, it is a precondition for production employees to communicate preferred working time ranges in advance that are included in the workforce planning processes or to allocate working time operationally and flexibly just in time, similar to flextime for office workers.

In summary, flexible working hours and individual workforce coordination within DCPS offer several advantages to both sides. Figure 14.2 illustrates some benefits of time-flexible working arrangements.

One of the advantages of time-flexible working arrangements for the company is the higher attractiveness as an employer. Such arrangement plays a strategic role, especially in an emerging or existing skilled worker shortage. Under these conditions, from the employer's point of view, a competitive situation for well-trained and motivated employees arises, which is an aspect not only in terms of recruiting but also in terms of employee retention. In countries such as Germany, signs of a shortage of skilled workers are currently emerging: For example, in the German STEM (science, technology, engineering, and mathematics) area in April 2022, there were around 499,600 listed vacancies compared to approximately seven million employees in this area (Anger et al., 2022). Suppose this number is reduced by job seekers (regardless of any possible mismatching attributes such as specific qualifications within the area). In that case, there are about 320,600 vacancies, which corresponds to around 2.7 times more than in 2014, i.e., about 117,300 (Anger et al., 2022).

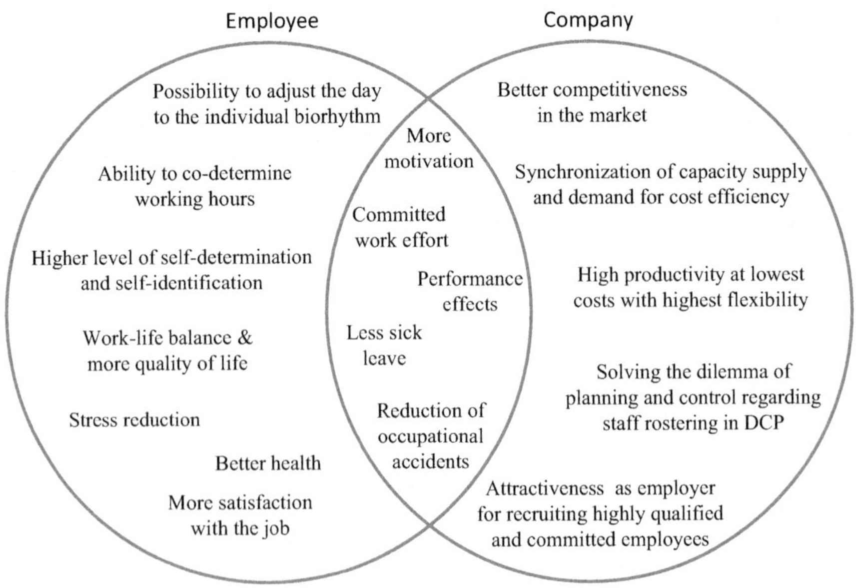

FIGURE 14.2 Advantages of time-flexible working.

From the employee's point of view, an advantage of time-flexible working hours is that a rhythm of working hours according to the own biorhythm is possible, or at least easier to realize. Humans have different biorhythms and can be divided into chronotypes according to their natural sleeping times (Wieden, 2016). Within this classification, a 'lark' is a person who usually wakes up early and goes to bed early. An 'owl' is the opposite. For example, at 6:00 am, larks may already be in a natural phase of high creativity (with high productivity at work). At the same time, an owl would still be in a sleeping phase without external compulsion (with low productivity at work) (Wieden, 2016). A study in five organization for economic cooperation and development (OECD) countries from 2016 shows that

> workers who sleep less than six hours per day report on average about a 2.4 percentage point higher productivity loss [...] than workers sleeping between seven to nine hours per day. [...] To put these numbers into perspective, assuming 250 working days in a given year, a worker sleeping less than six hours loses around six working days [...] more than a worker sleeping seven to nine hours per year.
>
> (Hafner et al., 2016, p. x)

In connection with sleeping habits, it is to emphasize that not only predetermined working hours but also other factors such as social coexistence, societal conventions, unhealthy lifestyle, physical activity, stress, anxiety, concerns, etc., can have a strong positive or negative effect concerning the circadian rhythm (Wieden, 2016; Hafner et al., 2016). However, the disruption of an individual's natural daily routine—for example, due to not-influenceable shift working times—can contribute to the loss of labor productivity and lead to illness and discomfort (Wong et al., 2019).

This chapter also relates to legal framework conditions regarding flexible working hours and working models. The aim of these laws is the health protection of the employees. Frequently too long or 'wrong' (e.g., nightly) working hours can lead to physical and mental harm (German Federal Institute for Occupational Safety and Health, 2019; Austrian Society for Occupational Medicine, 2019; Wong et al., 2019). Also, working long hours does not mean working efficiently. Performance and attention decrease with the duration of working time. Fatigue increases, and the risk of accidents increases, too (German Federal Institute for Occupational Safety and Health, 2019; Austrian Society for Occupational Medicine, 2019). The above effect can be seen across different occupational groups (Austrian Society for Occupational Medicine, 2019). The risk of an accident grows significantly after the seventh hour of work. After the ninth hour of work, it increases gravely, and after the twelfth hour of work, the risk of accident is approximately 80%–100% higher than during a normal eight-hour-working-day (German Federal Institute for Occupational Safety and Health, 2019; Austrian Society for Occupational Medicine, 2019). As a result, the risk of accidents increases, and individual productivity per hour decreases when working more than 40 hours per week. Working (time) laws specify a range within which employers, labor unions, or individual employees can negotiate. For example, there are working time limitations, minimum requirements for breaks, or compensatory times to be observed (see EU-Working Time Directive

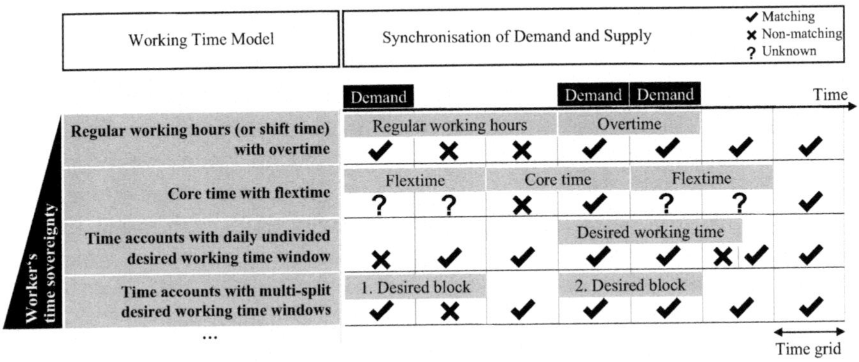

FIGURE 14.3 Time sovereignty in various working time models.

2003/88/EC[1] with the subordinate laws for the single EU-countries[2] or the United States labor law[3]).

Generally, the applicable working models are only as flexible as the legal framework allows. Due to the changing world of work, such a development can lead to new flexible working time models and will benefit all sides. The working time model significantly influences workers' working time sovereignty (see Figure 14.3). However, the employee must also be able to deal with the freedoms offered by working time sovereignty and bring along a certain degree of self-organization. Furthermore, time sovereignty should also fit the company's needs, like work organization and organizational culture. Figure 14.3 displays different working time models according to their timely co-determination possibilities of the workers and the synchronization of demands and supplies. Moreover, suppose the company needs to apply shift models, and there are no eight-hour shifts to plan; it is easier to integrate employees working in part-time models, as they may have a shorter daily working time and commonly do not match the rigid shift times.

It is essential to note that the desired time flexibility of the employee and the employer can be different. The production system likely requires some workers even when no worker would like to work at that particular moment (or the other way around). For example, a 24-hour running production system demands a manual work task between 3:00 am and 4:00 am. At this time, it is improbable to achieve coverage with the preferred working hours of an employee. Especially in this regard, it is essential to note that one-sided company-related flexibility requirements without respect to worker's needs and wishes leads to less satisfaction and motivation and more stress and health complaints amongst employees (German Federal Institute for Occupational Safety and Health, 2019). Also, globalization, digitalization, demographic development, and changing values of society have led to an increase in employees with atypical working hours (7:00 pm – 7:00 am or weekend) (German Federal Institute for Occupational Safety and Health, 2019). By considering flexible working hours within Industry 4.0, the issue is about finding a satisfactory compromise for both sides. However, the arrangement of such a compromise would still be the subject of further investigation.

14.4 PROBLEM FORMULATION AND CASE STUDIES

14.4.1 Problem Description and Modelling

Several requirements can be derived from the problem scenario described at the beginning, which must be included in modelling. In Schwemmer et al. (2022), the authors have elaborated on the most important model attributes qualitatively.

14.4.1.1 Problem Classification

Problems are categorized to facilitate the recognition and allocation of research endeavors. Currently, there is no commonly known problem classification determining the planning task described (except for the qualitative descriptions, e.g., in Schwemmer et al. (2022)). The problem scenario is primarily about machine-dominated, highly automated production systems. Accordingly, classifications with many machines are closest, such as job shop scheduling problems (JSSP) or line balancing problems (LBP). However, these existing problem classes do not include (medium-term) workforce rostering. On the other hand, simple employee-timetabling problems (e.g., general employee scheduling problem [GES] or minimization personal task scheduling problem [SMPTSP]) do not capture the complexity of task assignment regarding machines and the short work periods of humans in DCPS.

14.4.1.2 Decentralized Production Control

DPC is a centerpiece when considering the dilemma of workforce rostering in smart factories, as described in Section 14.2. There are no known baseline schedules in advance. DPC can make decisions flexibly, on the lowest shop floor level, as they arise. Therefore, the exact demand times of workers have yet to be discovered. However, classical rostering approaches need this information and are no longer appropriate.

14.4.1.3 Dual-Resource Constraint

There should be an overall view of the discussed scheduling and rostering problem. Humans and machines should be assigned to the shop tasks with the same optimization/priority level. The term is borrowed from a subclassification of the JSSP, in which the capacities of both types of resources are constrained, as they are usually in reality.

14.4.1.4 Short Discrete Worker Requirement Operation

As explained, increasing automation eliminates the need for many manual continuous work tasks. In the problem formulation, staff demand times can be modeled in terms of concise tasks that arise mainly discretely in time. The manual tasks can be shorter than related machine operations, which is the main difference from the Dual-Resource Constraint (DRC) JSSP-class, where machine tasks and the corresponding manual task are the same length. The assignment of a worker to a machine occurs at short selective intervals. Examples are collaborative subtasks, interim inspections, troubleshooting, or inserting and removing parts.

14.4.1.5 Individual, Heterogeneous Qualification Profiles

As described, the future production systems under consideration will be highly complex CPPS. The associated personnel are well-trained, highly specialized, and differ in their qualifications. Each employee can have their own training or knowledge, but the workers' skillsets can also coincide. Therefore, skillsets can contain single or multiple qualifications since, usually, only some workers are able to execute every task.

14.4.1.6 'Shiftless' Working Hours

'Shiftless' in this context means avoiding rigid traditional shift systems, such as the classic model of a two- or three-shift system characterized by an early shift, a late shift, and potentially a night shift. Nevertheless, at the end of the rostering process, each worker has their working times, which could be described as their 'shift'. This notion is mainly intended to refer to the fact that each worker has individual working hours, which can be completely detached from any previous time scheme. It is open whether a work assignment begins at 6 o'clock, 9 o'clock, 11 o'clock, or not before 14 o'clock. The working time of each worker may differ.

14.4.1.7 Integration of Working Time Preferences of the Employee

One-sided company-related flexibility requirements should not be the aspiration of a modern rostering and scheduling process. As described in section 14.3, employees should also have a right to co-determine their working time arrangements. Buzzwords are 'Work 4.0' and 'New Work' as well as labor shortage. For this purpose, asking for these preferences before rostering or allowing an operational leeway at the beginning or end of the work deployment is necessary. Thus, employee-related flexibility requirements should be included.

14.4.1.8 Stochastic Influencing Variables

There is only a rough expectation value and an approximate fitting distribution function for some parameters. On the one hand, due to the many individual tasks that may not be performed very often, the exact performance times are usually unknown. On the other hand, unpredictable uncertainties arise mainly due to the human individual performance variations of a worker over the day and between different workers. Moreover, production disruptions can be modeled this way, and other stochastic effects are also conceivable. This attribute usually makes finding a solution significantly more complex, especially in the form of computational expense. Furthermore, it can strongly impact the solution itself, primarily if solutions are intended to be more robust to stochastic effects.

14.4.2 Essential Thoughts Toward a Solution Approach

At present, there are few statements from practice and science on the discussed dilemma. To our knowledge, no quantitative models deal with this dilemma (Schwemmer et al., 2022, 2020). When considering the world of work in Industry 4.0, there is a significant divergence compared to, for example, the time-flexible workhours-centered research around the year 2000, as mentioned in Section 14.3, and the subsequent literature. In these 'old' models, humans are mainly still only a second-level resource in the optimization process (e.g., see Grabot & Letouzey, 2000). The demand is derived from an existing production plan and the effort is to serve it in the best possible way on that subordinated optimization level. In the CPPS scenario, however, one should go one step further, regarding the changed conditions. As a high-specialized production resource, the workers should be raised to the same optimization level as other resources (e.g., machines), and they should become a 'first-level resource' in the optimization process. Such a notion will change decisions regarding resource allocation and processing sequence. Consequently, baseline schedules optimized by a method that handles workforce deployment on the first level will differ from ones with workforce deployment at the second level.

However, by bundling work tasks of the human worker, the worker will probably receive continuous working time blocks again. Frequent idle times will almost be avoided. Nevertheless, the idea is not to optimize decisions for human resources but to strive for overall optimization. The processing of job and machine assignments continues to play an equally important role which will complicate the optimization accordingly (Schwemmer et al., 2020).

The second significant difference from already known workforce planning approaches is that in DCPS, decisions are made ad hoc. On the one hand, there is no baseline schedule to derive a demand. On the other hand, the assignment of tasks and workers weeks in advance destroys the decision-making autonomy of the DPC. Suppose human resource planning has already made the assignment. In that case, there is no longer any need for a decentralized decision system which leads to the conclusion that staff scheduling in a DCPS only determines the presence of workers (and possibly an on-call duty). In this way, not only is the human resource made available to the production system, but the decentralized decision system can still initiate the real-time-driven allocation and processing sequence of tasks and resources—at least within the scope of the predefined presences. Consequently, the goal is to predict attendance in such a manner that the appropriately qualified workers are already on-site at the exact times when they are needed (Schwemmer et al., 2020). When no workers are required, none have to be provided. Such decisions ensure effective capacity utilization and, thus, resource synchronization. Figure 14.4 represents the relationship between capacity demand and supply for the number of workers to be rostered. In the example, there are both over- and under-coverage and precisely coordinated requirements. Due to volatile variations in the demand for workers related to the short manual operating times, it is not always possible to ensure exact coverage.

Even though workforce planning in DCPS is a novel workforce planning problem, the general complexity of workforce planning problems remains. For the rostering and scheduling process, there are many regulations to comply with (e.g., regulations of labor law, collective labor agreements, or employment contracts) which are challenging boundary conditions for the model. Therefore, the derived planning and scheduling models are often highly constrained and very complex (Ernst et al., 2004) with a discontinuous solution space consisting of many infeasible solutions (due to restrictions regarding labor laws). The related problems of personnel rostering and scheduling problems can be classified in the complexity class of non-deterministic polynomial-time (NP)-hardness (Özder et al., 2020).

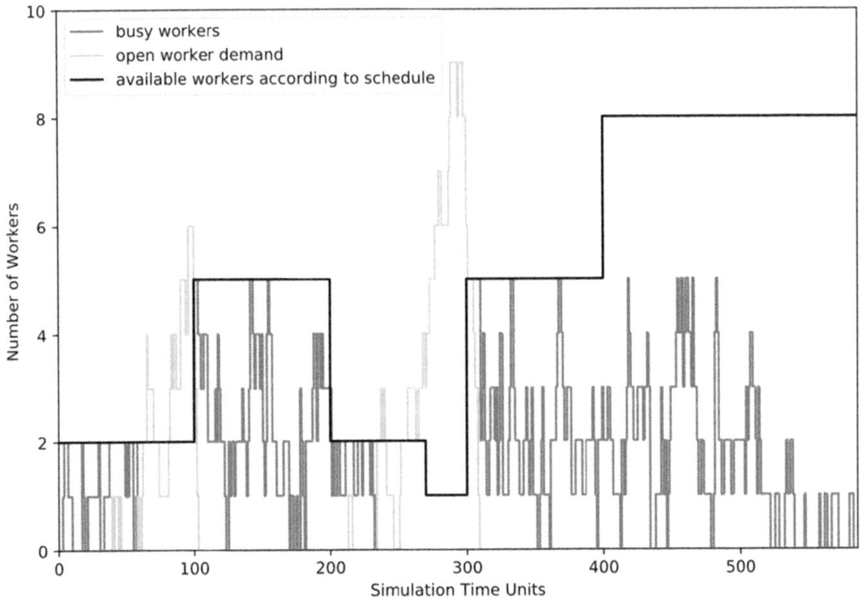

FIGURE 14.4 Exemplary representation of the relationship between worker supply according workforce roster, satisfied worker requests and open worker requests.

14.4.3 The Research Project: SIM4PEP

In the research project SIM4PEP (Schwemmer et al., 2020), a methodology is currently being developed to solve the planning and scheduling dilemma of workforce presence rostering based on the assumptions and restrictions discussed in Sections 14.4.1 and 14.4.2. The purpose is to conduct primary research for workforce attendance rostering in DCPS with a short- to medium-term planning horizon (up to six weeks). The project also intends to push to free production workers from rigid shift times and proactively incorporate individual working time requests. The approach uses a simulation-based forecasting method to predict the exact requirement times as well as the requested amount and qualifications of workers. Possible demand patterns are evaluated under different conditions to find desirable scenarios to support and undesirable bottlenecks to avoid. Furthermore, to apply the approach on a larger scale (like an application in the industrial sector), it is combined with simulation-based optimization (i.e., black-box optimization based on a genetic algorithm).

Prognostic inconsistencies for future forecasts are included via stochastically influenced parameters. For example, as mentioned before, humans are not machines. Variations in processing times will likely occur due to manual activities and individual skills or physical conditions. Although this complicates the solution approach and the calculation runs, it can be an essential issue in the topic under consideration, with at least an examination related to whether the fuzziness of the values based on stochasticity has a significant influence on the solution.

The above approach further indicates that the experiments serve to find a methodology to solve the dilemma and clarify fundamental research questions such as the exact use cases, limits of the method, and any particular effects that can apply in general.

As stated, the chapter is not only focusing on the workforce in isolation but also on the entire production environment. Therefore, the model within the research project also incorporates various optimization subjects for overall optimization and not just workforce-related objective values (e.g., delay costs of the scheduled jobs). Moreover, with particular regard in this chapter to the workers, the following optimization triggers for workforce deployment planning are identified:

- Minimizing the total attendance time of all workers (and on-call duty, if necessary) by avoiding idle times
- Minimizing work surcharges; thus, related work science aspects can be considered as well, such as avoiding night work
- Maximizing the scheduled working time, which is within each worker's individual desired working time.

Preliminary results of the project show that a time-flexible roster is superior compared to rigid two- or three-shift systems. However, the framework is still under development at the time of writing. The most critical aspects (except for stochastic influences) are already included in the model with these promising results. The evaluation of the preliminary results is based on three objective terms: the cost for worker attendance including night surcharges, surcharges for working outside the desired working hours, and delay costs. The researchers achieved resource synchronization through a targeted supply of workers. Moreover, the worker supply does not take place uniformly irrespective of the demand situation, as in two- or three-shift systems. Therefore, very high working time efficiency of the workers occurs. Delay costs can also be kept under control by taking advantage of the forecasts on the basis of the simulation model and some predefined slack times of the jobs. Slack times can be used to bundle manual tasks for the workers, making them an essential prerequisite for the methodology developed in this research project.

14.4.4 The Research Project: KapaflexCy

In contrast to the current project SIM4PEP, KapaflexCy (see Bauer, 2015) project is already completed but covers only some parts of the discussed issue. The project aims to introduce working time flexibilization for the production worker into companies quickly and in the spirit of Industry 4.0 and Work 4.0. The methodology involves both the demand and supply synchronization from the company's point of view and the employees' working time preferences—at least at a certain level.

Concerning the topic of this chapter, it is worth looking at one key element of the methodology: the KapaflexCy-App. This application is available for all production workers and coordinates shift assignments on the basis of direct requests to the workers where workers can accept and reject the requests. However, within this 'shift-doodle'-application, the assignments of workers to their working times still base on the time grid of rigid shift systems. The algorithm filters suitable employees in the following steps:

- Check of appropriate skill: Who can?
- Check of legal requirements: Who must not?
- Check of actual flextime wage: Who may?
- Check of individual preferences/requests via the app: Who wants to?

This way allows suitable employees to decide whether to work additional shifts according to their personal preferences, which increases self-determination concerning working hours, at least within the shift time grid and for additional shifts. Accordingly, flexible individual working hours are not possible in this system.

The methodology of KapaflexCy differs from the solution approach described in this chapter in one further aspect, as it does not deal with the dilemma of workforce rostering and scheduling in DCPS. The project covers workforce requirements based on predefined baseline schedules. Nevertheless, this user-oriented system offers the possibility to query employee preferences and is a worker-centered approach to react to short-term deviations in the production schedule, such as additional shifts.

14.5 CONCLUSION

In DCPS, a dilemma arises regarding the rostering and scheduling of the workforce. While the DPC no longer requires a (detailed) baseline schedule by deciding the allocation and sequencing in real-time, the workforce requirement and deployment planning require a specific lead time to provide workers. The changing world of work aggravates this conflict. Work 4.0 involves not only changes in the type and content of activities but also changes in the work models and working hours. The individual organization of working hours will play an important role now, including the working time preferences of the employees to improve work-life balance. Discussing workforce rostering for production environments of the future industry for both views, employee and company, a higher degree of co-determination for the workers regarding the daily working hours are feasible and preferable.

There needs to be more scientific or practical knowledge on dealing with the discussed dilemma. However, researchers are already studying effective methods to ensure efficient workforce planning and to close the scheduling gap. The focus is not only on the workforce as a subject of optimization but also on the overall optimization of jobs, machines, and workers. Precise synchronization of workforce demand and supply achieves optimality in terms of personnel cost, other production costs, and production flexibility without any cancellations and delays. Such a solution approach ensures a high level of competitiveness for the company on the market and, simultaneously, attractiveness as an employer.

Conclusively, with the complete implementation of DPC, the discussed dilemma is a challenge and an opportunity. By overcoming this obstacle, the high-tech strategy will work better.

ACKNOWLEDGMENTS

We would like to thank the German Research Foundation/Deutsche Forschungsgemeinschaft (DFG), which is funding our project with the title 'A simulation-based and flexi-time applying prediction model for scheduling personnel deployment times in the production planning process of cyber-physical systems' (Project-ID: 439188616).

NOTES

1 http://data.europa.eu/eli/dir/2003/88/oj
2 E.g., in Germany: Arbeitszeitgesetz: https://www.gesetze-im-internet.de/arbzg/
3 https://www.dol.gov/general/aboutdol/majorlaws

REFERENCES

Anger, C., Kohlisch, E., Koppel, O., Plünnecke, A. 2022. MINT-Frühjahrsreport 2022. Demografie, Dekarbonisierung Und Digitalisierung Erhöhen MINT-Bedarf – Zuwanderung Stärkt MINT-Fachkräfteangebot Und Innovationskraft. *Gutachten für BDA. Institut der deutschen Wirtschaft Köln e. V.*

Austrian Society for Occupational Medicine. 2019. Leitfaden Zur Arbeitsmedizinischen Beurteilung Langer Arbeitszeiten. Version 1.4. ÖGA.

Bauer, W. 2015. Selbstorganisierte Kapazitätsflexibilität in Cyber-Physical-Systems: Abschlussbericht. *Stuttgart: Fraunhofer Verl.*

Bergmann, F. 2019. *New Work, New Culture: Work We Want and a Culture that Strengthens Us*. Winchester, UK; Washington, USA: Zero Books.

Brauner, C., Anne, M. W., Alexandra, M. 2018. BAuA-Arbeitszeitbefragung: Arbeitszeitwünsche von Beschäftigten in Deutschland: Forschung Projekt F 2398. *1. Auflage. baua: Bericht.* Dortmund Berlin Dresden: Bundesanstalt für Arbeitsschutz und Arbeitsmedizin (BAuA).

Ernst, A. T., Jiang, H., Krishnamoorthy, M., Sier, D. 2004. Staff Scheduling and Rostering: A Review of Applications, Methods and Models. *European Journal of Operational Research*, 153(1): 3–27.

Eurofound, ed. 2016. *Working Time Developments in the 21st Century: Work Duration and Its Regulation in the EU*. Luxembourg: Publications office of the European Union.

Ganschar, O., Gerlach, S., Hämmerle, M., Krause, T., Schlund, S. 2013. Produktionsarbeit der Zukunft - Industrie 4.0: Studie. *Edited by Dieter Spath and Fraunhofer-Institut für Arbeitswirtschaft und Organisation.* Stuttgart: Fraunhofer-Verl.

German Federal Institute for Occupational Safety and Health, ed. 2019. Flexible Arbeitszeitmodelle - Überblick und Umsetzung. 44149 Dortmund.

German Federal Ministry of Education and Research. 2017. Industrie 4.0 | Innovationen für die Produktion von morgen. 11055 Berlin.

German Federal Ministry of Labour and Social Affairs. 2017. *Weißbuch Arbeiten 4.0: Arbeit Weiter Denken.* Berlin: BMAS.

Grabot, B., Letouzey, A. 2000. Short-Term Manpower Management in Manufacturing Systems: New Requirements and DSS Prototyping. *Computers in Industry*, 43: 11–29.

Hafner, M., Stepanek, M., Taylor, J., Troxel, W., Stolk, C. 2016. Why Sleep Matters – The Economic Costs of Insufficient Sleep: A Cross-Country Comparative Analysis. *RAND Corporation.* 10.7249/RR1791.

Hung, R. 1999. Scheduling a Workforce under Annualized Hours. *International Journal of Production Research*, 37 (11): 2419–2427.

Kagermann, H., Lukas, W.-D., Wahlster, W. 2011. Industrie 4.0: Mit Dem Internet Der Dinge Auf Dem Weg Zur 4. Industriellen Revolution. *VDI-Nachrichten*, April. Internet Archive. http://www.vdi-nachrichten.com/artikel/Industrie-4-0-Mit-dem-Internet-der-Dinge-auf-dem-Weg-zur-4-industriellen-Revolution/52570/1

Neumann, W. P., Winkelhaus, S., Grosse, E. H., Glock, C. H. 2021. Industry 4.0 and the Human Factor – A Systems Framework and Analysis Methodology for Successful Development. *International Journal of Production Economics*, 233: 107992.

Özder, E. H., Özcan, E., Eren, T. 2020. A Systematic Literature Review for Personnel Scheduling Problems. *International Journal of Information Technology & Decision Making*, 19(6): 1695–1735.

Pinzone, M., Albè, F., Orlandelli, D., Barletta, I., Berlin, C., Johansson, B., Taisch, M. 2020. A Framework for Operative and Social Sustainability Functionalities in Human-Centric Cyber-Physical Production Systems. *Computers & Industrial Engineering*, 139: 105132.

Schwemmer, J., Schmidt, T., Völker, M. 2020. A New Simulation-Based Approach to Schedule Personnel Deployment Times in Decentrally Controlled Production Systems. In *SIMUL*, 2020: 19–23.

Schwemmer, J., Kühn, M., Völker, M., Schmidt, T. 2022. Scheduling Workforce in Decentrally Controlled Production Systems: A Literature Review. In *Dynamics in Logistics*, Lecture Notes in Logistics. Cham: Springer International Publishing: 396–408, 10.1007/978-3-031-05359-7_32

Stutzer, E. 2012. Familienfreundlichkeit als Zukunftsfrage in Unternehmen. Text/html. *GMS Zeitschrift für Medizinische Ausbildung*, 29(2): Doc34; *ISSN 1860-3572*, 10.3205/ZMA000804.

Van den Bergh, J., Beliën, J., Bruecker, P., Demeulemeester, E., Boeck, L. 2013. Personnel Scheduling: A Literature Review. *European Journal of Operational Research*, 226 (3): 367–385.

Wieden, M. 2016. *Chronobiologie im Personalmanagement*. Wiesbaden: Springer Fachmedien Wiesbaden. 10.1007/978-3-658-09355-6.

Wong, K., Chan, A. H. S., Ngan, S. C. 2019. The Effect of Long Working Hours and Overtime on Occupational Health: A Meta-Analysis of Evidence from 1998 to 2018. *International Journal of Environmental Research and Public Health*, 16 (12): 2102.

15 Managing Risks in Product Development and Manufacturing Processes

Jelena Petronijevic
Arts et Métiers Institute of Technology, Campus de Metz, France

Stéphane Hubac
STMicroelectronics, Crolles, France

Marc Lassagne and Alain Etienne
Arts et Métiers Institute of Technology, Campus de Metz, France

15.1 INTRODUCTION

Industry 4.0 has introduced many changes in manufacturing systems and their design. Be it through Internet of things (IoT), numerical twins, big data, or cloud computing; it corresponds to a technological leap that increased the connectivity of manufacturing systems. In that context, since manufacturing systems are increasingly dependent on other systems and processes inside the company (design, production, quality assurance, maintenance, marketing, finance) and outside of the company (suppliers, customers, markets), the notions of connectivity and networking become essential. Consequently, risk management of manufacturing systems for Industry 4.0 uses adequate tools to contextualize and implement a multi-dimensional approach. In that sense, a systemic point of view (Helbing, 2013) is needed to guide proper decision-making in this complex environment.

This chapter focuses on product and process risks and their interactions. It discusses standardized risk approaches, their advantages, drawbacks, and decision-making methods. The first section of the chapter introduces the main concepts used in risk management and their interactions. The second and third sections consider the process and product risk management. The fourth section introduces a framework for process-product risk management, followed by the challenges raised by decision-making in practice in the fifth section. The sixth section presents the application of the proposed framework in three case studies.

DOI: 10.1201/9781003327523-18

15.2 MAIN CONCEPTS AND THEIR INTERACTIONS

Risk can be defined as the effect of uncertainty on objectives (31000: Risk Management - Principles and Guidelines, 2009). This effect can be observed through various scenarios and measured with a combination of their likelihood and consequences (Larson & Kusiak, 1996). In a slightly different but not incompatible view, Renn (1998) considers risk as a "possibility that human actions or events lead to consequences that affect aspects of what humans value". When managing risks in the technical domain, one first needs to determine the scope of the analysis (i.e., the risks under consideration), the nature of the conceptualization of risk (qualitative or quantitative), and the aggregation rule of risks (Renn, 1998). However, institutions, procedures, and social contexts also have an impact on risk assessment and need, therefore, to be considered. In that sense, even in the technical context of Industry 4.0, risk must be regarded as a social construct. Four metaphors can illustrate this idea (Renn, 1998): risk can be seen as pending danger (Damocles' sword), as slow killers (Pandora's box), through the lens of cost-benefit ratios (Athena's scale), and as an avocational thrill (Hercules' image). The following subsections further clarify different aspects of risks, keeping in mind such metaphors during the risk management process for Industry 4.0.

15.2.1 Risk and Uncertainty Taxonomy

"Risk" is often associated with a pending danger, an event that can potentially lead to severe adverse consequences. The likelihood of occurrence of these events is generally measured using probabilities. When historical data about a manufacturing process is available, it is possible to adopt a frequentist approach, where the likelihood of occurrence is inferred from past events. However, quite often, and mainly when dealing with innovative processes (as is the case with Industry 4.0), where no data is available, it is necessary to resort to subjective probabilities (Hacking, 2006; Wallsten & Budescu, 1983; Aven & Zio, 2011; O'Hagan, 2019).

The notion of risk is more complex than that of danger, in any case. The measurement of the occurrence of adverse events can be fraught with difficulties for at least two reasons, especially again in the context of Industry 4.0: on the one hand, the knowledge of these events may not be perfect; on the other hand, these events and their causes can be linked, which may lead to feedback loops and amplifications. Risk can be represented as a chain of connecting risk events or actions with their effects on the objectives, as shown in Figure 15.1.

This causal chain shows that understanding risk depends on the knowledge of uncertainty, which needs to be defined more precisely. International Organization for Standardization (ISO) 31000 defines uncertainty as a lack of information, experience, and knowledge related to events, consequences, or likelihoods. Although broad, this definition needs to address different types of uncertainty that can play a vital role in the selection of the appropriate method for risk assessment. Thunnissen (2003) proposes an uncertainty taxonomy for complex systems that includes epistemic uncertainty, aleatory uncertainty, interaction uncertainty, and ambiguity. Epistemic uncertainty represents the lack of knowledge related to the system's

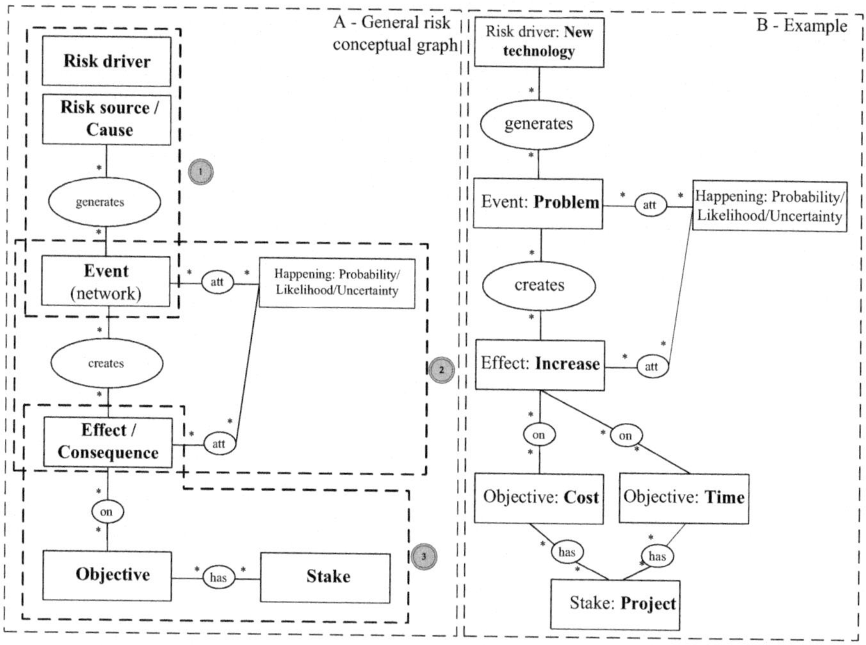

FIGURE 15.1 (a) Risk definition as a causal chain; (b) example of a causal chain (adapted from Petronijevic, 2020).

characteristics at stake and includes model, phenomenological, and behavioural uncertainties. Aleatory uncertainty considers the variations of the physical system. Interaction uncertainty is the interaction of events originating from the same or different areas but whose interactions should be predictable. Finally, ambiguity may be related to linguistic imprecision. Even if all these types of uncertainty may be encountered in the context of Industry 4.0, interaction uncertainty is undoubtedly the most crucial aspect that characterizes it.

15.2.2 Risk Management and Risk Interactions

ISO 31000 defines risk management as a series of steps: communicating and consulting, establishing the context, risk assessment including identification, analysis and evaluation, risk treatment, and monitoring and review. In that perspective, if the risk is mostly characterized as a "pending danger" (Damocles' sword), it can also have a delayed effect, and an adverse event can generate unexpected consequences (Pandora's box). The ISO 31010 standard (ISO, 2019) lists different techniques for representing consequences, likelihood, dependencies, and interactions (see Table 15.1).

Several techniques can be used to represent likelihood and consequences qualitatively or quantitatively, like fault tree analysis (FTA) and event tree analysis (ETA). However, these techniques (as is the case with Bayesian Networks) cannot explicitly consider feedback loops: they can only represent a specific scenario

TABLE 15.1
Techniques for Risk Consequence, Likelihood, Dependencies, and Interactions Analysis Based on ISO 31010 (Adapted from Petronijevic, 2020)

Approach	ISO Category	Description/Mathematical Formulation
Bayesian analysis and Bayesian network	Consequence and likelihood	Bayesian analysis is used to analyse likelihood on events based on prior judgements and empirical data. Bayesian network is used for risk estimation. It includes (conditional) variables and their cause and effect relationships. Bayes' theorem: $P(A\|B) = \frac{P(B\|A)P(A)}{P(B)}$ P(A|B) – probability of A given that B has occurred (the posterior assessment) P(B|A) – probability of B given A has occurred P(A) – the prior assessment of the probability of A P(B) – the prior assessment of the probability of B Bayes theorem can be extended to consider multiple events. Bayesian networks are based on Bayes inference and they posses the same properties as Bayesian analysis
Business impact analysis	Consequence and likelihood	Business impact analysis consider influence of incidents and events on organization and the means to manage them. It is performed using questionnaires, interviews and/or workshops.
Event tree analysis (ETA)	Consequence and likelihood	Event tree is the analysis of consequence and controls that give both quantitative and qualitative results. It is used to model and analyse the probability of outcomes scenarios based on the given event and controls. Qualitative results include descriptions of potential outcomes.
Fault tree analysis (FTA)	Consequence and likelihood	Fault tree is used for analysis of likelihood and causes. It analyses causes of events based on the faults and Boolean logic. It provides both qualitative and quantitative results. Qualitative result includes graphical representation and a list of minimal cut sets.
Cause – consequence analysis	Consequence and likelihood	Cause-consequence analysis approach provides quantitative results for cause and consequences. It is a combination of Event and Fault trees.
Markov analysis	Consequence and likelihood	Markov analysis is used to analyse likelihood. Calculates the probability that the system will be in one of its states at defined time in the future.
Monte Carlo simulation	Consequence and likelihood	Monte Carlo simulation is used to calculate probability of outcomes through multiple iterations and random variables.
Causal mapping	Dependencies and interactions	Causal mapping is used to analyse cause qualitatively. It represents events, causes, effect and their relationships.
Cross impact analysis	Dependencies and interactions	Cross impact analysis provides quantitative results on likelihood and cause. It considers the change in the probability of an events based on the occurrence of one of them.

comprising a series of events. Causal mapping and cross-impact analysis can be used to analyse risk interaction and dependencies. Still, while causal mapping is a qualitative method, cross-impact analysis is a quantitative method that uses Monte Carlo simulation based on the previously defined causal probabilities. These methods can be suitable for assessing risks in Industry 4.0. Still, they need to address the question of how to combine quantitative and qualitative data: more than a solely qualitative analysis is required to enable rational decision-making, but a quantitative analysis is also often not possible due to a lack of data. To sum up, the most widely used methods in industry are only sometimes suitable for Industry 4.0 as they either need to accommodate its complexity or make it possible to combine qualitative and quantitative data, an important issue when dealing with novel systems. The framework in section 15.2 aims at proposing possible solutions.

15.2.3 Cost-Benefit-Based Risk Analysis

Risk assessment is not simply an intellectual exercise that can lead to the determination of a specific risk level. Instead, it also must include an economic dimension to help decision-making among risk control measures, especially in the context of Industry 4.0, where the promise of improving production efficiency is paramount. In fact, it is only possible to argue for risk control measures on a solely technical basis by showing their benefit to the profitability of a company. Moreover, as limited resources are available in the business world, it is necessary to prioritize the areas where these resources will be allocated to achieve the highest level of safety for a given amount of money. The whole purpose of the economic evaluation of risk is to enable a company to achieve this rational prioritization, which Renn (1998) deems "Athena's scale".

Two perspectives can be considered: cost-benefit analysis (CBA) and cost-effectiveness analysis (CEA) (Persky, 2001; Cellini & Kee, 2015; Sunstein, 2018; Boardman et al., 2017). CBA relies on the computation of the Benefits (expressed in a monetary unit) associated with a risk control measure minus the costs (described as a monetary unit) it entails to implement any risk control measure where the benefits are higher than the costs. These beneficial risk control measures can also establish a ranking depending on the available budget.

Furthermore, the computation of the above implies that some non-monetary benefits need to be monetized, such as, for instance, the number of avoided fatalities or improved environmental protection. Several approaches have been proposed to perform this monetization (Hanemann, 1994; Diamond & Hausman, 1994; Portney, 1994; Kniesner & Viscusi, 2002), but it may encounter many difficulties: biases may exist in the results, and there may be a reluctance to monetize some benefits; for instance, there are many controversies surrounding the use of the Value of a Statistical Life (Viscusi, 2003; Viscusi, 2010), even if the most fundamental evidence shows that such a value is, albeit often implicitly, inherently present in social and business decisions.

Circumventing these difficulties can be achieved through CEA. Once a certain desired level of safety has been determined, the goal of CEA is to select the least expensive risk control option among those that can achieve this level. Using this approach corresponds to determining implicit values for non-monetary benefits: they do not appear as pronounced as with CBA, which is why CEA may be seen as more ethically acceptable.

Finally, CEA and CBA require the computation of the Present Net Value for the costs associated with the risk control measures. Several difficulties may also arise on that occasion: first of all, the choice of a discount rate may be disputable (the higher the rate, the lower the future costs will be); secondly, if some future monetary benefits may also be discounted, it may be ethically unacceptable to discount some non-monetary elements: this would mean for instance that a future life saved is less valued than a present one. Thus, comparing discounted costs and non-discounted benefits may sometimes lead to inconsistencies.

Overall, CBA and CEA may be useful tools for further technical risk analyses. In current industrial practices, they may be often used informally but are a required step to achieve rational decisions.

15.2.4 Decision-Making in Risk Management

Beyond the various limits outlined in the previous section, it is necessary to delve deeper into the issue of decision-making. Indeed, CEA and CBA may need to be improved, considering critical elements that are difficult to monetize. For instance, it is hard, if not altogether impossible, to monetize the familiarity of an operator with a new process, as the case would often be in the context of Industry 4.0. Another critical issue associated with CBA and CEA is that they need to account for risk attitudes and trade-offs explicitly.

Regarding risk attitude, CBA and CEA rely on expected value: the computation of expected benefits is based on the multiplication of their consequences by their probability of occurrence. In other words, a risk control measure would generate a benefit of €100 with a probability of 0.5. Similarly, a benefit of €0 with a probability of 0.5 would be considered equivalent to a risk control measure that generates with certainty a benefit of €50. However, from a decision maker's point of view, these two risk control measures may not be equal. As far as trade-offs are concerned, CBA and CEA add the benefits and the costs without any regard for the fact that a decision-maker may want to put more emphasis on one benefit or one cost versus another.

Addressing these issues corresponds to considering risk through Renn (1998)'s Hercules metaphor, where risk is seen as an "avocational thrill": risk cannot only conform to an expected value calculation, and trade-offs have to be made between various dimensions. Decision analysis, a set of techniques derived from economics and management science, deals explicitly with issues critical for uncertain contexts like the one of Industry 4.0. Although many other decision-making methods exist, this section explains about Multi-attribute utility theory (MAUT) (Keeney, 1982; Keeney & Raiffa, 1993), European school of multiple-criteria decision analysis (Roy & Vanderpooten, 1996), and analytical hierarchy process (AHP) (Saaty, 1990; Saaty & Vargas, 2012).

MAUT can make a choice based on evaluating the alternatives to various attributes (attributes being the measures of satisfaction concerning the various dimensions of the problem) and with partial utility functions. These evaluations are then combined into a single measure (multi-attribute utility function), which helps the decision-maker to make a decision. The decision maker's preferences are encoded in the partial utility and multi-attributes functions. The partial utility function enables the analyst to model the decision maker's risk attitude and the marginal decrease in user satisfaction. For instance, the same increase in safety would be valued differently in the case where protection is very low, and safety is already at a good level. Aggregating these partial utility functions into a multi-attribute, one allows for modelling the trade-offs between attributes. Once these functions have been determined, it becomes possible to choose among risk control measures by computing a score that considers the probability of occurrence of the events and the relative desirability of the multi-dimensional consequences of these events.

MAUT has been applied in various fields (Corner & Kirkwood, 1991; Keefer et al., 2004), such as presenting a survey of some studies. Specific applications to Industry 4.0 remain to be documented, but this approach could be fruitfully applied. This technique has various advantages, among them the fact that it enables a decision-maker to define their preferences regarding a given problem in a much more sophisticated manner than that which CBA or CEA allows. It dispenses with many problems regarding the monetary valuation of human life or the monetization of non-monetary aspects (e.g., environmental safety or other relevant decision parameters like technical familiarity and public image). Moreover, the partial utility functions, and hence the global utility function, are expressed in "utiles" (a unit that only measures the decision-makers satisfaction). Here, the interest is the decision maker's preferences over outcomes, regardless of what they are or in what unit they are measured.

However, as attractive as they may sound, decision analysis techniques present a certain number of difficulties, which may impede their implementation: they require a set of (conceptual and software) tools which may at first seem harder for the decision-maker to master than CBA or CEA; the expression of the results is less intuitive than CBA (where the outcome is directly expressed in monetary units); finally, it requires a decision-maker to make choices which seem to be more "binding" than those made for CBA-based techniques, in that they explicitly involve their preferences (in CBA, preferences and modeling choices are implicit, which makes it easier to justify the results, as they seem to be "objective" to the untrained eye).

15.2.5 Conclusion

This first section presented a first outlook on the tools and approaches that could be used for risk management in Industry 4.0. It showed that risk should be taken as a multi-dimensional construct and that addressing it requires both performing a risk assessment and using formal methods to make decisions among risk control measures. The following section now more specifically focuses on process and product risk management by suggesting an approach to alleviate the challenges associated with risk management in Industry 4.0.

15.3 PROCESS-RELATED RISK MANAGEMENT

15.3.1 On the Problem of Process-Related Risk Management

Manufacturing systems, especially in the context of Industry 4.0, are directly or indirectly related to numerous processes. This section focuses on risk management in the development process. However, the solutions discussed could be extended to other manufacturing processes. The development process relies on a temporary organization with resources and goals to deliver a change (Turner, 2007). This temporary endeavour (process, project) creates a product, a service, or a result (Project Management Institute, 2013). The development process has several phases: planning, concept development, system-level design, detail design, testing and refinement, and production ramp-up. Each phase can be further decomposed into activities with process attributes (goals) and variables (controls). Several domains of the process need to be synchronized to achieve these goals. These domains include process, organization, tool, product, and goal systems with objectives, requirements, and constraints. In other words, to successfully develop a product or a system, one should consider the set of activities to be performed together with the available resources and their goals.

Development phases are impacted by uncertainty (Morse et al., 2018): Ambiguity is present the most in the planning phase. Epistemic uncertainty impacts significantly conceptual development and system-level design. Aleatory uncertainty has the most significant influence on detailed design, testing and refinement, and production ramp-up. Finally, system-level design and detailed design are characterized by interaction uncertainty. It should be noted, however, that all types of uncertainties are present in all the phases of the development process. The most influential uncertainty type can help choose the appropriate risk assessment solution considering the targeted phase. Apart from their nature, uncertainties can be further categorized according to their impact on the process. It can be distinguished according to (Browning, 1999) as performance, schedule, development cost, technology, market, and business risks. Cause–effect relations are established between these uncertainties, meaning that risk in one domain can cause changes in another. As Browning (1999) states, since uncertainty and consequence lead to risk, one should focus on risk drivers and their relationships to improve risk assessment and mitigation.

Consequently, the risk can be observed in the form of a network. It could be wise to aim at the quantitative representation of risk that can be adapted to a qualitative description of risk in cases when the quantitative data are unavailable. This risk characteristic is essential in the early stages of development when risk identification is mainly based on expert opinions. Lastly, the representation of risk should include likelihood, causalities, and interactions.

Another vital aspect of relation establishment in the domain of the development process is the representation of the process itself. It has already been stated that the development process is made of activities and the (precedence) relations between them. When choosing an appropriate risk management solution, one should be able to consider the connectivity between activities and their associated risks.

The networked risks should be observed from the systemic perspective (Helbing, 2013). This problem focuses on consequences analysis and interdependencies and spill-overs between clusters (Renn, 2008). Large-scale data mining, network analysis, systems dynamics, scenario modelling, sensitivity analysis, and agent-based modelling are some techniques that can be used to resolve this problem (Helbing, 2010). Similarly, Morse et al. (2018) state that solutions for interaction uncertainty include simulation, multidisciplinary design optimization, and complexity science.

From the perspective of risk networking, a fuzzy cognitive map (FCM) can represent a good candidate to deal with risk's qualitative and quantitative nature and different risk effects. FCM (Axelrod, 2015; Kosko, 1986) is a cyclic-directed graph that can define relationships between risk drivers/factors. Cyclic graphs respond well to one of the problems in risk management – they can consider loops. Since FCM is fuzzy, it suits qualitative problems and manages uncertainties. Due to its similarity with neural networks (Carvalho, 2013), it can be used for quantitative problems and achieve learning. The central limit of this approach is that it cannot calculate the likelihood. In that sense, it needs to be combined with other methods. Apart from FCM, other network-based solutions can also be good candidates for representations of risk interactions. One could turn towards matrix-based or value-chain approaches to process representation and relationships between activities.

The final choice that one should consider is related to the simulation method. Different simulation types (discrete event, system dynamics, agent-based simulation) are used in risk management, often in combination with Monte Carlo simulation, to handle stochastic events and behaviours. According to Helbing (2010), system dynamics and agent-based simulation are suitable for tackling the problem of systemic risks. The main difference between these two solutions is in their scale. System dynamic is used as a top-down approach that provides a global overview of the problem. An agent-based simulation is more flexible and can be used for top-down to bottom-up problems. Barbati et al. (2012) underline that agents are particularly adequate for large-scale problems, modular domains, and frequent changes. Finally, the problem of networked risks could also be addressed using data mining techniques. The problem with these solutions lies in the need for more data in the early stages of the development process. If one would like to apply these techniques, one should first ensure proper data collection and transition between the existing risk assessment techniques and the new ones.

Based on the presented discussions, the following subsection presents one possible framework to deal with risk management in development processes.

15.3.2 A Framework for Risk Assessment in Development Processes

Section 15.2.1 has shown some of the main aspects that should be considered when proposing the risk assessment solution for the development process. The proposed solution should enable adequate representation of risks in their quantitative and qualitative nature, likelihood, causalities, and interactions. The risk management process should be naturally connected to the development process and decomposed into different activities with their resources, goals, and relations. The complexity of this problem calls for integrating different methods to enable risk assessment. In this section, we present one possible framework that can tackle the problem of risks and their interactions in development processes.

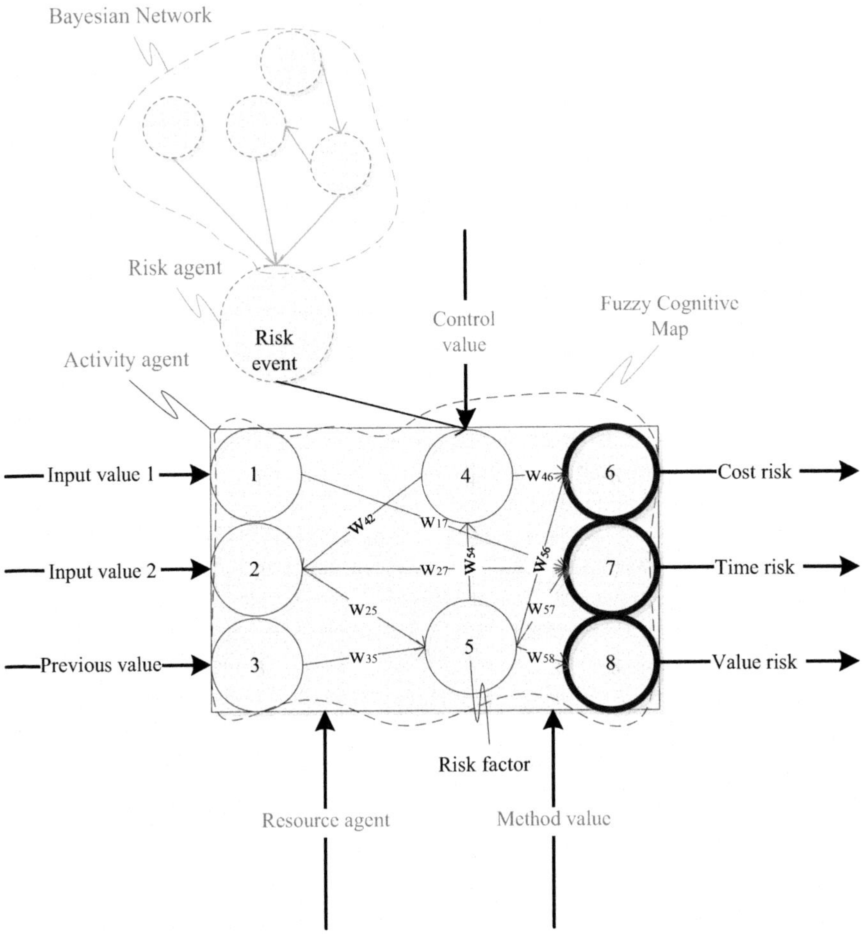

FIGURE 15.2 Framework for risk assessment in development process (adapted from Petronijevic, 2020).

The framework (Petronijevic, 2020; Petronijevic et al., 2022) is illustrated in Figure 15.2. It includes an activity agent with inputs, outputs, resources, and controls. The process is modelled as a chain of activities with precedence relationships established through input, output, and control connections. In development processes, different activity types can occur with their main risk drivers or factors. The risk factors and their relations are represented using a FCM. Inputs and outputs are risk factors and part of FCM. Other factors can be related to the activities' resources, methods, or controls. Relationships between factors are known and represent the behaviour of the activity. They should be defined with experts and actors involved in the development process. Risk events and other factors influence the values of the risk factors. If data are available, events that impact risk factors can be modelled using Bayesian networks. Otherwise, experts can propose values of risk factors. Once there is a change in one or several risk factors, FCM can evaluate new values of all the risk factors in the network. This way, the risk assessment on the activity level is performed. Through the relations between activities, the risk propagates from activity to activity and forms the risk on the process level. Keeping in mind that the framework lies on agent-based simulation and stochastic behaviour of the events and risk values, Monte Carlo simulation can provide an overview of the risk effects.

The framework's application is based on several steps presented in Figure 15.3. Firstly, the risk factors are identified and formed into networks based on the risk register data (Project Management Institute, 2013). These steps can be performed manually by several experts during brainstorming sessions. The obtained networks are then analysed using the proposed framework that evaluates risk on the activity and process levels. This framework is one of many that can be applied in the development process. However, it presents the advantage of addressing the problems of risk networking and the qualitative-quantitative nature of risk in the development process.

15.4 PRODUCT-RELATED RISK MANAGEMENT

This section focuses on product risk management and the advances in this domain needed to reply to the changes brought by Industry 4.0. Since the main points are based on the work of Cabanes et al. (2021), the reader is invited for further clarifications to consult the reference and its corresponding literature.

To assess and categorize failures according to their influence on mission success and equipment safety, the US Armed Forces created the failure mode and effects analysis methodology (FMEA) in 1949 (Military Procedures paper MIL-P-1629). Since then, the industry has seen a significant increase in the application of the FMEA technique, including the Apollo space program in the 1960s, followed by different industrial domains starting in the 1980s. The primary goals of FMEA in large sectors are problem avoidance and design/process improvement. Also, preventing issues is preferable to correcting them in terms of cost, quality, and dependability. FMEA is a tool that enables problem prevention before the testing phase and can improve design and process. The key goals are to produce secure, reliable, and trouble-free designs and have error-proof manufacturing procedures. FMEA is one of the most fundamental procedures for determining the scope of risk as a necessary step before risk reduction, according to the Quality System Requirements QS-9000. Organizations should examine and approve FMEAs before production phases as this approach attempts to prevent defects rather than fault identification. Industrial companies should record their processes for managing product safety-related products and manufacturing processes, including FMEA, following the international automotive task force (IATF) 16949:2016 standard. FMEA is also used to enhance test plans and process controls because they are essential components of successful product development. The capacity to choose alternatives, define prospects for establishing fundamental distinction, improve the company's reputation and competitiveness, and boost customer satisfaction are additional advantages of doing FMEA (AIAG, 2008).

System FMEA, design FMEA, and process FMEA are the three subtypes of FMEA. The highest-level analysis of a complete system made up of various

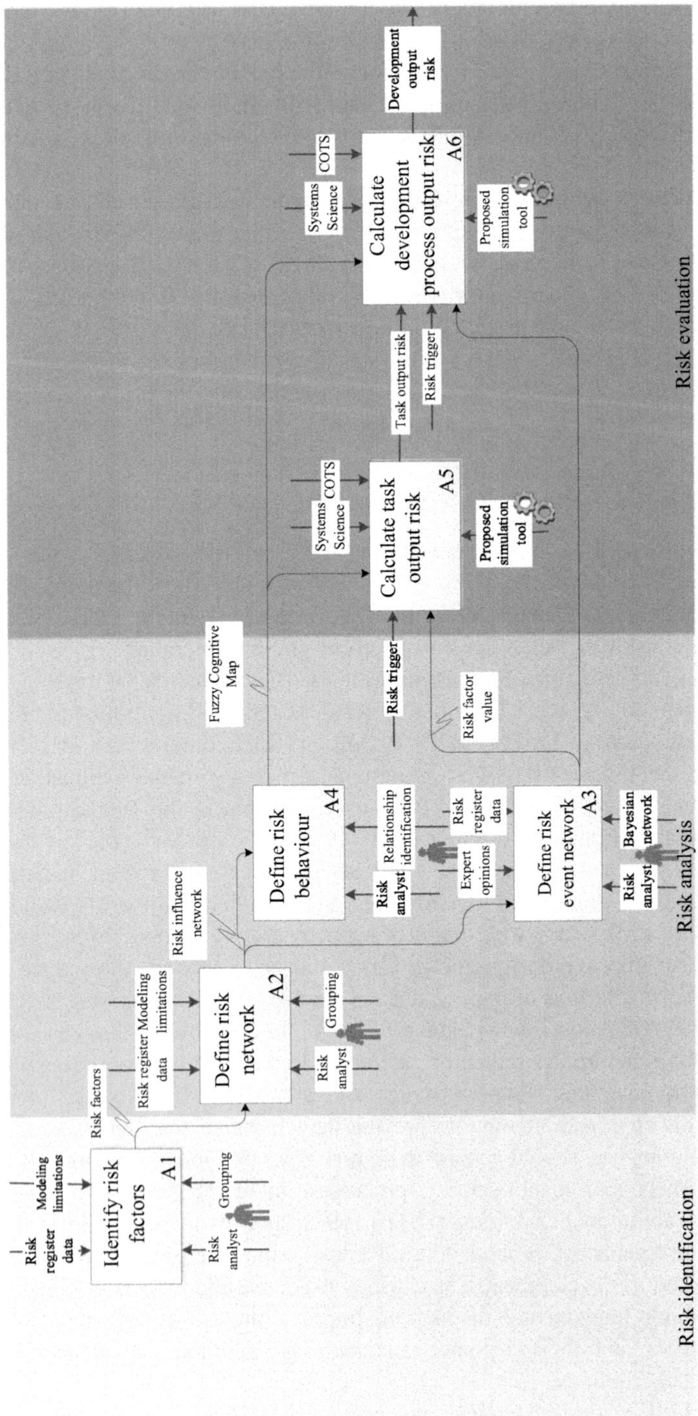

FIGURE 15.3 Risk assessment process for development process (adapted from Petronijevic, 2020).

subsystems is a system or concept FMEA. Design FMEA seeks to discover and demonstrate engineering solutions corresponding to system FMEA standards and customer criteria. Product design engineers often manage it. Process FMEA covers manufacturing processes. To ensure that products or technologies are constructed per design specifications while maximizing the quality, reliability, productivity, and efficiency of the various processes, the focus is on defining how manufacturing and assembly processes might be designed. These types of FMEA are intended to support the product development process and are also intended to be interdependent, as shown in Figure 15.4.

FMEA is, therefore, one of the primary methods used in the industry for reliability management in product design and development. The scholarly literature, however, points out several problems with the FMEA methodology. As a result, the main goals of this section are an analysis of the method's shortcomings, its improvement, and a brief presentation of an alternative implementation methodology capable of supporting quality and reliability management more effectively. This methodology extends the traditional FMEA approach through C-K design theory (Cabanes et al., 2021), which enables considering a more comprehensive framework that addresses the knowledge gap of FMEA.

The tabular technique of presenting data forms the foundation of the FMEA methodology. A series of spreadsheet rows and columns visually represent the information from the analysis (see Figure 15.5). Industrial handbooks state that the FMEA process is built on three key steps: (1) potential failures and effects analysis

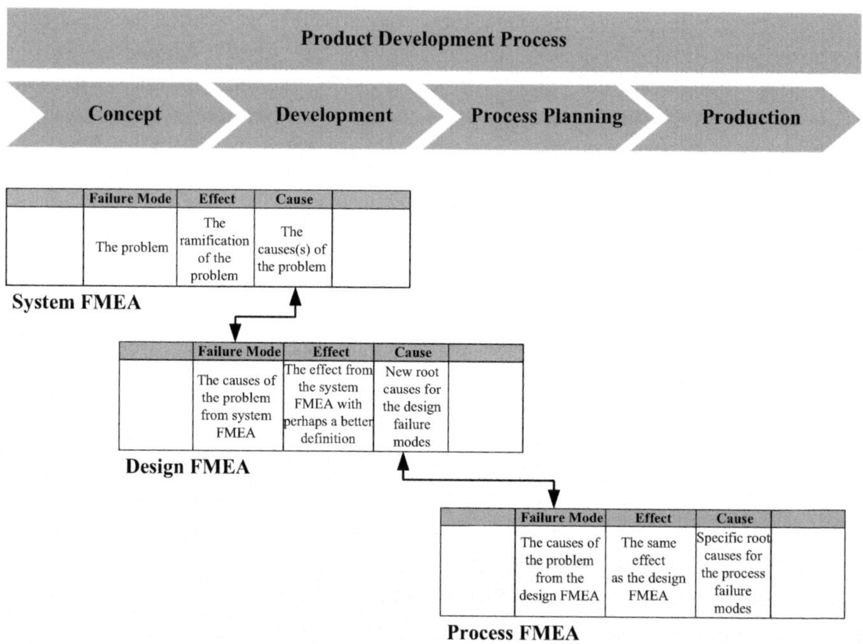

FIGURE 15.4 Relationship of system, design, and process FMEA (adapted from Cabanes et al., 2021).

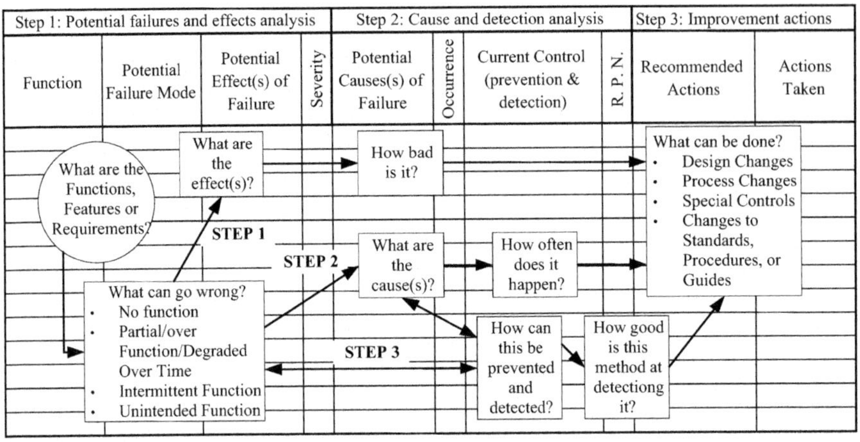

FIGURE 15.5 Generic FMEA worksheet (adapted from AIAG (2001), and Cabanes et al., 2021).

(Step 1 in Figure 15.5), (2) cause and detection analysis (Step 2 in Figure 15.5), and (3) improvement actions (Step 3 in Figure 15.5).

A function must represent the task, system, design, and process. A function is typically described with an active verb. The potential failure mode of a product or process is how it might not meet the design requirements. The potential failure modes for each function may vary, and they should be defined in "technical terms, not as a symptom necessarily apparent by the client" (AIAG, 2008). The result and impacts of the failure on the system, design, and procedure are potential effects. Failure's potential impacts must be examined from two viewpoints: local effects and global effects. Local effects imply that the failure may be isolated and has no external effects. Global effects imply that the failure could impact other functions. The potential cause, often known as the failure's underlying cause, is the cause of the failure. A potential cause of failure could indicate a design flaw, with the failure mode as a result (AIAG, 2008). FMEA methods are advised in most industries for developing new products or processes, changing current ones, accommodating client requests, improving quality, and managing reliability.

FMEA uses the following three-step process:

- Step 1: potential failures and effects analysis.
- Step 2: cause and detection analysis, including current controls regarding prevention and detection.
- Step 3: improvement actions: this can be done by identifying preventive actions that reduce or eliminate potential failure modes or detective actions (e.g., testing) aimed at helping to identify a weakness.

The first two steps are the critical part of an FMEA process, as the performance of improvement actions depends on these two steps. Prevention, detection analysis, and improvement activities are no longer meaningful if potential failure modes, effects, and causes are wrongly or not identified.

Although FMEA is a widespread method used across industries, practitioners and academic researchers highlight many challenges in efficiently implementing it. A general challenge (that FMEA shares with other risk assessment methods) is that it may take much work to demonstrate its actual contribution to the improvement in the quality of product and process design. This may be due to several reasons that are particularly prevalent in the context of Industry 4.0: the complexity and novelty of the systems it encompasses, characterization by many interfaces, and integration of multiple domains. Because of these features, FMEA often fails to identify all failure modes and does not allow the discovery of unexpected potential failures, especially when multiple technical teams develop their FMEA without considering the connections between them. Finally, updating in a dynamic manner FMEA reports is difficult and often leads to out-of-date information.

FMEA Guidelines (AIAG, 2008; Mode, 2011) provide recommendations on the objects that must be reviewed in the analysis but provide scarce guidance on how to proceed. Some tools may be used to enhance the analysis, such as boundary diagrams, p-diagram, interfacing diagrams, FTA, or brainstorming (more techniques are available in the ISO 31010 standard (ISO, 2019). These tools can help better understand the system and ensure a more robust analysis of the system. For example, FTA consists in developing scenarios that can lead to an adverse event by considering the various logical combinations of other prior events: using it can help increase the quality of the content generated by the FMEA. From a more general standpoint, enhancing FMEA with other tools allows a better understanding of complex systems and can help improve information sharing between teams across an organization.

However, although they can help collect data, information, and knowledge on the product/system under consideration and support collective learning and knowledge sharing, they are designed to only partially develop an FMEA. The recent AIAG and VDA handbook (2019) proposes a new approach for FMEA development: the 7-step approach (see Figure 15.6). This new procedure emphasizes system analysis; however, it only includes a slight improvement in analyzing potential failures, effects, and causes (Step 4 in Figure 15.6).

A standard answer to the previously outlined difficulties is often said to be addressed using brainstorming. However, it is unlikely to systematically tackle the complexity challenge, primarily because of its unstructured nature, which may fail to identify risks efficiently. Another way to deal with this challenge is the four-step failure mode avoidance (FMA). It consists of a structured framework based on four steps: (1) function analysis, (2) function failure analysis, (3) robust countermeasure

System Analysis			Failure Analysis & Risk Mitigation			Risk Communication
1st Step	**2nd Step**	**3rd Step**	**4th Step**	**5th Step**	**6th Step**	**7th Step**
Planning & Preparation	Structure Analysis	Function Analysis	Failure Analysis	Risk Analysis	Optimization	Results Documentation

FIGURE 15.6 The 7-step FMEA approach (adapted from Cabanes et al., 2021).

development, and (4) robust design verification. This framework is mainly focused on the development of design FMEA. Although it is possible to use a similar approach for the process FMEA, this framework implies a clear separation between the development of the different types of FMEA (system, design, process FMEA), which in turn fails to address the question of complexity and interactions. Many products or processes are so complex that correcting one type of FMEA can cause new mistakes in the same type or another. Since a system FMEA is carried out before a design FMEA, a mistake discussed during the design FMEA is not considered anymore in the system FMEA. This approach carries the risk that potential sources of error may be overlooked in the development of the FMEA; it would be more reasonable to carry out all three methods simultaneously, but it is rarely the case. Finally, even if a structured framework allows more rigour than a brainstorming approach, it may impede creativity in problem identification. The framework suggested subsequently does not aim to fully replace the tools presented with an alternative process but offers a complementary approach combining creativity and robustness and allowing the development of several types of FMEAs simultaneously. The risk management framework method depicted in Figure 15.7 redefines reliability management using FMEA from a traditional approach based on a hierarchical and linear model of reliability to a chain-linked model of reliability approach known as C-K oriented FMEA.

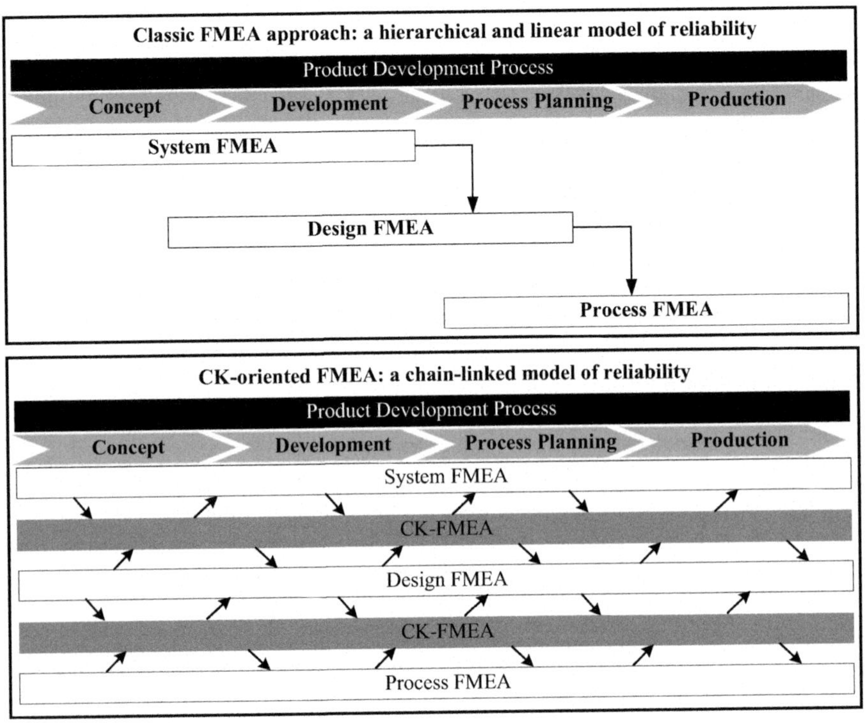

FIGURE 15.7 Classic FMEA approach vs. CK-oriented FMEA approach (adapted from Cabanes et al., 2021).

A functional framework is used in the C-K approach to alter FMEA to allow new collective learning, better social cohesion, and strong coordination between FMEA teams. However, it also extends FMEA usage from a problem-solving to a design-oriented approach. From a managerial perspective:

- Creating new concepts becomes an essential aspect of the design process.
- Collective learning determines problem generation and is regarded as a design area.
- Social interaction is both a design resource and an area that may be designed.

15.5 PROCESS-PRODUCT RISK MANAGEMENT

In the development process, the product is the goal composed of many connected activity goals. The relation between the product and its process(es) is bidirectional. Change in the process influences the product. Product error demands actions on the process level. In theory, process and product risk management are connected (see discussion on FMEA in Section 15.3). In reality, the problems still appear due to the need for more awareness of the connection between the process and the product. A couple of years ago, Takata recalled over 41.6 million vehicles due to the airbag issue (Consumer Reports, 2019). The problem, on the product side, was related to an unstable chemical, while on the process side, defective manufacturing and control processes provoked the issue (Tabuchi, 2016). This is one of many situations where it is hard to categorize the problem. In the context of a highly connected Industry 4.0 world, it is time to go beyond one domain and focus on the impact of the links between the areas. The question that arises is how to establish this connection. In this section, we are proposing one possible way to do it.

For years, researchers have been discussing the connection between different domains. Considering solutions in change management, Eckert et al. (2015) observed that even when the integrated process-product solution is proposed, the solution is usually centred around one of the domains while the other needs to be adapted. Browning actively discusses the link between the product and the process (Browning, 1998), defining performance risk as an "uncertainty in the ability of a design in its current state to meet desired quality criteria (along any one or more dimensions of merit) and the consequences thereof". The author further proposes a causal map that links different products (e.g., product complexity) and process characteristics (e.g., development process iterations, schedule risks, and cost risks) to performance risks. As seen in Section 15.3, the knowledge enables the creation of new concepts, which is essential for adequate risk assessment.

In addition, one could turn towards the design domain to enable interconnected process-product risk management. The design domain is less enclosed than the risk management domain regarding the product or process orientation. Hence, it can be a good starting point for the new risk management perspective. In this domain, design activity theory (Weber, 2007) connects the designer's activities to the product characteristics. Another design approach, FBS(-linkage), has already been used in the context of risk management (Haley et al., 2014; Haley et al., 2016; Hamraz et al.,

2012a; Lough et al., 2009; Tumer & Stone, 2003). FBS framework (Gero, 1990; Hamraz et al., 2012b) decomposes the system into its functions, behaviors, and structures. These two design frameworks can be combined to enable a neutral (no pure process or product orientation) process-product risk management perspective.

The main idea of the process-product view is to enable interactions between the risk and the appropriate knowledge integration. At the same time, this new perspective should be introduced subtly and follow well-established practices. For this reason, as the first step of introducing a new risk management perspective, it is possible to extend the FMEA approach. Hence, the idea is to introduce concepts from Sections 15.2, 15.3, and the beginning of the current section into the FMEA. The FMEA table would still represent the resulting table of risk assessment. At the same time, other (networking) approaches would be used to update the table according to the current knowledge and state of the system. The framework for the FMEA table update is given in Figure 15.8.

The framework is based on the active unit. The unit of action/activity is the essential element of every function. The process uses human and technical resources. Human resources are also characterized by their knowledge of the system's elements (process, product, technical resources, and human resources). The output of the process is its product or a goal. All system elements have their functions and consequently perform actions or activities. Consequently, no matter the element of the system, they can all be represented through their unit of action. The deficiency of the action unit is the failure mode of the FMEA. It is identified by

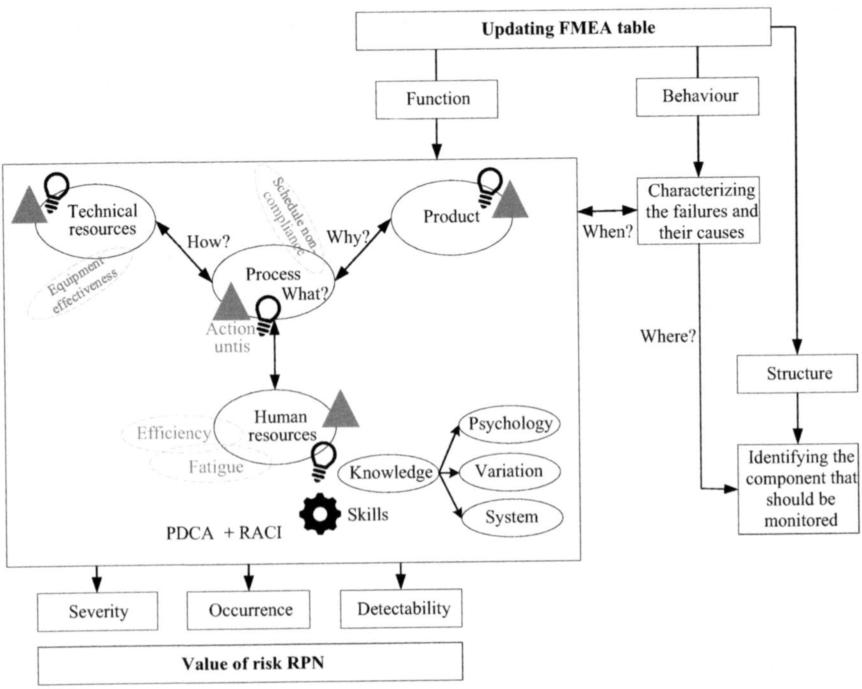

FIGURE 15.8 The framework for the process-product linkage integration into the FMEA.

the question "What?". To be performed, the activity uses resources identified by the question "How?". The goal/product of the activity is identified by the question "Why?". Hence, it is possible to establish a systemic network between process, product, and resource action units. This function network is the core of process-product connections. When uncertainty is introduced in one action unit, it propagates through the network and updates the risk of the connected action units. The uncertainty of each function is the result of its risk behaviour causal network, as presented in Section 15.2. This risk behaviour sets "When?" uncertainty will be introduced in the system. Finally, every risk behaviour is made of risk drivers or factors that define its structure. They define "Where?" the risk appears and what factors should be observed to prevent the risk effect. The knowledge of the process-product network grows based on the C-K framework (see Section 15.3).

As can be seen, the network can be used to connect all the interactions between different elements of the manufacturing systems (process, product, resources). The elements are connected on their functional level (F – Function) through their activities. Every activity is defined through its risk behaviour network (B – Behaviour) composed of risk drivers/factors (S – Structure). Consequently, FBS (-linkage) and design activity frameworks are used as a base for this solution. The solution is compatible with the C-K framework since new concepts and their links can be added to the network. Finally, this framework is the first stage of developing the process-product risk management view. It aims to provide a perspective for the paradigm shift in risk management towards a systemic view of risk. As a result, it should be improved in the future with further academic and real-world applications.

15.6 DEBIASING DECISION-MAKING

Implementing the techniques and approaches we described may suffer from various biases that derive from the workings of the human mind and from collective phenomena, which may impede both the elicitation of expert opinion and rational decision-making. We will develop in this section some examples of individual and collective biases that may be relevant to Industry 4.0 and suggest ways to debias decision-making.

The heuristics and biases research program has been developed since the seminal work (Tversky & Kahneman, 1974). Initially, the idea behind this research programme was to examine the deviations of the human mind from standards of rationality. For instance, a rational decision should only be based on an optimization calculation. It should be independent of how the alternatives the decision-maker has to choose are presented. Since then, many cognitive biases have been identified, and the list is ever-growing. Recently, Kahneman (2013) has also identified two ways in which the human mind works in forming thoughts, depending on the nature of the task to be performed: "System 1" corresponds to an instinctive, automatic, and unconscious way of reasoning, which occurs for simple tasks, and "System 2" reflects a conscious, calculated way of thinking, appropriate for more complex tasks.

On a collective level, the pioneering work of Janis (1982) has shed some light on various phenomena that may occur when conducting collective thinking, which he labelled under the term "groupthink". Groupthink corresponds to the fact that, out of necessity for a group to maintain its cohesion when trying to reach a decision or to perform an assessment, dissenting voices will tend to be silenced, critical thinking may decrease, and overall, the consensus that may be reached can be unsatisfactory to all participants in the discussion.

The existence of the two ways of thinking, as well as groupthink, has several implications for risk management: Risk assessment may encounter some difficulties, especially when dealing with complex phenomena, such as those that are present in Industry 4.0, as it has been shown earlier. For instance, biases such as anchoring (when a base value influences a judgement) or availability (favouring the elements that most immediately come to mind when expressing a judgement) may alter both the capacity to adequately identify the feedback and amplifications in networked risk as well as most the ability to express non-biased subjective probabilities when resorting to expert opinion. In the same fashion, overconfidence in innovative technical systems may impede a proper risk assessment and be associated with an optimism bias. Finally, since managing risk in Industry 4.0 is often based on a joint exercise due to its innovative and complex nature, it may be subject to groupthink.

Despite the sobering account, cognitive biases may further impede the already complex exercise of risk management in Industry 4.0, and some of the technologies involved may help alleviate them. For instance, digital technologies help automate certain decisions that would otherwise be fraught with biases. However, this only corresponds to a transfer of the influence of the biases. Eventually, human judgement may come into play when making high-level decisions based on the information systems provide. In practice, two strategies can be envisioned to counter cognitive biases' deleterious effects. At the individual level, Morewedge et al. (2015) and Sellier et al. (2019) show that, besides being conscious of their existence to avoid them, training may improve performance in situations where the decision-maker would be subject to cognitive biases. At the collective level, Janis (1982) suggests ways to avoid groupthink. For instance, they submitted the group to an external audit or had a group member play devil's advocate. Other techniques may also be used, such as the Delphi technique (Rosin et al., 2022), to aggregate expert opinion regarding evaluating risks.

15.7 CASE STUDIES

The primary goal of design for manufacturing (DFM) is to enlarge process yield, cost, cycle time, safety, and environmental product boundaries. In comparison, the primary goal of manufacturing for design (MFD) is to keep the manufacturing inside these boundaries and, thus, to prevent, control, and manage the risk of unscheduled or abnormal behaviours. In the industry, the prevention, control, and management of these abnormal behaviours are performed by using different quality management systems (QMS) and methodological processes like FMEA, statistical process control (SPC), and, more recently, automated technology like robots (3.0) or human-tools interfaces (4.0) allowing the evolution of SPC to advanced process/equipment control (APC/AEC). This section presents three French case studies from the semiconductor domain where the high automation level and usage of state-of-the-art IT applications are very high. The main issue in these case studies is based on the so-called 4.0

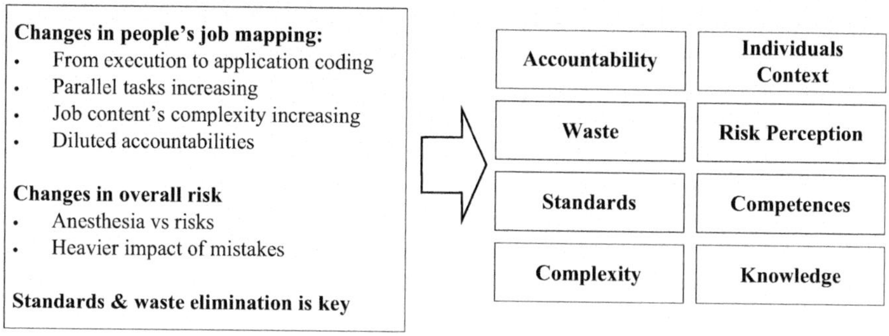

FIGURE 15.9 The 4.0 Automation's paradox: tangible benefits, but new side effects must be integrated.

automation paradox (see Figure 15.9), which introduces complex risk management challenges. These challenges must be taken into account when developing every 4.0 tool. Consequently, to address the side-effect brought by the 4.0 automation paradox, the approach presented in Section 15.4 will be used to develop different risk management solutions. Through different standpoints, the three examples summarize the necessary management evolution of industrial methodological processes in the context of Industry 4.0 with consideration of human aspects.

Use case 1 focuses on the out of control action plan (OCAP). An OCAP is used when SPC approaches are implemented to identify abnormal variability caused by identifiable causes to make the process more stable, minimize the variability (cost, cycle time, yield, safety, and environmental impact), and improve the process performance. When an out-of-control occurs, an OCAP is performed as a problem-solving tool. In a multi-products and technology-manufacturing context, and even more when products and technologies are frequently evolving, managing OCAPs can become a nightmare in terms of knowledge and competency management, and, as a consequence, manufacturing line overall performances can be at stake.

Use case 2 is called "Maintenance Tablet". Maintenance procedure design and sustainability may be quite complex when the procedures are numerous and require regular updates. Weak preventive and corrective maintenance procedures significantly impact manufacturing cost and cycle time, product yield, safety, and the environment. Ensuring sustainable design, update, and competency transfer of these procedures to the end-user in a functional, dynamic, and efficient manner has strategic importance for high-tech industrial environments.

Use case 3 is called "equipment installation process (EIP) competency sustainability". The more the industrial facility grows and evolves, its management becomes more complex. The process of installing new equipment is no exception. In many cases, this process has been carried out since each company's very creation of the manufacturing line. Over the years, the capabilities of a manufacturing line and its installation process are supposed to improve. However, their complexity may also increase if more 4.0 tools are used (as described in Figure 15.9).

In the case of EIP, the process (function of the process – figure 15.8) begins with the identification of equipment and infrastructure needs (functions of technical resources) by users (function human resources). It then proceeds to an arbitration process in which the investments to be made are defined based on the available budgets and the relevance of the users' requirements. Afterward, these investments must be defended and approved. Once approved, the entire planning process begins to prepare the acquisition of the new equipment - to purchase the equipment and finally to proceed to its installation and validation. Although the previous description is a general summary of the process, its complexity is evident due to the involvement of many teams, users, the finance department, suppliers, top management of the company, etc. Therefore, this process also includes documents and procedures at different times (when) and places (where).

Similarly, as in use cases 1 and 2, EIP's major issue is that the entire process is dynamic. Also, with each installation of a new piece of equipment, there is new knowledge or actions that were not considered and must be added to the process for future installations. Keeping knowledge and know-how updated in that kind of context makes the entire process a real concern. Another major issue is how to organize the documentation in a concise and useful way for the end users. Since the very purpose of any documentation or tool is to serve as a guide for the daily work of those involved in the process, it is imperative to prevent risk from spreading due to a lack of competency.

Here, what does it signify to perform 4.0 tools' implementation in the context of use cases? It adds as a reply to 4.0 Automation's paradox wherein tangible added values, and new side-effects must be integrated, as shown in Figure 15.9.

C-K framework (figure 15.8, section 15.4) methodological process was used to design and implement a 4.0 tool (i.e. technical resource):

- For use case 1, the developed tool is called e-cap (electronic control action plan). For use case 2, the tool is called a "maintenance tablet". For use case 3, it is called "IEP C-K TABLE & Competency Checklist".

All these 4.0 tools were implemented in order to link better the units of actions/activity elements of the system function network formed by:

- Process Functions (What?):
  - **Use case 1**: The OCAP flow actions units are performed WHEN non-compliance (i.e., system behaviour drift) occurs someWHERE.
  - **Use case 2**: The Maintenance actions units are performed WHEN preventive or corrective actions are performed someWHERE.
  - **Use case 3**: The Equipment Installation Process actions units are performed WHEN a piece of equipment needs to be installed somewhere.

- Resources Functions (How?)
  - Technical: The existing functional tools and procedures associated with each actions units (steps) of the OCAP (use case 1), Maintenance (use case 2), or IEP (uses case 3) flows, which need to be linked and/or

replaced and/or sustained by the new 4.0 automated application – e-Cap, Maintenance tablet, IEP C-K Table/Checklist

  - Human: The RACI (Realize, Accountable, Consulted, Informed) action unit associated with use cases 1, 2, and 3 flow action units (steps).

- Product Functions (Why?): Represent the effectiveness of use cases actions units/flows (What we Do, Plan, and/or Check or study) and the efficiency of the wisdom used (How Human / Technical resources link) in case of system behaviour drift When:
  - For example
    - Use case 1:
      - A manufacturing Control detection characterizing a known failure and causes occurs.
  - Highlight manufacturing system structure that should be monitored and/or adjusted.
    - When an unscheduled crisis occurs and requires an update of the OCAP System Functions action units
      - ability to update and transfer knowledge in real-time to prevent from lack and spread of incompetency

All this process uses cases to represent know-how that has been developed over the years through continuous learning. Due to the dynamic documentation, the process is robust and can serve as an essential guide for users. Also, if a team member leaves the company, the functions and tasks performed by this member can be assured with (more) ease by the new members since the process is well documented and up to date. For these reasons, a C-K Framework table (see Figure 15.10) was used to design the 4.0 tools (i.e., Technical Resource).

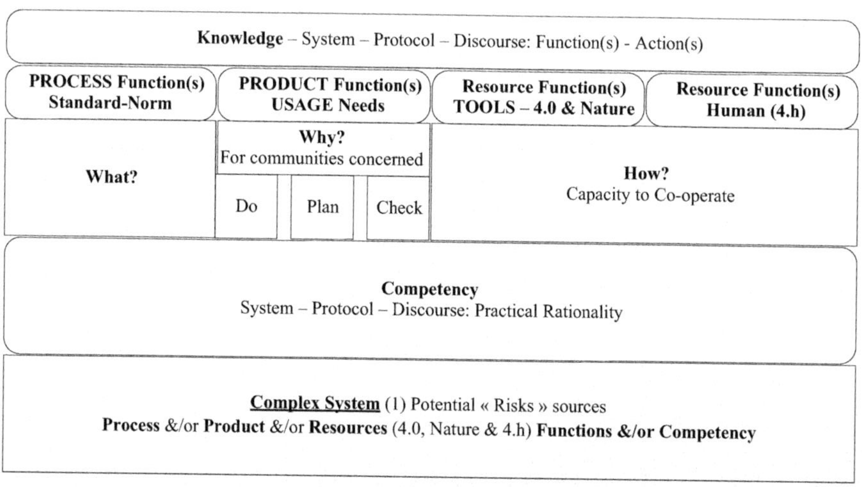

FIGURE 15.10 The C-K Framework table used to collect knowledge in complex systems and used to design the e-cap 4.0 tool from the existing manufacturing context.

- The C-K Framework table describes the different specific actions of each Functional structure of the systems (Figure 15.8). These actions, in turn, involve the interaction of different functional components for their realization. On the one hand, there are human resources functions since each task has persons with a specific function (RACI) that cooperate in carrying out the task. On the other hand, these people use Technical resources functions (i.e., documents or tools) to do their work. The general idea is to have a single document (i.e., C-K Framework Table or a relational database) and a visualization system linked to this document (i.e., Power BI or automated Excel checklist) which allows sustaining the knowledge of each function's families associated with action units and their interrelation. This solution supports each person who needs to perform a task within the process by answering questions such as What?, Who?, Why?, How?, When?, and Where?

The C-K Framework table is used to design and implement the mentioned 4.0 tools (i.e., Technical Resources) from the existing manufacturing context. This table helps to collect functional actions unit knowledge from the existing competency, i.e., the way human resources are doing the job from the existing complex systems. It allows designing the 4.0 technical resource tool by considering human resources social interaction (RACI) and not just a design resource.

Based on the needs of specific actors and the associated quality norms, for every unit in the function network and their competencies, the efficiency and effectiveness are to be associated:

- Severity: process yield, equipment cost & uptime, 8D, safety, etc.
- Occurrence: enforcement of the reaction – speed – accuracy, cycle time, productivity gain, etc.
- Detection: OCAP, maintenance procedures, EIP traceability, improvement, rationalization, implementation, and competency propagation of new standards in each workshop of the manufacturing area.

It is to be noticed here that the same methodological process applied to three different use cases has shown similar results:

- Complex Systems can coherently prevent the systemic potential of "Risks" related to Process and/or Product and/or Resources (4.0, Nature & Human) Functions and/or (lack or spread of) Competency.
- The conducted operational studies have shown up to a 30% reduction of functional blocks, operational cost, and cycle time directly related to improved competency sharing.

15.8 CONCLUSION AND FUTURE WORK

This chapter discussed the challenges in risk management for product development and manufacturing processes. While risk management solutions are evolving, the lack of performance we might face lies in the absence of a global view of the problem. To achieve this global vision, one should consider all aspects of the development or manufacturing process. Product, process, and resource characteristics need to be taken into account, along with their interactions. With the growing number of technologies in Industry 4.0, the solution should be adaptable to introducing new concepts. Finally, we must ensure that the decisions made are unbiased. The framework proposed in this chapter addresses different aspects (product, process, resources) and their interactions while considering the inclusion of new concepts. The solutions have been applied to several technical case studies with promising results. While the solution is general, conducting case studies in different engineering domains would be beneficial. To achieve its systemic potential, further work is needed in applying the data modelling and process solutions, which raises the issue of how to include these solutions in the risk management process. Some of the questions that should be answered are: What aspects need to be treated automatically? How should qualitative data be considered? How are the results to be integrated into the global process-product-resource network? While we proposed several directions in our framework, further work is needed to obtain optimal results. Finally, all these solutions should be part of the debiased decision-making strategy. Consequently, developing a systemic risk management approach should include an appropriate decision-making process. In this sense, the solution proposed in this chapter should be considered a first but not the final step towards systemic risk management in product development and manufacturing processes.

REFERENCES

AIAG. 2001. *Potential failure mode and effects analysis (FMEA): Reference manual.* Third Edition. Southfield, MI: Chrysler Corporation, Ford Motor Company, General Motors Corporation.

AIAG. 2008. *Potential failure mode and effects analysis (FMEA): Reference manual.* Fourth Edition. Southfield, MI: Chrysler Corporation, Ford Motor Company, General Motors Corporation.

AIAG, VDA. 2019. *Failure mode and effects analysis – FMEA Handbook: Design FMEA, process FMEA, supplement FMEA for monitoring and system response.* Southfield, MI: Automotive Industry Action Group.

Aven, T., Zio, E. 2011. Some considerations on the treatment of uncertainties in risk assessment for practical decision making. *Reliability Engineering & System Safety*, 96(1): 64–74.

Axelrod, R. 2015. *Structure of decision: The cognitive maps of political elites.* Princeton University Press.

Barbati, M., Bruno, G., Genovese, A. 2012. Applications of agent-based models for optimization problems: a literature review. *Expert Systems with Applications*, 39(5): 6020–6028.

Boardman, A.E., Greenberg, D.H., Vining, A.R., Weimer, D.L. 2017. *Cost-benefit analysis: Concepts and practice.* Cambridge University Press.

Browning, T.R. 1998. Modeling and analyzing cost, schedule, and performance in complex system product development (Doctoral dissertation, Massachusetts Institute of Technology, Sloan School of Management, Technology and Policy Program).

Browning, T.R. 1999. Sources of performance risk in complex system development 1. In *INCOSE International Symposium*, 9(1): 611–618.

Cabanes, B., Hubac, S., Le Masson, P., Weil, B. 2021. Improving reliability engineering in product development based on design theory: The case of FMEA in the semiconductor industry. *Research in Engineering Design*, 32(3): 309–329.

Carvalho, J.P. 2013. On the semantics and the use of fuzzy cognitive maps and dynamic cognitive maps in social sciences. *Fuzzy Sets and Systems*, 214: 6–19.

Cellini, S. R., Kee, J. E. 2015. Cost-effectiveness and cost-benefit analysis. In J. S. Wholey, H. P. Hatry, & K. E. Newcomer (Éds.), *Handbook of Practical Program Evaluation* Fourth Edition: 636–672.

Consumer Reports. 2019. Takata airbag recall: Everything you need to know. Consumer Reports. https://www.consumerreports.org/car-recalls-defects/takata-airbag-recall-everything-you-need-to-know/

Corner, J.L., Kirkwood, C.W. 1991. Decision analysis applications in the operations research literature, 1970–1989. *Operations Research*, 39(2): 206–219.

Diamond, P.A., Hausman, J.A. 1994. Contingent valuation: Is some number better than no number? *Journal of Economic Perspectives*, 8(4): 45–64.

Eckert, C., Albers, A., Bursac, N., Chen, H.X., Clarkson, J., Gericke, K., Gladysz, B., Maier, J., Rachenkova, G., Shapiro, D., Wynn, D. 2015. Integrated product and process models: Towards an integrated framework and review, ISBN: 978-1-904670-65-0.

Gero, J.S. 1990. Design prototypes: A knowledge representation schema for design. *AI Magazine*, 11(4), pp. 26–26.

Hacking, I. 2006. *The emergence of probability: A philosophical study of early ideas about probability, induction and statistical inference*. Cambridge University Press.

Haley, B.M., Dong, A., Turner, I.Y. 2014. Creating faultable network models of complex engineered systems. In *International Design Engineering Technical Conferences and Computers and Information in Engineering Conference*, 46315: V02AT03A051.

Haley, B. M., Dong, A., Turner, I. Y. 2016. A comparison of network-based metrics of behavioral degradation in complex engineered systems. *Journal of Mechanical Design*, 138(12): 121405.

Hamraz, B., Caldwell, N.H., John Clarkson, P. 2012a. A multidomain engineering change propagation model to support uncertainty reduction and risk management in design. *ASME Journal of Mechanical Design*, 134(10): 100905 (14 pages).

Hamraz, B., Caldwell, N.H., Clarkson, P.J. 2012b. A matrix-calculation-based algorithm for numerical change propagation analysis. *IEEE Transactions on Engineering Management*, 60(1): 186–198.

Hanemann, W.M. 1994. Valuing the environment through contingent valuation. *Journal of economic perspectives*, 8(4): 19–43.

Helbing, D. 2010. Systemic risks in society and economics. IRGC–The Emergence of Risks: Contributing Factors.

Helbing, D. 2013. Globally networked risks and how to respond. *Nature*, 497(7447): 51–59.

Janis, I.L. 1982. *Groupthink: Psychological studies of policy decisions and fiascoes*. Houghton Mifflin School.

31000: Risk management—Principles and guidelines, 36 (2009) (testimony of ISO).

ISO. 2019. *IEC 31010 Risk management—Risk assessment techniques*. ISO.

Kahneman, D. 2013. *Thinking, fast and slow* (1st pbk. ed.). New York: Farrar, Straus and Giroux, 499.

Keefer, D.L., Kirkwood, C.W., Corner, J.L. 2004. Perspective on decision analysis applications, 1990–2001. *Decision analysis*, 1(1): 4–22.

Keeney, R.L. 1982. Decision analysis: An overview. *Operations research*, 30(5): 803–838.

Keeney, R.L., Raiffa, H., Meyer, R.F. 1993. *Decisions with multiple objectives: Preferences and value trade-offs*. Cambridge University Press.

Kniesner, T.J., Viscusi, W.K. 2002. Cost-benefit analysis: Why relative economic position does not matter. Cost-Benefit Analysis: Why Relative Economic Position Does Not Matter (March 2002).

Kosko, B. 1986. Fuzzy cognitive maps. *International Journal of Man-Machine Studies*, 24(1): 65–75.

Larson, N., Kusiak, A. 1996. Managing design processes: A risk assessment approach. *IEEE Transactions on systems, man, and cybernetics-part A: systems and humans*, 26(6): 749–759.

Lough, K.G., Stone, R., Tumer, I.Y. 2009. The risk in early design method. *Journal of Engineering Design*, 20(2): 155–173.

Mode, F. 2011. Effect Analysis, FMEA Handbook (with Robustness Linkages). FMEA Handbook Version 4, Ford Motor Company.

Morewedge, C.K., Yoon, H., Scopelliti, I., Symborski, C.W., Korris, J.H., Kassam, K.S. 2015. Debiasing decisions: Improved decision making with a single training intervention. *Policy Insights from the Behavioral and Brain Sciences*, 2(1): 129–140.

Morse, E., Dantan, J.-Y., Anwer, N., Söderberg, R., Moroni, G., Qureshi, A., Jiang, X., Mathieu, L. 2018. Tolerancing: Managing uncertainty from conceptual design to final product. *CIRP Annals*, 67(2): 695–717.

O'Hagan, A. 2019. Expert knowledge elicitation: Subjective but scientific. *The American Statistician*, 73(sup1): 69–81.

Persky, J. 2001. Cost-benefit analysis and the classical creed. *Journal of Economic Perspectives*, 15(4): 199–208.

Petronijevic, J. 2020. Global risk management in a product development project (Doctoral dissertation, HESAM Université).

Petronijevic, J., Etienne, A., Siadat, A. 2022. Global risk assessment for development processes: From framework to simulation. *International Journal of Production Research*: 1–25.

Project Management Institute (Ed.). 2013. *A guide to the project management body of knowledge (PMBOK guide)*, 5th edition, Project Management Institute.

Portney, P.R. 1994. Contributions to a symposium on contingent valuation. *Journal of Economic Perspectives*, 8(4): 3–64.

Renn, O. 1998. Three decades of risk research: Accomplishments and new challenges. *Journal of risk research*, 1(1): 49–71.

Renn, O. 2008. White paper on risk governance: Toward an integrative framework. In *Global risk governance*: 3–73. Springer, Dordrecht.

Rose, K.H. 2013. A guide to the project management body of knowledge (PMBOK® Guide)—Fifth Edition. *Project Management Journal*, 3(44): e1–e1.

Rosin, F., Forget, P., Lamouri, S., Pellerin, R. 2022. Enhancing the decision-making process through industry 4.0 technologies. *Sustainability*, 14(1): 461.

Roy, R. B., Vanderpooten, D. 1996. The European School of MCDA: A historical review. In *Proceedings of EURO XIV Conference*: 3–6.

Saaty, T.L. 1990. Multicriteria decision making: The analytic hierarchy process: Planning, priority setting resource allocation, 1: 287. McGraw Hill.

Saaty, T.L., Vargas, L.G. 2012. The seven pillars of the analytic hierarchy process. In *Models, methods, concepts & applications of the analytic hierarchy process*: 23–40. Springer, Boston, MA.

Sellier, A.L., Scopelliti, I., Morewedge, C.K. 2019. Debiasing training improves decision making in the field. *Psychological Science*, 30(9): 1371–1379.

Sunstein, C. R. 2018. *The cost-benefit revolution*. The MIT Press.

Tabuchi, H. 2016. *As Takata Costs Soar in Airbag Recall, Files Show Early Worries on Financial Toll*. The New York Times.

Thunnissen, D.P. 2003. Uncertainty classification for the design and development of complex systems. In *3rd Annual Predictive Methods Conference*, 16. CA: Newport Beach.

Tumer, I.Y., Stone, R.B. 2003. Mapping function to failure mode during component development. *Research in Engineering Design*, 14(1): 25–33.

Turner, R. 2007. *Gower handbook of project management*. Routledge.

Tversky, A., Kahneman, D. 1974. Kahneman, judgment under uncertainty, 1974.pdf. *Science*, 185: 1124-1131.

Viscusi, W.K. 2010. The heterogeneity of the value of statistical life: Introduction and overview. *Journal of Risk and Uncertainty*, 40(1): 1–13.

Viscusi, W.K., Aldy, J.E. 2003. The value of a statistical life: a critical review of market estimates throughout the world. *Journal of Risk and Uncertainty*, 27(1): 5–76.

Wallsten, T.S., Budescu, D.V. 1983. State of the art—Encoding subjective probabilities: A psychological and psychometric review. *Management Science*, 29(2): 151–173.

Weber, C. 2007. Looking at "DFX" and "product maturity" from the perspective of a new approach to modeling product and product development processes. In *The Future of Product Development*: 85–104. Springer, Berlin, Heidelberg.

16 Reverse Engineering

Past, Present, and Future Prospects

Anas Bin Aqeel
Department of Mechatronics Engineering, College of Electrical and Mechanical Engineering, National University of Sciences and Technology, Islamabad, Pakistan

Muhammad Irfan Aziz
College of Aeronautical Engineering, National University of Sciences and Technology, Risalpur, Pakistan

Uzair Khaleeq uz Zaman
Department of Mechatronics Engineering, College of Electrical and Mechanical Engineering, National University of Sciences and Technology, Islamabad, Pakistan

Nayyer Aafaq
College of Aeronautical Engineering, National University of Sciences and Technology, Risalpur, Pakistan

16.1 ENGINEERING DESIGN PROCESS

An engineering design process is a tool that engineers use to guide the development of products and processes to make one's dreams into reality. Going through its steps helps engineers to empathize with the people who use their finished products. Moreover, engineering design is an iterative process developed by learning from failures and following problem-solving steps to get the best possible solution. The measures may include, for example, determining objectives and constraints, prototyping, testing, and evaluation. The process starts with empathy. Engineers must clearly define the problem to be solved and the underlying constraints, often requiring communicating with the clients or a group of people for whom the solution is designed. Problem definition also requires conducting background research on the issue to get insight into the technologies that might be adaptable to meet their needs. After clearly defining the constraints, the design team brainstorms to enlist many new and creative ideas that may need

DOI: 10.1201/9781003327523-19

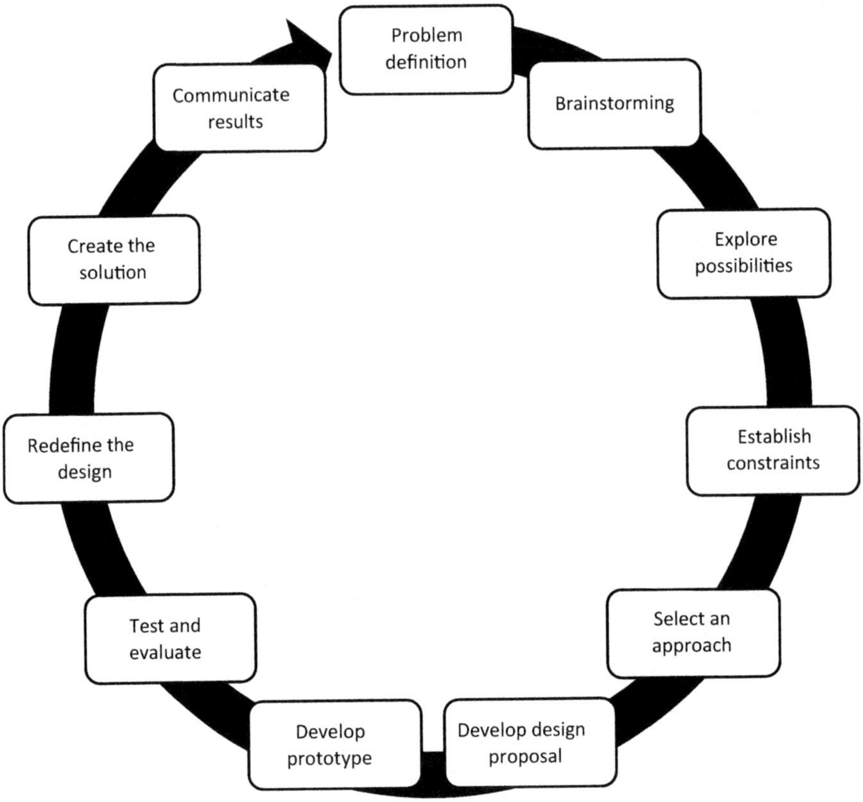

FIGURE 16.1 Schematic for engineering design process (adapted from Eunsang & Dongkuk, 2016).

to be more realistic. After having enough ideas, engineers focus on promising ideas, work toward their solutions, and plan the final product. The engineers then build, test, and evaluate multiple prototypes. With each prototype, they learn from the failures and return to the earlier steps of the design process to improve the product or method itself. They help shape the future by using design thinking to solve complicated problems.

The design process can be conventionally divided into multiple predetermined steps which are not required to be followed in sequence. A few of the steps may need to be repeated before moving on to the next step, or the sequence may change altogether. The schematic representing the steps for the engineering design process is shown in Figure 16.1.

The engineering design process steps include the following:

a. **Problem definition:** This step involves formulating the user needs into a definite engineering problem which includes defining the limitations and constraints of the design process.

b. **Brainstorming:** Engineers brainstorm to enlist possible solutions and ideas before the design process starts. It is better to let all the design ideas flow at this stage instead of judging the designs.
c. **Research ideas/explore possibilities:** Researching past projects and current technological advancements will help improve the design process and will result in avoiding the problems faced by past design teams. Additionally, engaging with the users may help find solutions or design aspects that were not considered by the design team.
d. **Establish criteria/constraints:** After due diligence to enlist the project requirements and possible solutions, this step requires revisiting the same and pondering upon any potential factors that may constrain the possible solution to the problem.
e. **Select an approach:** Once the possible solutions have been listed, alongside constraints, limitations, and technology at hand, engineers choose the best solution, meeting the requirements. This may also include consideration for alternative solutions and may involve repetition of some of the earlier steps.
f. **Develop a design proposal:** After the finalization of the design approach, the proposed solution is refined and improved, and a design proposal is prepared, which is an iterative process and may continue throughout the project duration.
g. **Develop prototype:** The refined design proposal is ready for prototype development allowing the design engineers to see how the product functions. The primary focus of the prototype product is to see the performance/functionality of the product. The robustification, system-level sustainability, and lifetime support aspects are incorporated at a later stage when the design and operational requirements of the project are fulfilled.
h. **Test and evaluate:** The prototype is then subjected to the test and evaluation phase to ascertain whether the developed product meets the design requirements. Again, this is an iterative process involving testing, evaluation, and improvement.
i. **Refine the design:** If the product is not meeting some of the test's requirements, changes are made to improve the design that can incorporate the testing requirements while maintaining other essential design aspects.
j. **Create the solution:** After the design and testing requirements have been met, the product may be improved to create the final product. Developing a product with such a process is sometimes called a 'field prototype' that is ready for serial production/delivery if there is no observation/feedback from the user.
k. **Communicate the results:** This step includes the final closure steps of the project. In this stage, the final product and its documentation are presented to the user/customer. The documentation allows for further manufacturing of the product.

16.2 INTRODUCTION TO REVERSE ENGINEERING

The engineering process consists of designing, assembling, and manufacturing the products and their maintenance. Moreover, engineering is classified into forward

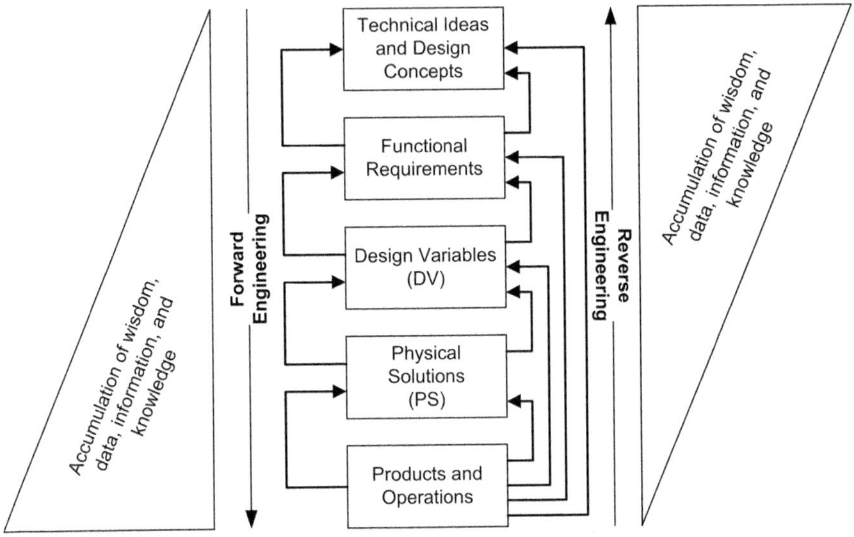

FIGURE 16.2 Forward and reverse engineering schematic (modified from Bi & Wang, 2020).

engineering (FE) and reverse engineering (RE), as shown in Figure 16.2. FE is a conventional technique for transforming the conceptual design into a logical design and finally implementing it physically. In some positions, the designed part/product needs more technical details such as engineering drawings, bill-of-materials, and manufacturing processes. For such situations, processes are replicated to achieve existing parts, subassemblies, and products without the available engineering drawings, data sets, and computational models, thereby inviting the concept of RE. Yau et al. (1995) defined RE as a process of retrieving new geometry using a manufactured product and modifying the previous computer-aided design (CAD) model through digitization. Numerous researchers described RE as capturing a component's physical entities through digital analysis. It is a process of obtaining a CAD model through 3D points captured using scanning and digitizing the existing products (Motavalli & Shamsaasef, 1996). RE can also be termed a learning process and implementing how an object works. It includes and is not limited to software, physical machines, and military technologies. Even biological functions related to how genes work can be reverse-engineered. Furthermore, RE software requires many essential tools to reverse engineer a product properly. RE is primarily used to replicate a product more inexpensively or because the original product is no longer available. It can also help improve your product quality. Generally, RE is divided into three main steps:

a. **Information extraction:** The product or its design is studied, and all the essential information is extracted.
b. **Modeling:** Modeling includes abstracting the extracted information into a conceptual model, which specifies actual information to a standardized model.

c. **Review:** Testing the model to see if it has been successfully reversed-engineered.

The points to address before and during the RE strategy is applied include:

- The logic for RE a part/product
- Analyzing the number of parts to be scanned
- The size of parts to be analyzed (large or small)
- The complexity level of parts to be analyzed (simple or complex)
- Material study of parts to be analyzed (stiff or flexible)
- Required accuracy (linear or volumetric)

The generic process of RE is divided into three phases, as shown in Figure 16.3. These generic phases of geometric model development include scanning, point processing, and applications (Raja et al., 2008a).

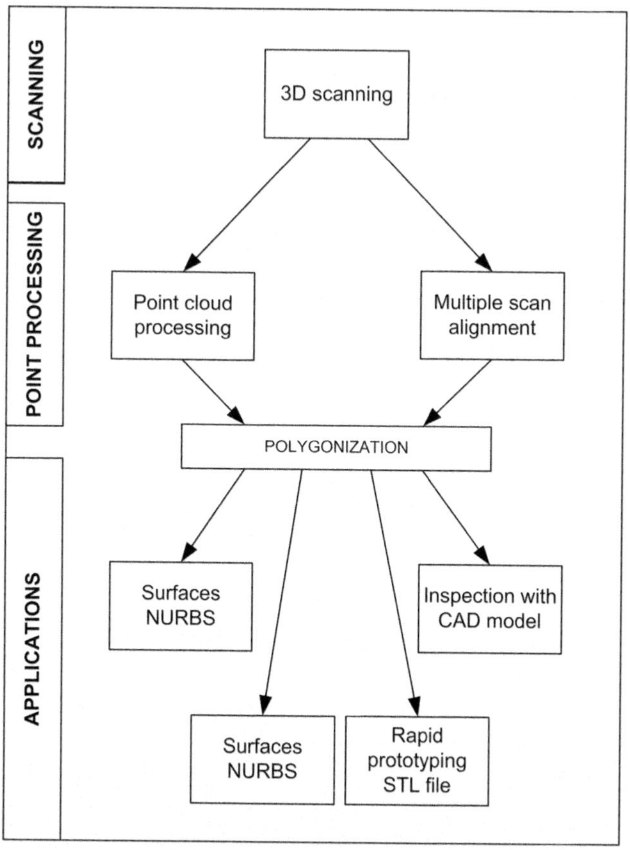

FIGURE 16.3 Phase categorization of reverse engineering as a generic process (modified from Raja et al. 2008a).

16.2.1 Scanning Phase

The scanning phase involves the strategies used to scan the part/product to be reverse-engineered through the proper selection of scanning techniques. The selection is pivotal to performing and capturing actual geometric features, including holes, slots, steps, and edges. Surface geometry is extracted for 3D models, and a cloud of points is obtained through 3D scanners. The scanning tools are mostly accompanied or added to computer numerically controlled (CNC) tools, and their scanning abilities are categorized into two different types: contact and non-contact scanners.

16.2.1.1 Contact Scanners

Contact scanners are contact-measuring machine technology (CMM)-based systems having a tolerance range of 0.01 to 0.02 mm. It consists of contact probes that allow it to follow the surface along the part contour. A few advantages include great accuracy, low cost, not being sensitive to color, and the ability to evaluate deep slots. The drawback of contact scanners depends on the size as it generates the data sequentially at the probe's tip. Another problem occurs due to the contact pressure requirement during scanning, as the tactile probe uses deflection to register data points. This process limits its use for soft materials such as rubber.

16.2.1.2 Non-Contact Scanners

Non-contact scanners are no physical part contact-based systems having a tolerance range of 0.025 to 0.2 mm. They use a laser, charge-coupled device (CCD), and optical sensors to collect data points. This faster technology can capture a large amount of data quickly. The disadvantages of non-contact scanning systems include light impingement on shiny surfaces due to light within the data-capturing process which requires the extra effort of fine powder coating on such surfaces before scanning. Another disadvantage is seen with the surfaces parallel to the axis, as shown in Figure 16.4. These issues could be improved in applications where precision is required compared to speed. The scanning techniques allow the cloud data sets to be stored in an accessible format with RE software providing it in the X, Y, and Z coordinates.

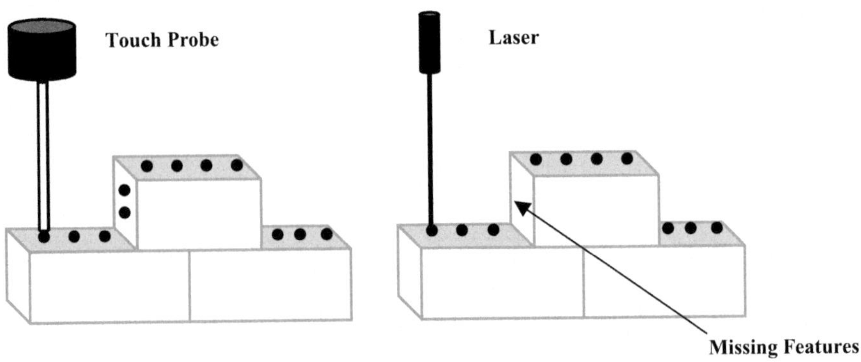

FIGURE 16.4 Presentation of missing features at vertical faces – touch probe vs laser source (modified from Raja et al. 2008a).

16.2.2 Point Processing Phase

The point processing phase consists of importing the cloud data points and reducing both data noise and the number of data points. This reduction is accumulated by using the predefined filters used during the process—the selection of predefined filters and its understanding of whether to use a relevant filter for a specific task. During the scanning process, it is observed that multiple scanning processes may be required for better results, and datum selection becomes an important aspect that helps the point processing system to reduce errors.

16.2.3 Application Geometric Model Development Phase

Rapid prototyping and tooling techniques significantly shorten the time to create physical representations of CAD models. However, RE is similarly creating CAD models from actual physical representations.

16.3 WHY REVERSE ENGINEER A PRODUCT?

RE is a multi-disciplinary generic science that can be applied across various fields universally. There are multiple reasons for RE products. One is that the OEM must manufacture the product, and the user needs lifelong support. Moreover, inadequate documentation or loss of original documents of the product leads to RE of the product to get the insight details. Another reason could be that the original supplier is unable to or unwilling to supply the required parts. Also, sometimes the user needs to update obsolete equipment or antiquated processes with modern or less expensive techniques. Last but not least, it is also used to analyze the market competitor's product design features to develop a better product. There are several more reasons for RE a product, and a few of them are as listed:

a. High cost of procurement from Foreign OEMs
b. Export Restrictions
c. Control over hardware/software for future upgrades
d. To address operational limitations
e. To enhance operational capabilities
f. To mitigate possible 'Backdoors' in sensitive/ops critical systems

16.4 FORWARD/REVERSE AND RE-ENGINEERING RELATIONSHIP

FE primarily deals with developing a product from the concept and design requirements. The FE results in a new system developed following recurrent design and development stages. In contrast, RE, also known as backward engineering, follows the reverse path. In RE, the existing system/application is broken down into sub-systems to retrieve information for system understanding and transformation. In comparison to FE, RE involves low-proficiency skills. Additionally, RE is adaptive compared to FE and is known to be prescriptive. The foremost challenge faced by RE is the non-availability of prior knowledge of the system/application that needs to be extracted without any technical drawings/schematics. It further involves investigating the relationship among various components/parts/modules of the system/application, which often becomes very time and resource intensive. Lastly, re-engineering deals with redesigning the aspects of the existing system/component to improve the design or add any component to the existing design.

16.5 RE AS DESIGN METHODOLOGY

The evolution of human civilization significantly depends on discoveries, innovation, knowledge, experiences, and technologies. RE is a design process of physical objects that works on current principles, how it can be improved, and how it can be improvised to other technological advancement areas. Since the start of humanity, we have known that nature is the best source of inspiration and knowledge for our technological development. In early 1900, the Wright brothers mimicked birds' flying behavior and reverse-engineered it to create the first-ever wings for flying an aircraft. Likewise, a robotic manipulator with parts like links, joints, sensors, controllers, actuators, etc., mimics our body arm. A few examples that highlight the RE of nature are shown in Figure 16.5.

Another essential concept in RE is digital modeling, a digital representation of a part/object using computational power like 3D scanning. It's an important tool to summarize the object into a digital model in useful formats of a point cloud or parametric surfacing (Bernard et al., 2010). Figure 16.6 shows the digital modeling and re-conception of knowledge, and Figure 16.7 shows the role of digital modeling in RE using an example.

Return on investment (ROI) must be provided by the object being reverse-engineered. In this case, cost and life-cycle analysis are required to provide evidence of cost-effectiveness. This chapter hence enlightens mostly the mechanical parts during RE with a special focus on computer-aided RE (CARE) on virtual solid model

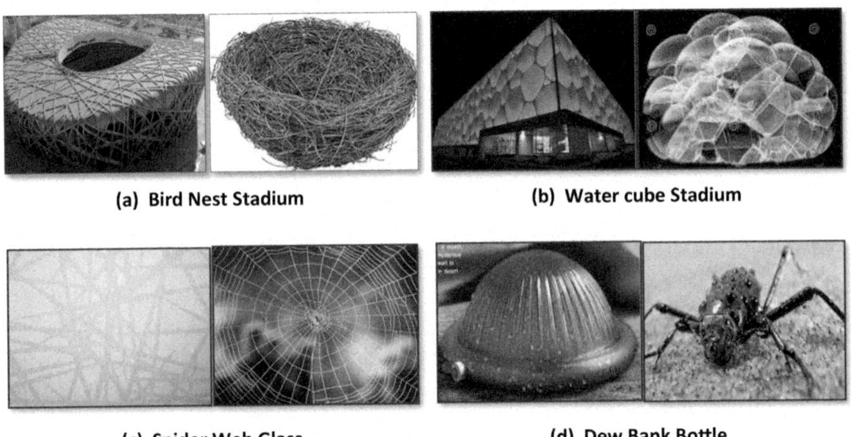

FIGURE 16.5 Nature-invented products (a) bird nest stadium, (b) water cube stadium, (c) spider web glass, and (d) dew bank bottle (modified from Hennighausen & Roston, 2015).

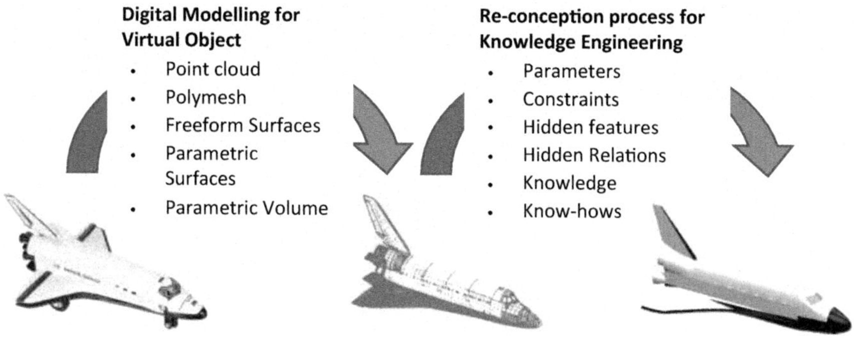

FIGURE 16.6 Digital modelling and re-conception process in reverse engineering (modified from Bernard et al., 2010).

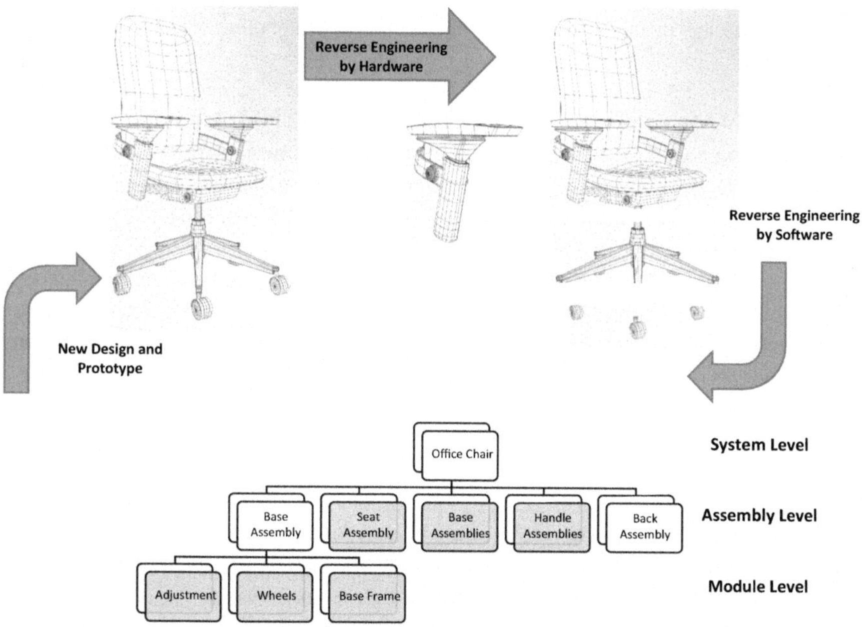

FIGURE 16.7 Role of digital modelling in reverse engineering using chair example (modified from Bi & Wang, 2020).

generation and identification of constraints and associated parameters throughout the process.

16.6 RE PROCEDURE

The RE project begins with collecting the point cloud using 3D sensors such as CMM, computer tomography (CT), or cameras. After the point cloud collection, a polymesh is created consisting of numerous parametric surface patches, which are later cleaned

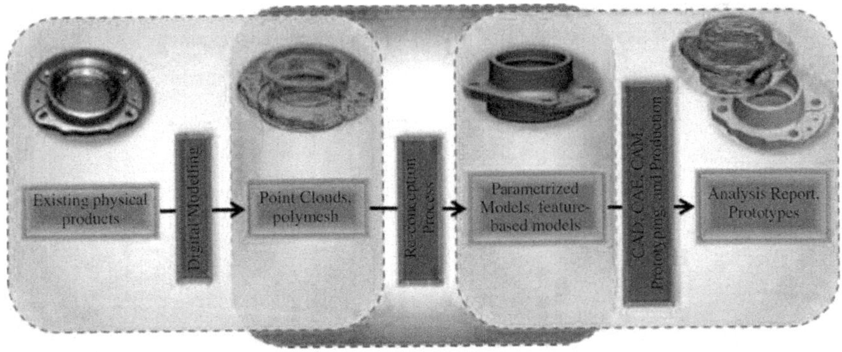

FIGURE 16.8 Three phases of reverse engineering (a) creation of polymesh or point cloud model using existing objects, (b) creation of parameterized models, and (c) utilization of reverse engineering model for CAD, CAM, CAE, and prototyping (modified from Bi & Wang, 2020).

and exported for parametric modeling. The process is divided into three stages, as shown in Figure 16.8., i.e., digital modeling, re-conception, CAD/CAE/CAM, and prototyping. Digital modeling is the computer model creation of an object to its nearest original. Data acquisition systems such as scanners are used to capture 3D data and transfer it to computer processes such as cleaning, filtering, and processing for polymesh model creation. The re-conception process uses knowledge-based feature identification for parametrized model development. The last phase allows the re-conception process to utilize in CAD/CAE/CAM and prototyping applications.

16.7 DIGITAL MODELING

From the computational capabilities, the representation of objects and using it is the first pivotal step for its advantage. In numerous circumstances, parametric models are only sometimes available and need digital modeling to capture data for the physical representation of objects. This process allows designers to understand the object's functionality, operations, and governing principles. A few examples include archeological objects, natural objects, and used products. In archeological objects, products such as pottery, weapons, and ancient infrastructure from history lack parametric models. Likewise, natural objects or creatures such as human or animal organs and their development with time cannot be found naturally and needs digital modeling to fasten the designing process.

16.7.1 Types of Digital Modeling

Based on different stages of reverse engineering, digital modeling can be categorized into different types, such as point clouds, meshes, and polygon modeling, which are explained next.

Point clouds – is a cleaned data form of data acquisition. It is a collection of data points collected for an object along the surface using a coordinate system. For the said purpose, the points are captured across X, Y, and Z coordinates using a reference coordinate system. Point clouds can further be categorized into organized and unorganized data points, with an organized system affecting the reduction in time for the modeling process. Unorganized point clouds represent the dispersed data set across a surface. On the contrary, an organized data set is in the shape of an organized grid or lines across the cross-section of the surface. It is observed that organized data can easily be converted to a mesh, whereas unorganized data needs organization through the division of volume/surface to cells and mapping those cells in the form of nodes or elements (Bi & Kang, 2014).

Polygon models – the process of scanning to acquire data from visible surfaces. Initially, the unstructured 3D data is represented as a point cloud, which is later converted to polygon models. A Polygon model is also termed a mesh which is a faceted model and a combination of numerous triangles. Mesh modeling is somewhat limited to visualizations, verifications, and inspections.

NURBS model – is a conversion of mesh model using point cloud to parametric model. The parametric model is termed a non-uniform rational basis spline (NURBS) surface. These surfaces are wrapped around polygon wireframes and are smoother than polygon models. In some scenarios, NURBS uses a set of boundary surfaces to form a watertight volume and a dumb solid. This is a knitted process of surfaces and conforms to no history of parametrization.

16.7.2 Surface Reconstruction

In a RE process, if a geometric representation of an object is acquired from data on boundary surfaces, the process is called surface reconstruction. Hoppe et al. (1992) defined surface reconstruction as a compact representation of the surface through construction as much as possible, given partial information about the unknown surface.

Surface reconstruction was traditionally used for scientific heritage visualization, cartography, and medical imaging. An era of architectural reconstruction research started in early 2000 when Rocchini et al. (2001a) developed a toolbox for heritage image reconstruction. Afterward, Milajlovic et al. (2004) analyzed three-dimensional images using unorganized point clouds through a developed MATLAB toolbox. Later, Rebolj et al. (2008) developed a monitoring system to differentiate scheduled and as-build construction models. Like building reconstruction, surface reconstruction of physical manufacturing parts was done in detail (Thompson et al., 1999; Fisher, 2002; and Pernkopf, 2005).

16.7.3 Algorithms for Surface Reconstruction

Surface reconstruction using point clouds is essential to RE. Azernikov et al. (2003) classified surface reconstruction into computer graphic and computational geometry approaches. They defined computational geometry as a piecewise-linear interpolation of unorganized points using the Delaunay triangulation system, which was later explored in detail (Cazals & Giesen, 2004). Amenta et al. (2001) approximated surfaces with points using medial axis transformation, a technique representing an object as an infinite union of balls and surface obtained termed as power crust. Previously, a ball-pivoting algorithm (a three-point triangulation technique) for surface reconstruction through point clouds was also developed (Bernardini et al., 1999). Abdel-Wahab et al. (2005) compared crust, power crust, tight cocone, and ball-pivoting algorithms to quantify their reconstruction quality, memory occupancy, and computational time. They devised crust and power crust algorithms to balance memory and timely execution. On the contrary, the ball-pivoting algorithm outcast the cocone algorithm by minimizing memory and time execution. It was observed that effective usage of Delaunay triangulation defined the computational cost.

The computer graphic approach aims for visualization quality, not the interpolated surface's closeness. A signed distance function was developed in which the surface was considered as a zero set of projected signed distance function resulting in surface topology with boundary curves (Hoppe et al., 1992; Neugebauer & Klein, 1997). Zhao et al. (2001) reconstructed surfaces using variational and partial differential methods.

16.7.4 Data Flow in Surface Reconstruction

Stereo cameras or laser sensors are employed to gather 3D range data for reconstructing a surface. The process from acquired data to the identification of geometric changes is complex.

Data Acquisition – initially, the point clouds are captured by scanning the surface. For large and complex objects, the scanning to capture points is done through different positions and views for proper data acquisition.

- Data filtering – the raw data, such as noise, distortion, and irrelevant data caused by either hardware system or environment, is filtered to be removed.
- Data registration and integration – laser scanner acquires data from a visible surface. Multiple views (using different references) are used to acquire data, and different layers/views data points are integrated to model the entire surface. Registration is termed as the system that allows the integration of multiple view data points using the same coordinate system. In comparison, integration is the merger of creating a single surface from multiple views.
- Surface reconstruction – a process through which raw point cloud or volumetric data from an approximate surface is formed.
- Data simplification and smoothing – a process of reduction in the amount of data obtained without affecting the quality of the reconstruction surface. It also allows faster engineering analysis, such as finite element, shape recognition, and collision detection. Moreover, it aims to achieve better computation and storage capacity.
- Data comparison – allows the data to be compared between the virtual and original object.

- Feature detection – allows the detection of features in CAD models. It provides the information if the relevant feature (such as edges, slots, holes, dimensions, etc.) is in place (data collection).

16.8 RE HARDWARE

From an engineering point of view, RE is defined as a process of analyzing an object (hardware and software) to recognize components and investigate how to reproduce it without original designs. It is a technology for reproducing 3D models in different geometric forms. RE hardware is used for RE data acquisition, which in 3D model representation is the collection of geometric data that signifies physical part/product. The technologies related to RE are categorized into three divisions: contact, non-contact and destructive, with the first two, briefly explained in the chapter earlier. The data acquisition output of reverse engineering is acquired to generate geometry from 2D images and point cloud data.

16.8.1 Contact Methods

Digitization of surfaces in contact method constitutes of two data collection techniques:

- Point-to-point sensing using touch-trigger probes
- Analogue sensing using scanning probes.

For the point-to-point sensing technique, a touch-trigger probe is utilized with a CMM or an articulated mechanical arm for point collection along the surface. A manual articulated mechanical arm can be used for a higher degree of freedom (DOF) movement for data collection. On the contrary, a programmed CMM system with a lower DOF can also be used. CMM technology provides better data collection accuracy than an articulated mechanical arm. However, its low DOF is a significant disadvantage for digitizing complex surfaces compared to mechanical arms. In the analog sensing technique, the scanning probe offers uninterrupted deflection data merged with machine position for surface location. These probes are attached to CMM or a CNC machine. The probe stylus tip continuously moves across the surface, gathering data. The analog sensing technique is faster than the point-to-point technique by three folds. Table 16.1 represents the list of commercial contact-based hardware systems for data acquisition (Pham & Hieu, 2008).

16.8.2 Non-contact Methods

Non-contact methods utilize 2D images and point clouds for geometrical representation of the associated surface. The transmitted energy source in the form of light, sound, or magnetic fields is captured, and this geometric data is calculated based on the algorithms. The algorithms introduced during the non-contact methods

TABLE 16.1
Commercial Hardware for Reverse Engineering Using Contact Method (Modified from Pham & Hieu, 2008)

Technology	Company	Model	Volume (mm)	Accuracy, Resolution, and Speed	Operation
Point-to-point sensing using a touch-trigger probe, mechanical arms	Faro Technologies	FaroArm Advantage	1200–3700	Acc: ±0.09 to 0.431 mm	Manual
		FaroArm Platinum	1200–3700	Acc: ±0.018 to 0.086 mm	
	Immersion Corp.	MicroScribe MX	1270	Acc: 0.1016 mm	Manual
		MicroScribe MLX	1670	Acc: 0.1270 mm	
Analogue sensing using scanning probe, CNC machine	Roland DGA Corp.	he Picza – PIX-30	305 × 203 × 60	Scan pitch in Y, Y, Z axis: +(X,Y): 0.05-5 mm in steps of 0.05 mm. +Z: 0.025 mm	Programmed
		MDX – 15	150 × 100 × 60		
		MDX – 20	200 × 150 × 60		
Point-to-point sensing using a touch-trigger probe, CMM	Mitutoyo	Euro – C – 121210	1205 × 1205 × 1005	Acc: 0.001 mm	Programmed
Analog sensing using scanning probe, CMM & CNC machine	Renishaw Inc.	Renscan 200	Based on CMM and CNC machine volume	+ Speed: 508 – 1016 mm/min + Max data rate: 70 points/s	Programmed

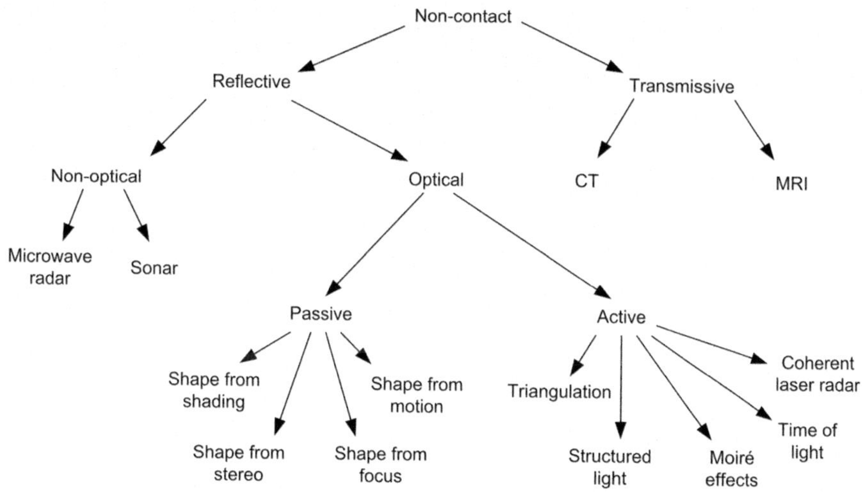

FIGURE 16.9 Classification of non-contact reverse engineering methods (modified from Pham & Hieu, 2008).

include triangulation, wave-interference information, image processing, and time-of-flight. RE can further be classified based on sensor technologies or data acquisition techniques (Alain, 1999; Rocchini et al., 2001b; Tamas et al., 1997). A classification of RE hardware systems based on non-contact methods is shown in Figure 16.9. The non-contact data acquisition techniques used in RE, such as optical, nonoptical, and transitive, are explained as follows.

16.8.2.1 Optical Techniques

16.8.2.1.1 Triangulation

Geometric triangulation is the most used optical technique for an object's coordinate point orientation. Triangulation involves distance measurement by angular calculation by a laser light and digital camera/charged-coupled device camera (CCD)-based method to measure surface coordinates. Figure 16.10 shows the two modifications of the triangulation technique with one and multiple CCD cameras employed. Using one CCD camera, a light through a light source is projected at an angle to the object, which is detected by the position of a reflected point (or line) using CCD camera. For multiple CCD camera systems, measuring functions are independent of light projection and may involve the movement of the spotlight or line.

The schematic using one CCD camera shows a light source projected at a specified angle (θ) on a surface object. As the baseline length (L) is known from calibration, different parameters, such as camera focal length (F), illuminated point (P), and fixed baseline length (L), can be calculated using the equation 17.1 and 17.2 (Park & DeSouza, 2005). Error in P and θ can be examined using equation 17.3.

$$Z = \frac{FL}{P + F\tan\theta} \tag{17.1}$$

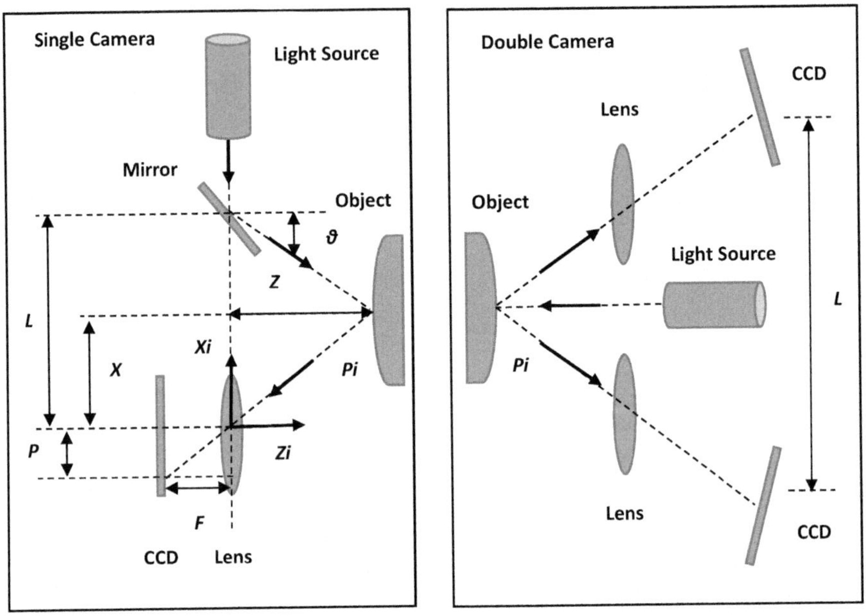

FIGURE 16.10 Triangulation method representing two variants of with single and double camera arrangements (modified and re-drawn from Wu et al., 2019).

$$X = L - Ztan\theta \tag{17.2}$$

$$\Delta Z = \frac{Z^2}{FL}\Delta P + \frac{Z^2\theta}{L}\Delta\theta \tag{17.3}$$

From the above formula, an increase in baseline length results in elevated measurement accuracy. Due to the scanner limitation of baseline, triangulation scanners are usually used for analyzing small-scale objects over shorter distances.

16.8.2.1.2 Structured Light

The structured light technique works with the projection of pattern lights using a scan head at a known angle and an image of the resulting pattern reflected by the surface. These images are analyzed to attain the surface data points. Bell et al. (1999) defined the light pattern into a single point, a sheet of light, a strip, a grid, and complex coded light, as seen in Figure 16.11. A sheet of light is the universally renowned used pattern that, when projected to an object, creates a line of light along the surface of the object. The detected line with *X, Y, and Z* coordinates are calculated using triangulation, and the images are stored along the X direction with index '*k*' assigned to each image. A light pattern with a series of different strips captures the contour along the surface. This technique is beneficial in digitizing human beings as it has higher data acquisition speed, doesn't use laser, and provides color texture information.

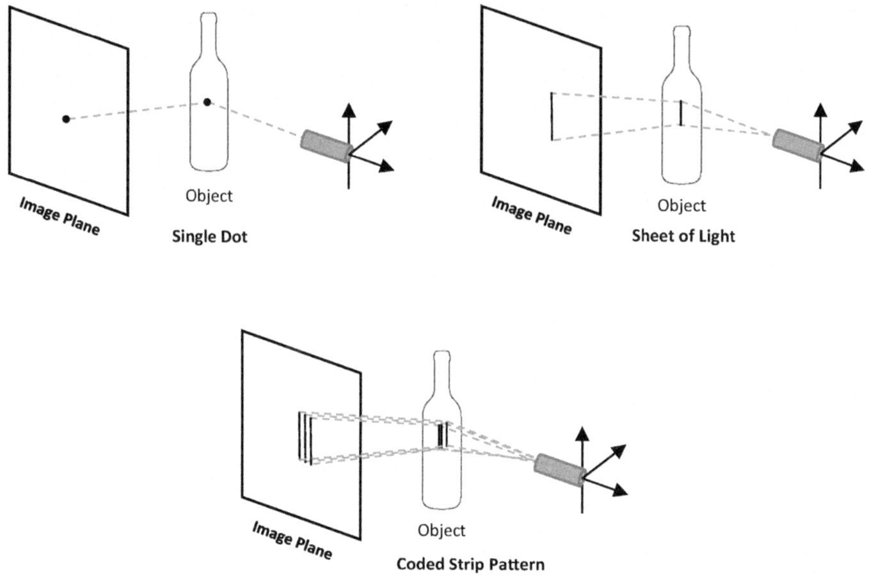

FIGURE 16.11 Structured-light technique showing different light patterns (modified from Bell et al., 1999).

16.8.2.1.3 Time of Flight

Time of flight (TOF) is the time a light pulse (laser) takes to reach the object and return. Distance, *D,* is equal to one-half the distance of laser pulse traveled and is given by $D = C * t/2$, where *C* is the speed of light, and 't' is the time of light pulse. For long-range scanners, TOF systems have an accuracy of a few mm to 2–3 cm. The accuracy of the distance depends on laser pulse width, detector speed, and timing resolution. TOF's disadvantage is its lack of object texture-capturing ability, and not beneficial for small- and medium-sized objects. A few other important non-contact methods are nonoptical and transitive techniques.

16.8.3 Destructive Method

The part which is small, complex, and requires scanning to measure internal and external features uses the destructive RE method. The process includes exposure of 2D cross-sectional sliced images through a CNC milling machine gathered by a CCD camera, which later is converted into edge points. This technology was introduced by CGI Inc. and is known as cross-sectional scanning (CGS) (Pham & Dimov, 2001). A representation of the destructive method can be seen in Figure 16.12. Moreover, RE data points are generally explored by 2D slicing of layers. However, the destructive way is a reverse process, as the 2D sliced images are collected by layer-by-layer destruction of the part. Its advantages include high speed, data capturing of internal structures, and easily workable with the machinable object. The disadvantage, however, is the destruction of the respective part.

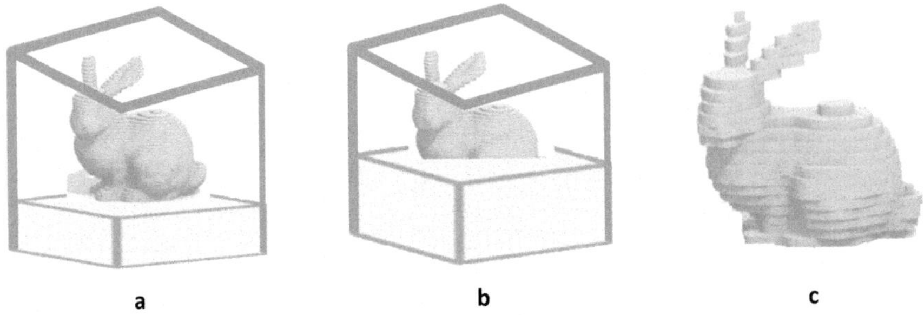

FIGURE 16.12 The data acquisition process for destructive reverse engineering (a) sliced rabbit part and matrix embedded together, (b) layer-by-layer machining for cross-sectional images, (c) and sliced reproduction of rabbit part.

16.9 RE SOFTWARE

16.9.1 RE Software Classification

RE can be classified into software requirements with no single software individually able to fulfill the data processing and geometric modeling requirements. A list is classified for RE software in Table 16.2 based on different engineering applications. This classification of applications is divided into numerous groups, as seen in the table.

16.9.2 RE Phases

Here, RE data processing is explained first to understand the RE software operation. Later, RE operations are described. According to researchers, the complete data processing is chained from scanning to NURBS modeling, as seen in Figure 16.13.

16.9.2.1 Points and Images Phase

Scanning data helps create 3D polygon models through registration, preparation, and optimization in this phase. As shown in Figure 16.14, a flowchart illustrates the transformation of scan data to 3D polygon models. Two flowcharts show the input data variation as either point clouds or images. Transitive techniques such as CT and magnetic resonance imaging (MRI) offer large sets of images as input data. On the contrary, cloud data is extracted for remaining RE techniques such as triangulation, ToF, structured light, etc.

16.9.2.2 Polygon Phase

Polygon models are built in this phase, where application requirements are achieved through controlling and manipulation, resulting in 3D polygon models. These models are utilized for the applications such as 3D animations, graphics, and prototyping. They can also generate CAD entities and construct NURBS surfaces for CAD, CAM, and CAE applications.

TABLE 16.2
Application-Based Classification of Reverse Engineering Software (Modified from Pham & Hieu, 2008)

No.	Application	Functionality	Software
1	Hardware control	Control of RE hardware for data acquisition. Also provides data processing operations and data conversion.	GSI Crystal Studio, Cyberware CyDir, Metris Scan and Mitutoyo Cosmos.
2	CAD entity manipulation	Extraction of point clouds and poly meshes for CAD entities manipulation such as points, contour lines, and CAD primitives (circle, rectangle, etc.).	Pro-E, SolidWorks, ICEM Surf, Imageware.
3	2D scan image processing and 3D modeling	Utilized to process 2D scan images (CT/MRI) and 3D reconstruction	Rapidform, Scan IP, 3D Doctors, Mimics, BioBuild Velocity2, Amira Scan IP.
4	3D inspection	Utilized for 3D inspection, error map creation and analysis, and creating inspection reports and documents.	COMET inspect, Power INSPECT, Geomagic qualify, Polyworks Inspector, and Metris Focus Inspection.
5	Polygon manipulation	Editing, modification, and optimization of 3D Polygon	Catia, Magics RP, Viscam RP, and Desk Artes.
6	Polygon and NURBS surface reconstruction	Provision of RE data processing toolset through point clouds and polygons to NURBS surfaces construction and 3D inspection.	CopyCAD, GSI Studio, Geomagics, Rapidform, Polyworks, and Paraform,
7	NURBS surface and solid modeling	Provision of NURBS modeling and editing tools through basic 3D CAD tools.	Pro-E, SolidWorks, Catia, UG, and Rhino.

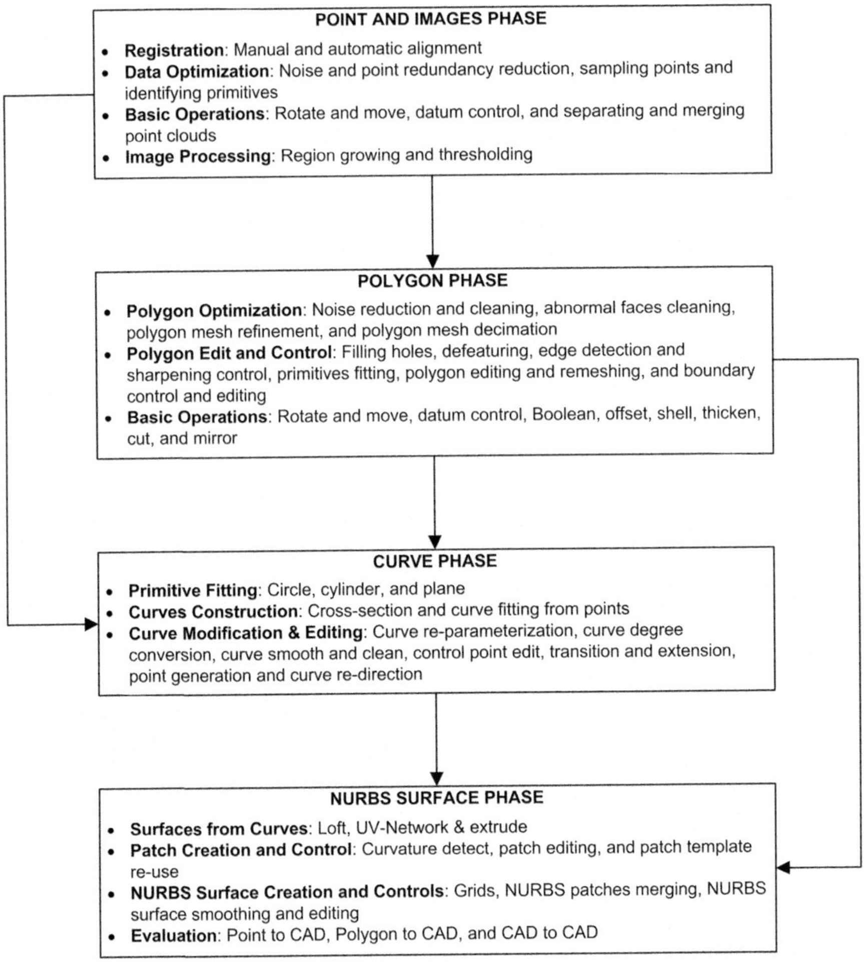

FIGURE 16.13 Phase representation of reverse engineering data processing chained with its fundamental operations.

16.9.2.3 Curve Phase

In numerous RE projects, particularly in RE of mechanical parts, CAD entities are the most essential for geometric modeling, either constructed through point clouds promptly or implicitly through editing, sectioning, and fitting processes of polygon models. It is known that simple geometries can be utilized to generate parts by extracting limited reference points using mechanical probes via the contact method. However, complex geometrical parts with free-form surfaces, such as curves, require larger data points. This data acquisition is generally obtained using non-contact methods.

16.9.2.4 NURBS Surface Phase

NURBS is the most exact method to define free-form curves and surfaces accurately. Its significance is due to points such as:

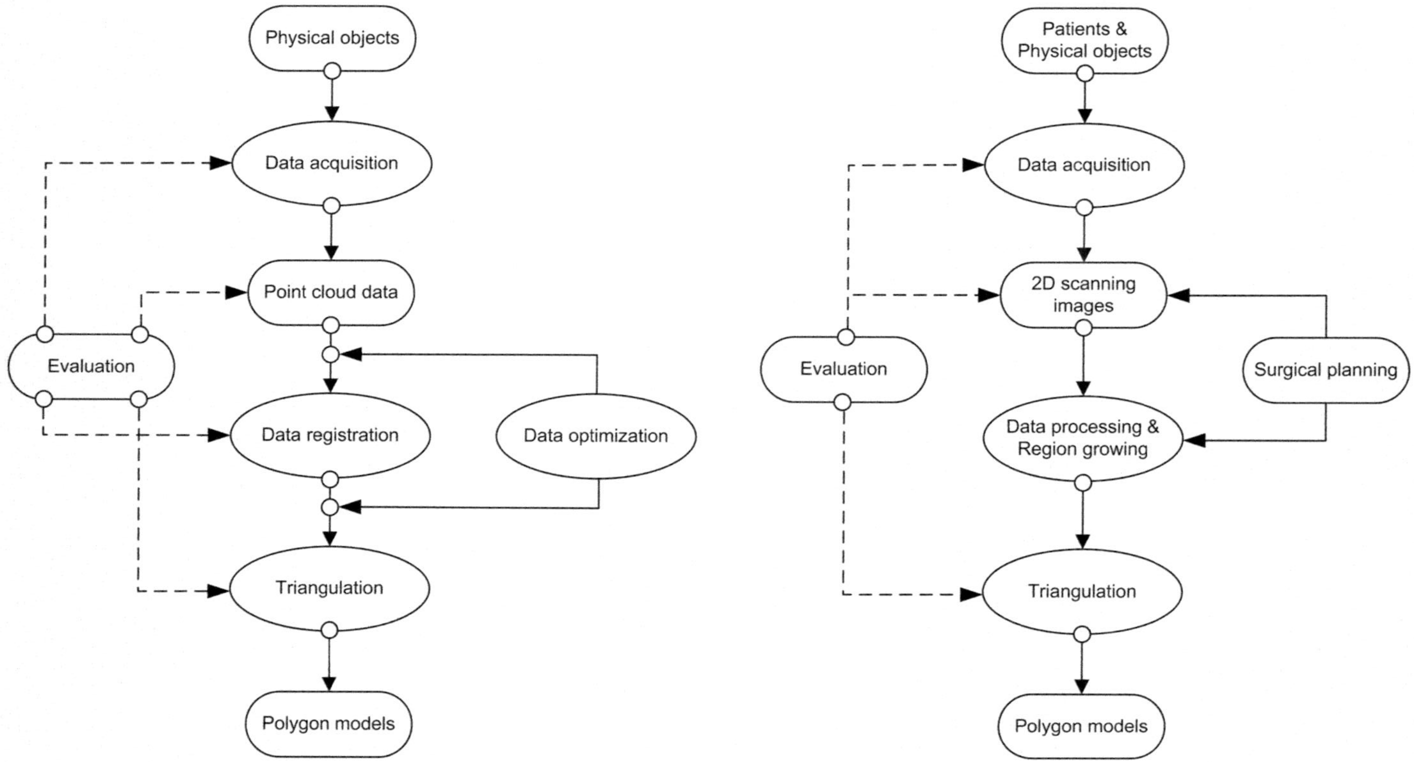

FIGURE 16.14 Transformation of reverse engineering scanned data into 3D triangular mesh models (a) flowchart with input as point cloud data, and (b) flowchart with input as 2D scan images (modified from Pham & Hieu, 2008).

- It offers a common mathematical model for free-form and basic analytical shapes.
- Memory consumption is less when storing shapes.
- NURBS offers flexibility to design a substantial number of shapes.
- It can be accessed through fast and numerically stable algorithms.
- NURBS is an overview of rotational Bezier, also nonrotational Bezier and B-Spline curves and surfaces (Rogers, 2001).

Figure 16.15 shows a NURBS construction flowchart in RE data processing chain. The three approaches used to generate NURBS surfaces includes manual creation using CAD entities, patches and automatic creation using polygon models. NURBS CAD solid model is formed through its import into CAD/CAM system.

Manual NURBS uses CAD entities like CAD modeling with models developed through basic entities like points, curves, etc. The only difference is with the construction using scan data points and images to form.

Manual NURBS using patches – NURBS patches are constructed through a patch structure utilized in this approach having a quadrangular shape. Figure 16.16

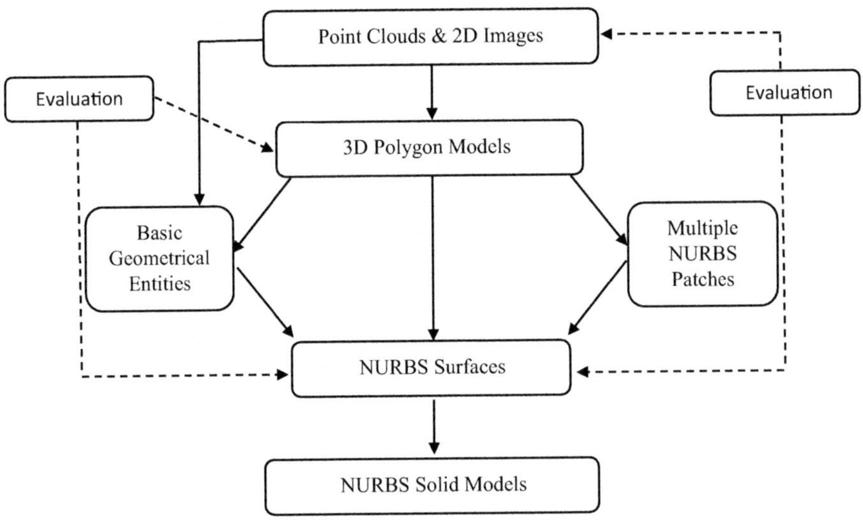

FIGURE 16.15 Different reverse engineering approaches for NURBS construction (modified from Pham & Hieu, 2008).

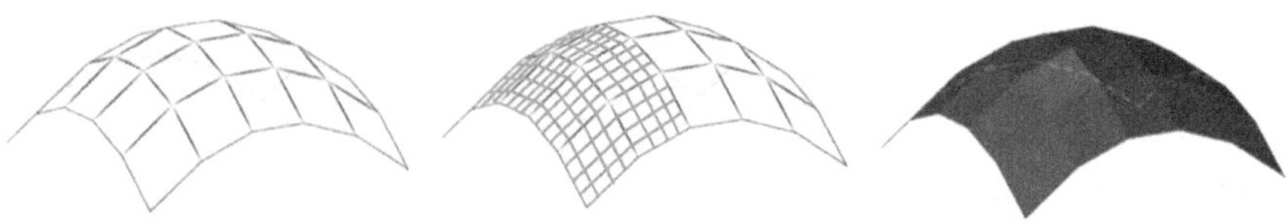

FIGURE 16.16 (a) A curved shape with NURBS patch layout, (b) UV grids having NURBS patches, and (c) NURBS solid model of curve surface.

shows a polygon surface defining a patch. This is the most effective surface fitting technique for complex geometrical (anatomical structures) models. With the increase in control points, the accurate model can be retrieved through patches.

Automatic NURBS creation using Polygon Model – NURBS surfaces using a polygon model is processed through a combination of repeating operations utilized in the NURBS patch approach. It is a fast outcome-oriented method with restrictions to straightforward geometries and specific applications using the draft NURBS surface model.

16.10 RE OF ELECTRICAL SYSTEMS

RE is required when one does not have the requisite schematic diagrams, and OEM is no more available or supporting the system. Additionally, it may be needed if one has the intellectual right to the system and wants to improve it but need the schematics, etc. Similarly, the track sizes, shape widths, placement, etc., have certain connotations, especially at higher frequencies or currents. Following paragraphs explain steps and methodologies in implementing RE of electrical systems.

16.10.1 RE at System Level

RE, at the electrical system level, can be explained based on the following:

a. PCB with electronic components connected to generate a definite response against definite inputs.
b. Single/double layer PCBs – Simplest form.
c. Multi-layer PCBs – Complex form.
d. Programmable Components.
e. Readable programs/software.
f. Locked programs.
g. Design & Production Process.
h. Performance Qualification.
i. Environmental Qualification.
j. EMI/EMC Qualification.

16.10.1.1 Types of Systems

Following are the types of systems used in the electrical domain:

a. Analog Systems
   - Analog components
   - Components/systems that are easiest to replicate
b. Digital Systems with Non-intelligent Devices
   - Digital ICs
   - Readable configurations
c. Digital Systems with Intelligent Devices
   - Micro-controllers
   - Micro-processors

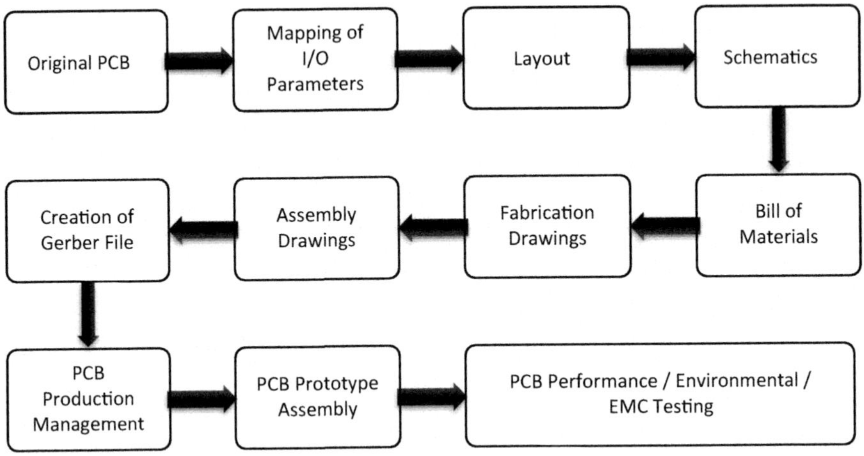

FIGURE 16.17 Schematic representing reverse engineering process flow for PCB electrical system.

- FPGAs
- Programs/software
- Locked programs

RE process flowchart for electrical systems is shown in Figure 16.17.

16.10.2 RE at Sub-level Systems

RE for electrical sub-level systems can be explained based on the following:

a. Input/output level
b. Proper bias levels
c. Signaling pulses
d. Clock frequency
e. Duty cycle
f. Input/output impedances
g. Functional parameters
h. Power consumption
i. Specialized functional parameters

16.11 MATERIAL ANALYSIS

Correct and thorough material analysis is fundamental in the RE process and requires precision and patience. Material analysis should be carried out on multiple points/cross sections based on part machine design and functionality. Furthermore, to ensure material equivalence (OEM and indigenous component), the following should be adhered to:

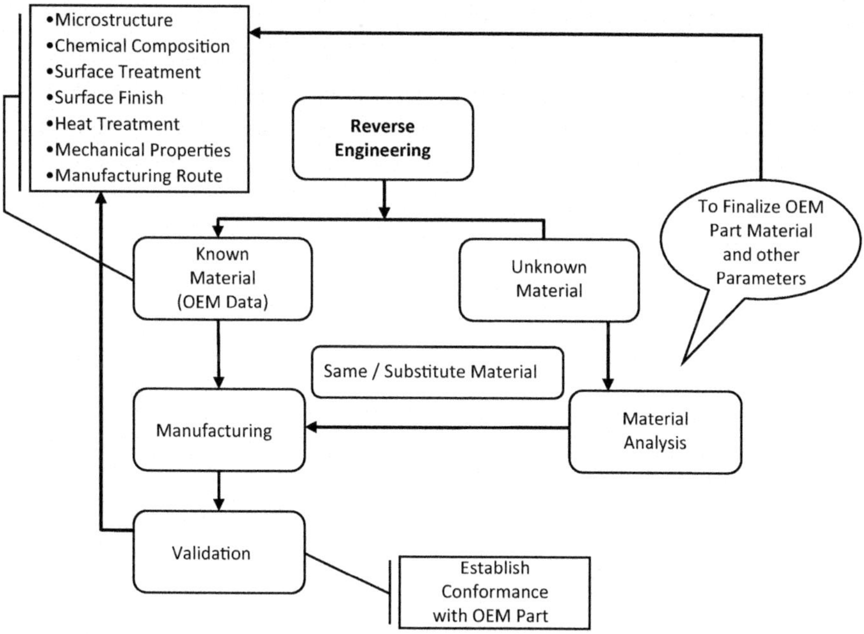

FIGURE 16.18 Process chart representing material analysis process for same or substitute materials.

a. Microstructure and chemical composition
b. Surface/heat treatment and surface finish
c. Mechanical properties and manufacturing routes

A process chart for the same/substitute material is shown in Figure 16.18.

16.11.1 Characterization Techniques

The following are the most used characterization techniques:

a. Microscopy
   - Optical microscopy (OM)
   - Scanning electron microscopy (SEM)
   - Transmission electron microscopy (TEM)
b. Spectroscopy
   - Scanning electron microscopy (energy dispersive spectroscopy)
   - X-ray diffraction
   - X-ray fluorescence
c. Mechanical testing
   - Compression testing
   - Tensile testing
   - Hardness testing

16.12 DIMENSIONAL ANALYSIS

Process of duplicating an item functionally and dimensionally by physically examining and measuring existing parts to develop the technical data (physical and material characteristics) required for competitive procurement. Both original and reverse-engineered parts should be compared to create geometric similarity.

16.12.1 Part Specifications

a. Manufactured parts can be specified by:
   - Mechanical properties
   - Electrical properties
   - Physical properties
b. Dimensions – linear or angular sizes of a part specified on the drawing
c. Tolerances – allowable variations from the specified part dimensions that are permitted in manufacturing

16.12.1.1 Dimensions

Types of dimensions include:

a. Basic size: A numerical value used to describe the exact theoretical size, actual profile, orientation, or location of a feature or datum target.
b. Variations in size may be due to the manufacturing process
c. Upper limit: Maximum limit of the size of a manufactured part
d. Lower limit: Minimum limit of the size of a manufactured part

16.12.1.2 Deviation

It is the algebraic difference between a size to its corresponding basic size.

a. Upper deviation
   - It is the algebraic difference between the upper limit of a part and its basic size.
   - (Maximum limit) – (Basic size)
b. Lower deviation
   - It is the algebraic difference between the lower limit of a part and its basic size.
   - (Minimum limit) – (Basic size)

16.12.1.3 Tolerance

The total amount a dimension may vary and is measured due to the difference between the maximum and minimum size limits concerning the basic size.

a. Effects of tolerance include high manufacturing costs, which may affect the system's proper functioning.
b. The three types of tolerances are unilateral, bilaterally symmetric, and bilateral non-symmetric.

16.13 RE – CASE STUDIES

16.13.1 Case Study – Integrated RE and Failure Analysis for Recovery of Mechanical Shafts

Numerous studies have shown the integration of the RE approach in the mechanical domain. Due to complex design constraints, uncontrolled failure, and wear conditions, remanufacturing needs to be standardized. This case study presented a recovery process of a rotary draw bending machine shaft through an integrated RE approach utilizing a non-contact hardware method along with analytical calculations and finite element analysis (Engel & Al-Maeeni, 2019).

The authors categorized the integration and failure analysis to recover the shaft into four steps. Firstly, a FARO Edge Scan Arm HD laser scanner, a non-contact system, was used for 3D point cloud data extraction to reconstruct a 3D model and identify the material type. Polyworks software was later used to evaluate the digital model through comparison with the manual part. For the chemical composition of the material, Spectro technology was used. The second step was the optical examination to inspect the manual shaft's surface characteristics, which is done through hardness and roughness tests. An optical microscope was also used to assess the machining process done on the manual shaft. Furthermore, the hardness of the surface was found using a Wolpert tester (a Brinell hardness testing system) with a 2.5 mm ball indenter and an applied load of 1830 N. In the third step, FEA analysis using Abaqus software was used to underpin the maximum stresses produced at maximum loading conditions and determine the safety factor. In the last step, identified design specifications were compared to existing specifications. Through this comparison, future surface treatments were recommended for the shaft.

A non-destructive material test was done to extract the metal's chemical composition and compare it with the standard metal database. The test results found the material as C15Pb. The nominal chemical composition, test results, and mechanical characteristics are shown in Tables 16.3–16.5.

After material identification, the surface hardness test was conducted and found the increase in hardness with an increase in its carbon concentration. To evaluate the permissible hardness value, assembly analysis was undertaken, as shown in Figure 16.19. Likewise, to find the type of machining process, the optical microscope was used to calculate standard surface roughness. Jamshidi et al. (2006) defined the standard shapes of textures and surface roughness as shown in Figure 16.20. It was found from optical imagery that the turning process was used

TABLE 16.3
Chemical Composition of C15Pb (Modified from Engel & Al-Maeeni, 2019)

C%	Si%	Mn%	P%	S%	Pb%
0.12–0.18	≤0.40	0.30–0.88	≤0.045	≤0.045	0.15–0.30

TABLE 16.4
Chemical Composition at Tested Locations (Modified from Engel & Al-Maeeni, 2019)

Sec.	C%	Si%	Mn%	P%	S%	Pb%
1	0.82	0.23	0.84	0.014	0.018	0.24
2	0.76	0.23	0.82	0.014	0.015	0.20
3	0.20	0.24	0.87	0.019	0.022	0.27
4	0.18	0.22	0.83	0.016	0.025	0.19
5	0.14	0.24	0.83	0.015	0.026	0.29

TABLE 16.5
Nominal Mechanical Properties of C15Pb (Modified from Engel & Al-Maeeni, 2019)

Tensile Strength	Yield Strength	Brinell Hardness	Poisson's Ratio	Rigidity's Modulus	Mass Density	Young's Modulus
500 MPa	385 MPa	143 HB	0.28 %	80 GPa	7.8 Mg/m^3	200 GPa

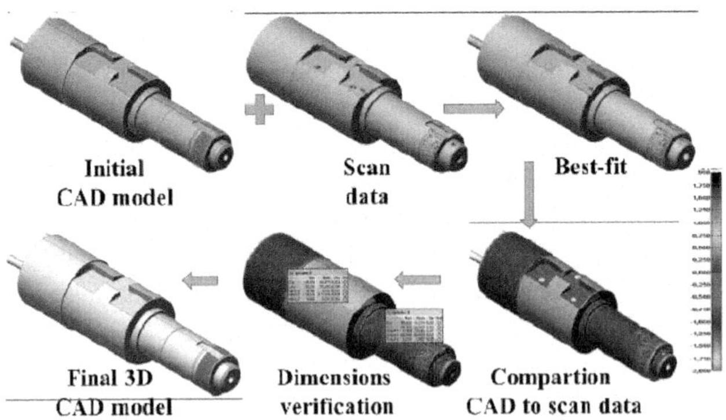

FIGURE 16.19 Machining processes and their standard roughness values (adapted from Engel & Al-Maeeni, 2019).

on the shaft having 0.03 ≤ Ra ≤ 25.0 and verified using assembly analysis. Also, the roller and needle bearing roughness of Ra ≤ 1.6 μm and Ra ≤ 0.2 μm was within the turning process range. The authors reported a comparison via Table 16.6 for the same specifications between extracted and measured values.

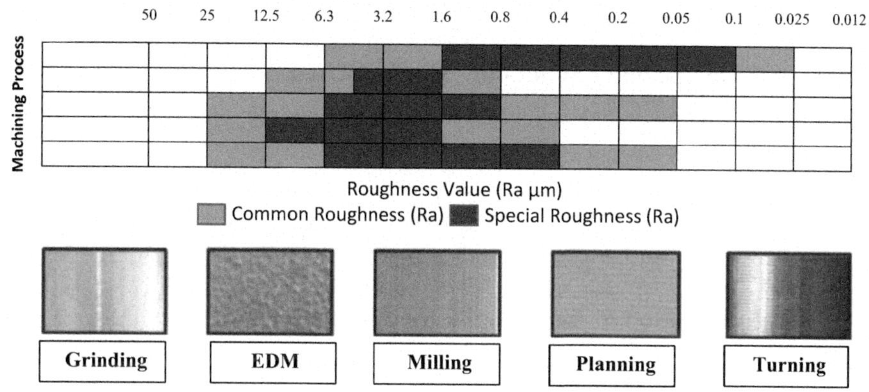

FIGURE 16.20 Reconstruction of CAD model for the shaft (modified from Jamshidi et al., 2006).

TABLE 16.6
Comparison of Nominal and Measured Design Specifications (Modified from Engel & Al-Maeeni, 2019)

Specifications	Original Design Characteristics	Measured Value
Surface hardness	613–681 HB	146–470 HB
Surface roughness	Ra ≤ 0.2 µm	0.073–4.09 µm
Bending stress	27.62 MPa	33.98 MPa
Material	C15Pb	–
Machining process	Turning process	–
Surface treatment	Carburizing	–
Design factor safety	3.05	–

16.13.2 Case Study – Analysis of Unknown Material

The unknown structure and design of the non-return valve was the main issue to be resolved in this case study. The following problems were to be addressed.

a. Establish equivalence
   i. Disassembly diagram
   ii. Microstructure analysis
   iii. Chemical composition (grade identification)
   iv. Heat treatment/temper state
   v. Surface treatment/case hardening
   vi. Mechanical properties
   vii. Manufacturing route
b. Validation(properties) of ingenious products with OEM

TABLE 16.7
Chemical Composition of Superalloy Material Used

Elements	Ni	Al	Co	Cr	Fe	Ti	Total
%Age	44	4	22	12	12	6	100

TABLE 16.8
Chemical composition of the material used in non-return valve

Elements	OEM	Indigenous	Coupon	Standard Composition
Si	0.54	0.5	0.46	0.5–1.2
Mn	0.68	0.65	0.55	0.40–1.2
Mg	0.56	0.61	0.57	0.20–0.80
Fe	0.4	0.31	0.3	≤ 0.70
Cu	4.16	4.52	4.12	3.90–5.0
Cr	–	–	–	≤ 0.10
Zn	–	–	–	≤ 0.25
Zr + Ti	–	–	–	≤ 0.20
Ti	–	–	–	≤ 0.15
Al	Balance	Balance	Balance	Balance

The casing of the non-return valve was the trickiest part. Since OEM's non-return valve was available for destructive testing, both OEM and indigenous casings were cut longitudinally using electric discharge machining (EDM). Material characterization on selected cross-sections was then carried out for both OEM and indigenous parts. Finally, validation of manufactured products (material/processes) was carried out to establish material equivalence.

Table 16.7 shows the chemical composition of the material used in the non-Return valve, while Table 16.8 shows a comparative analysis of the OEM and indigenously developed non-Return valve.

16.13.3 Case Study – Analysis of Another Unknown Material

The unknown material in this case study was to be analyzed, and the material was conventional superalloy. Since the same material was not readily available, RE was required to find the correct composition of the subject material. The following challenges needed to be handled:

a. Material (conventional superalloy) composition to be found
b. To assess the effect of heat treatment on microstructure and mechanical properties
c. To determine the impact of cooling methods

TABLE 16.9
Comparative Analysis of OEM and Indigenously Developed Non-Return Valve

Analysis	OEM	Indigenous	Coupon
Hardness, HV	139 ± 3	122 ± 2	120 ± 1
Microstructure	A typical microstructure of 2xxx series with Al-Mg-Si-Cu intermetallics	A typical microstructure of 2xxx series with Al-Mg-Si-Cu intermetallics	A typical microstructure of 2xxx series with Al-Mg-Si-Cu intermetallics
Surface treatment/ case hardening	Anodizing	Anodizing	Anodizing
Heat treatment condition	T4	T4	T4
Manufacturing route	Machined from extruded/rolled blank	Machined from extruded/rolled blank	–

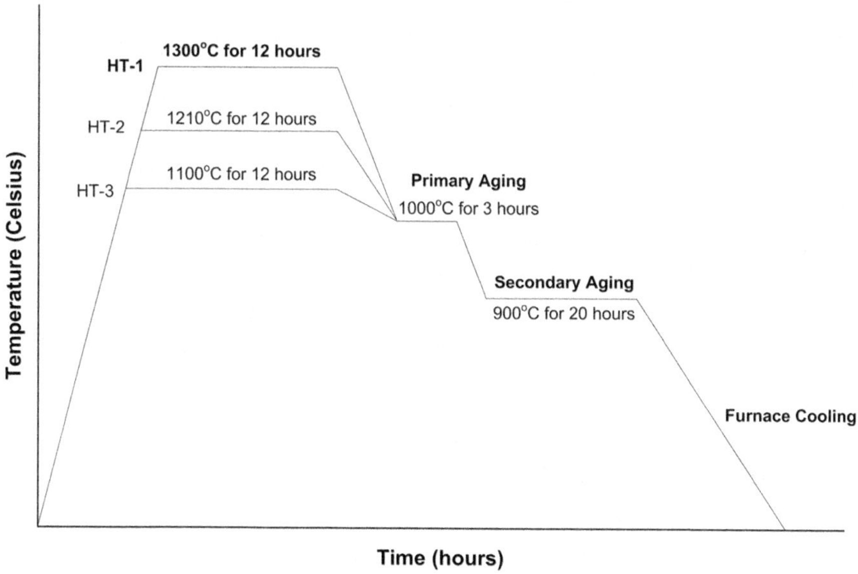

FIGURE 16.21 Composition of alloy and heat treatment cycles.

The chemical composition of the superalloy was found, as shown in Table 16.9. The composition of alloy and heat treatment cycles are shown in Figure 16.21. The microstructure-characterization and micro-hardness after all three heat treatment cycles are also shown in Figure 16.22. The cooling effects after both furnace-cooled and air-cooled are shown in Figure 16.23.

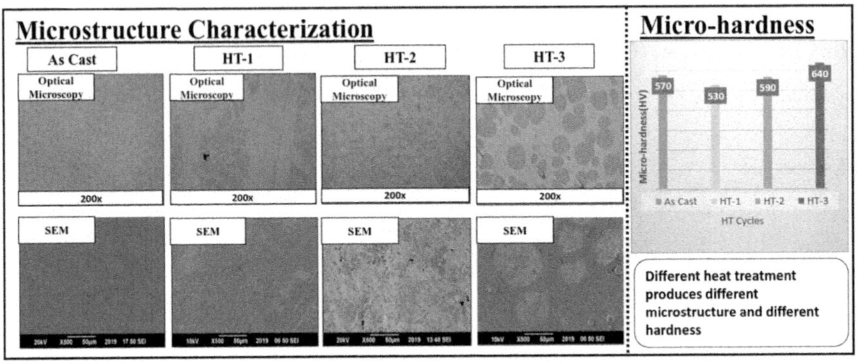

FIGURE 16.22 Microstructure-characterization and micro-hardness for all three heat treatment cycles.

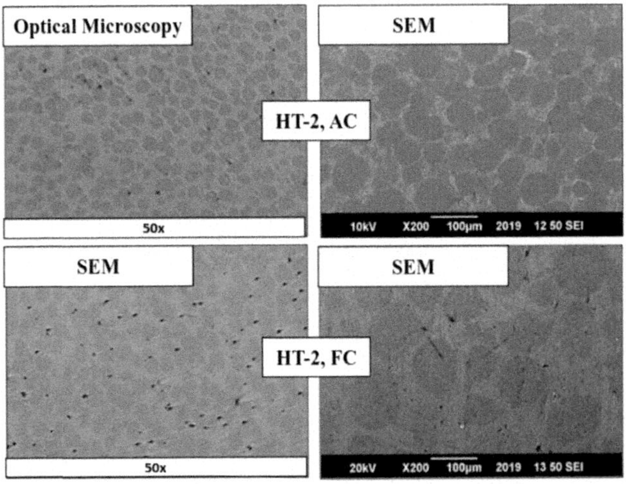

FIGURE 16.23 Effects of cooling after furnace-cooled and air-cooled processes.

16.14 CONCLUSION

Like developments in rapid prototyping and tooling techniques assist physical models in shortening time through CAD models, RE processes are helping to reduce the time for converting physical models to CAD models. One of the most complex actions in RE is the representation of CAD models from point data, as a 3D surface reconstruction requires complex surface fitting algorithms. The evolution of RE software is in place as most CAD software can't handle large amounts of point data and need discrete software packages for point processing. Feature-based algorithms are revolutionizing the reverse engineering system from point cloud to complete solid models.

This chapter started with the definition of RE, followed by why there is a need to reverse engineer products. Afterward, RE strategy, generic processes associated (with contact and non-contact scanning), point processing, and geometric model development were introduced. The RE concerning nature was also explained with examples. RE procedures and digital modeling were then detailed with a few algorithms that are generally used. Later the in-depth description of methodologies and techniques used in RE was explained.

This chapter also presented the computer-aided RE system using CMM technology. Computer vision solutions were presented as an alternative approach to CMM technology. Then RE hardware and software systems were introduced. RE hardware includes contact, non-contact and destructive methods. The essential non-contact methods with an optical technique, such as triangulation, structured light, and TOF, were explained in detail. The software section explained engineering phases such as point and images, polygon, curve, and NURBS surface phases. At the end of the chapter, RE in an electrical system, material analysis, and dimensional analysis were explored. The chapter finishes with three case studies for mechanical shaft RE, material RE, and non-Return valve RE.

REFERENCES

Abdel-Wahab, M.S., Hussein, A.S., Taha, I., Gaber, M.S. 2005. An enhanced algorithm for surface reconstruction from a cloud of points. In *GVIP05 Conference*, Cairo, Egypt: 19–21.

Alain, B. 1999. A review of state-of-the-art reverse engineering. *Proceedings of TCT (Time Compression Technol) European Conference 1999*, Nottingham Royal Moat House, United Kingdom: 177–188.

Amenta, N., Choi, S., Kolluri, R. K. 2001. The power crust, unions of balls, and the medial axis transform. *Computational Geometry*, 19(2-3): 127–153.

Azernikov, S., Miropolsky, A., Fischer, A. 2003. Surface reconstruction of freeform objects based on multiresolution volumetric method. In *Proceedings of the Eighth ACM Symposium on Solid Modeling and Applications*: 115–126.

Bell, T., Li, B., Zhang, S. 1999. Structured light techniques and applications. *Wiley Encyclopedia of Electrical and Electronics Engineering*: 1–24.

Bernard, A., Laroche, F., Remy, S., Durupt, A. A. 2010. New trends in reverse-engineering: Augmented-semantic models for redesign of existing objects. *Ouvrage Collectif Reverse-Engineering*: 27.

Bernardini, F., Mittleman, J., Rushmeier, H., Silva, C., Taubin, G. 1999. The ball-pivoting algorithm for surface reconstruction. *IEEE Transactions on Visualization and Computer Graphics*, 5(4): 349–359.

Bi, Z., Wang, X. 2020. *Computer aided design and manufacturing*. John Wiley & Sons.

Bi, Z.M., Kang, B. 2014. Sensing and responding to the changes of geometric surfaces in flexible manufacturing and assembly. *Enterprise Information Systems*, 8(2): 225–245.

Cazals, F., Giesen, J. 2004. Delaunay triangulation based surface reconstruction: A short survey (Doctoral dissertation, INRIA).

Engel, B., Al-Maeeni, S.S.H. 2019. An integrated reverse engineering and failure analysis approach for recovery of mechanical shafts. *Procedia CIRP*, 81: 1083–1088.

Eunsang, L., Dongkuk, L. 2016. Analyzing team based engineering design process in computer supported collaborative learning. *Eurasia Journal of Mathematics, Science and Technology Education*, 12(4): 767–782.

Fisher, R.B. 2002. Applying knowledge to reverse engineering problems. Proceedings of Geometric Modeling and Processing, Riken: 149–155.

Hennighausen, A., Roston, E. 2015. 14 Smart inventions inspired by nature: Biomimicry. https://www.bloomberg.com/news/photo-essays/2018-10-25/bloomberg-s-week-inpictures (accessed 10 January 2023).

Hoppe, H., DeRose, T., Duchamp, T., McDonald, J., Stuetzle, W. 1992. Surface reconstruction from unorganized points. In *Proceedings of the 19th Annual Conference on Computer Graphics and Interactive Techniques*: 71–78.

Jamshidi, J., Mileham, A. R., Owen, G. W. 2006. Dimensional tolerance approximation for reverse engineering applications. *International Design Conference-design*, 2006: 855–862.

Milajlovic, Z., Jovanovic, B., Maric, F. 2004. Matlab toolbox for analysis of 3D images. *Review of the National Center for Digitization*, (4): 70–77.

Motavalli, S., Shamsaasef, R. 1996. Object-oriented modelling of a feature based reverse engineering system. *International Journal of Computer Integrated Manufacturing*, 9(5): 354–368.

Neugebauer, P.J., Klein, K. 1997. Adaptive triangulation of objects reconstructed from multiple range images. *In IEEE Visualization*, 97(6): 12.

Park, J., N-DeSouza, G. 2005. 3-D modeling of real-world objects using range and intensity images. In: *Machine learning and robot perception*, Springer, Berlin, Heidelberg: 203–264.

Pernkopf, F. 2005. 3D surface acquisition and reconstruction for inspection of raw steel products. *Computers in Industry*, 56(8–9): 876–885.

Pham, D., Dimov, S. 2001. *Rapid manufacturing: the technologies and applications of rapid prototyping and rapid tooling*. London: Springer-Verlag, ISBN: 978-1-4471-0703-3.

Pham, D.T., Hieu, L.C. 2008. Reverse engineering–hardware and software. In: *Reverse Engineering*. Springer, London: 33–70.

Raja, V. 2008a. Introduction to reverse engineering. In: Raja, V., Fernandes, K. (eds) *Reverse Engineering. Springer Series in Advanced Manufacturing*, Springer, London.

Rebolj, D., Babič, N.Č., Magdič, A., Podbreznik, P., Pšunder, M. 2008. Automated construction activity monitoring system. *Advanced Engineering Informatics*, 22(4): 493–503.

Rocchini, C., Cignoni, P., Montani, C., Pingi, P., Scopigno, R. 2001a. A suite of tools for the management of 3d scanned data. *3D Digital Imaging and Modeling Applications of: Heritage, Industry, Medicine* & Land: 2.

Rocchini, C., Cignoni, P., Montani, C., Pingi, P., Scopigno, R. 2001b. A low-cost scanner based on structured light. *Computer Graphics Forum (Eurographics 2001 Conference Proceedings)*, 20(3): 299–308.

Rogers, D. F. 2001. *An introduction to NURBS: With historical perspective*. Morgan Kaufmann, ISBN: 978-1-55860-669-2.

Tamas, V., Ralph, R. M., Jordan, C. 1997. Reverse engineering of geometric models–an introduction. *Computer Aided Design*, 9(4): 255–268.

Thompson, W.B., Owen, J.C., Germain, H.J.S., Stark, S.R., Henderson, T.C. 1999. Feature-based reverse engineering of mechanical parts. *IEEE Transactions on Robotics and Automation*, 15(1): 57–66.

Wu, C., Baijin, C., Chunsheng, Y., Xiaopeng, Y. 2019. Modeling the influence of oil film, position and orientation parameters on the accuracy of a laser triangulation probe. *Sensors*, 19(8): 1844.

Yau, H., Haque, S., Menq, C. 1995. Reverse engineering in the design of engine intake and exhausts ports. *ASME prod eng div publ ped, ASME, New York, NY, (USA)*, 64:139–148.

Zhao, H.K., Osher, S., Fedkiw, R. 2001. Fast surface reconstruction using the level set method. *In Proceedings IEEE Workshop on Variational and Level Set Methods in Computer Vision*: 194–201.

17 Conclusion

Uzair Khaleeq uz Zaman, Ali Siadat, Aamer Ahmed Baqai, Kanwal Naveed, and Atal Anil Kumar

17.1 INTRODUCTION

Smart manufacturing (SM), by pure definition, is the integration of modern machinery, the internet of things (IoT), and artificial intelligence (AI) into industrial processes, thereby enabling a more efficient and error-free design, manufacturing, and customer delivery of products. Today, every new industrial setup is trying to incorporate the ideas of SM while competing with each other in innovation and customer-product relations, thereby increasing not only productivity but also helps in improving the product every day. Therefore, declaring that SM is the backbone of any progressive nation wouldn't be an understatement. Moreover, SM is a crucial player in the future of the industrial revolution. With the help of all the cutting-edge technologies, it is hoped to replace traditional products and processes with intelligent and customized products per customer requirements. This chapter summarizes all the key concepts explained in the preceding chapters one by one to present a complete overview of SM for a better understanding of the readers.

17.2 OVERVIEW OF HANDBOOK CHAPTERS

The focus of this Handbook is to provide the end user with a comprehensive overview of the state-of-the-art regarding all the critical factors in SM and what makes an industry worthy of being categorized as part of the Fourth Industrial Revolution. In each chapter, the future trends and frameworks are discussed, along with practical case studies re-emphasizing the core concepts present in the chapters. Starting from introduction to SM and Industry 4.0 (Chapter 1), computer-aided design/computer-aided process planning/and computer-aided manufacturing (CAD/CAM/CAPP) (Chapter 2), involvement of Robots in Industry 4.0 (Chapters 5 and 6), IoT-enabled systems (Chapter 11), material supply and logistics 4.0 (Chapter 13), the Handbook ends with reverse engineering (Chapter 16), thereby covering all the significant aspects of Industry 4.0 and SM.

Chapter 1 provides an introduction related to SM and Industry 4.0. The origins and key pillars of SM/IM are discussed along with the concept of the factory of the future (FoF) for manufacturing systems and their design. The chapter ends with the future perspectives of SM/IM in Industry 4.0 and lays the foundation for the rest of the chapters in the Handbook.

Chapter 2 discusses the basic understanding of CAD, CAPP, and CAM systems concerning their definition, classification, mathematical modeling, and software

DOI: 10.1201/9781003327523-20

implementation. Integrating CAD, CAPP, and CAM systems are also explored through state-of-the-art industrial examples to link the individual systems with Industry 4.0. At the end of this chapter, essential aspects of CAD, CAPP, and CAM have been detailed as well.

In Chapter 3, the focus is on developing an understanding of Quality 4.0 through background and definitions. It also discusses the dimensions of Quality 4.0 and various models being implemented worldwide. Industrial case studies are further presented, and a conceptual framework is proposed for Quality 4.0. Finally, the chapter concludes with the existing challenges associated with Quality 4.0.

In Chapter 4, the author presents the idea of a digital twin for smart product design, biomanufacturing, and IoT. All the concepts are explained with relevant case studies and the associated challenges. An application framework for digital twin design is also discussed at the end of the chapter.

In Chapter 5, the primary focus is on the application of robotic systems in the manufacturing industry and how their involvement can improve the overall manufacturing process. Various essential aspects, like sensor applications and ML in sensor fusion, are discussed. The chapter also introduces basic planning and control techniques for robotic systems and concludes with relevant case studies, current challenges, and future frameworks.

In Chapter 6, further to the concept of Chapter 5, a detailed discussion on collaborative robots (cobots) is presented. This chapter focuses on the challenges for SMEs when introducing cobots in their manufacturing setups to create a better understanding of identifying the suitable applications and approaches to make manufacturing smart. The importance of workers in manufacturing is also highlighted along with industrial case studies. The chapter concludes with a conceptual framework for the implementation of cobots, and the conclusions are drawn.

Chapter 7 discusses the concept of control strategies for product customization and the inclusion of big data analytics in product customization. The utilization of big data for efficient manufacturing and identifying faulty products is also discussed, especially concerning predictive maintenance. Finally, the chapter concludes with case studies and a framework for future application.

In Chapter 8, the concept of MR/AR as a new control strategy is introduced for manufacturing systems. The chapter explains the types of AR/VR devices along with their benefits and limitations. The environment preparation and adaptation in manufacturing systems are also discussed with the help of several case studies for both VR and MR systems.

In Chapter 9, soft sensor inferential models are presented with a focus on bioprocesses. A detailed discussion on already existing soft sensor technologies is presented along with the case studies in the bioprocesses domain as the chapter's primary focus. The chapter concludes with a perspective on the ML method's revolution in soft sensors.

In Chapter 10, SM is discussed incorporating energy harvesting concepts in Industry 4.0. The chapter discusses energy harvesting, its types, and the associated applied perspectives. The connection between SM, IoT, and Industry 4.0 is also discussed with the help of practical case studies. The chapter ends with future trends and possibilities in energy harvesting.

In Chapter 11, IoT for the manufacturing industry is discussed. The chapter starts with the enablers of IoT such as RFID and WSN, and then discusses the benefits and applications of IoT for the manufacturing industries. The manufacturing policies and strategies being employed by various countries are also elaborated. Finally, the chapter concludes with real-life case studies, challenges, and opportunities in manufacturing.

In Chapter 12, the integration of PLM with CM is presented. The chapter starts with sustainable production using PLM, discusses synergetic production through CM, offers various frameworks and methods in PLM, and explains Industry 4.0 technologies such as AI, big data, and blockchain with respect to PLM and CM. Different industrial methods with the inclusion of characteristics and challenges of CM for product life paradigms are also discussed. The chapter concludes with various case studies related to PLM and CM, along with the proposition of a conceptual framework.

In Chapter 13, the concepts of material supply and Logistics 4.0 are introduced for SM. PPS, such as the AACHEN model, is presented with established approaches like the LOOR approach, CONWIP, etc. Furthermore, types of DSS are discussed with the potential use of communication-driven systems in Logistics 4.0. The chapter ends with case studies focusing on decentralized supply chain systems and Logistics 4.0.

In Chapter 14, workforce engagement with DCPS is presented. The production system's flexibility is the primary goal of this chapter based on self-controlling resource allocation and sequencing. It was discussed how an efficient workforce roster results in the lowest possible labor costs while all production requirements are covered. Thus, the aim is to match capacity supply and demand in the most appropriate manner possible.

In Chapter 15, the authors advocate a systemic approach to managing risks in product development and manufacturing processes. The chapter defines risk, uncertainty, and decision-making in risk management and ends with various use cases.

Finally, Chapter 16 concludes with the introduction of the concept of RE and how it can be used for Industry 4.0. The chapter emphasizes why RE is critical and discusses the concepts of both FE/RE and re-engineering. The RE procedure is discussed in detail, along with the tools and hardware/software used. The chapter concludes with case studies focusing on RE and its integration with other domains.

Conclusively, with the detailed discussions, practical applications, case studies, and several future framework inclusions throughout the Handbook, it is evident that SM is the key to the industrial revolution and the design of future manufacturing systems. It is being implemented in almost all modern industrial setups and is becoming the founding stone for a new era of more efficient and productive industries. Moreover, now is the high time to align more research activities and efforts toward the advancements and application methods of modern machinery, control strategies, quality, and risk management methods. In this regard, this Handbook will be beneficial for the research community and the practitioners. It will be able to provide valuable insights into the current trends and future frameworks concerning manufacturing system design in Industry 4.0.

Index